SPATIAL PROCESSING IN NAVIGATION, IMAGERY AND PERCEPTION

SPATIAL PROCESSING IN NAVIGATION, IMAGERY AND PERCEPTION

Edited by

Fred Mast
University of Lausanne, Switzerland

and

Lutz Jäncke
University of Zürich, Switzerland

Springer

Fred Mast
University of Lausanne
Department of Psychology
Anthropole 5189
CH-1015 Lausanne, Switzerland

Lutz Jäncke
University of Zürich
Department of Psychology
Treichlerstrasse 10
CH-8032 Zürich, Switzerland

Library of Congress Control Number: 2007924155

ISBN 978-0-387-71977-1 e-ISBN 978-0-387-71978-8

Printed on acid-free paper.

9 8 7 6 5 4 3 2 1

springer.com

This book is dedicated to our sons Cedric, Yannick and Frederic

Contents

Contributing Authors

Marios Avraamides
Department of Psychology, University of Cyprus, Nicosia, Cyprus

Laura Bamert
Department of Psychology, University of Lausanne, Switzerland

Thomas Barkowsky
University of Bremen, Germany

Dominique Boisson
INSERM, Université Claude Bernard, and Hospices Civils de Lyon, Bron; Institut Fédératif des Neurosciences de Lyon; Hôpital Henry Gabrielle, St Genis Laval, France

Bruce Bridgeman
Department of Psychology, University of California, Santa Cruz, USA

Michael Corballis
Department of Psychology, University of Auckland, New Zealand

Sean Fannon
Center for Mind and Brain, University of California, Davis, USA

Vittorio Gallese
Department of Neuroscience, University of Parma, Italy

Reginald Golledge
Department of Geography, University of California, Santa Barbara, USA

Cecilia Guariglia
Department of Psychology, University of Rome "La Sapienza", Italy

Irina Harris
School of Psychology, University of Sydney, Australia

Bernhard Hess
University Hospital Zürich, University of Zürich, Switzerland

Marc Himmelbach
Section Neuropsychology, Center for Neurology, Hertie-Institute for Clinical Brain Research, University of Tübingen

Sophie Jacquin-Courtois
INSERM, Université Claude Bernard, and Hospices Civils de Lyon, Bron; Hôpital Henry Gabrielle, St Genis Laval, France

Lutz Jäncke
Department of Psychology, University of Zürich, Switzerland

Kirsten Jordan
Institute of Medical Psychology, University of Goettingen, Germany

Hans-Otto Karnath
Section Neuropsychology, Center for Neurology, Hertie-Institute for Clinical Brain Research, University of Tübingen, Germany

Roberta Klatzky
Department of Psychology, Carnegie-Mellon University, Pittsburgh, USA

Stephen Kosslyn
Department of Psychology, Harvard University, Cambridge, USA

Simon Lacey
School of Human Sciences, Southampton Solent University, Southampton, UK

Brian Lathrop
Department of Psychology, University of California, Santa Cruz, USA

Yvonne Lippa
Department of Psychology, Miami University, Oxford, USA

Jack Loomis
Department of Psychology, University of California, Santa Barbara, USA

George Mangun
Center for Mind and Brain, University of California, Davis, USA

Fred Mast
Department of Psychology, University of Lausanne, Switzerland

Christoph Maurer
Neurology, Neurocenter, University of Freiburg, Germany

Thomas Mergner
Neurology, Neurocenter, University of Freiburg, Germany

Branka Milivojevic
Department of Psychology, University of Auckland, New Zealand

Nathaniel Newby
Johnson Space Center, Houston, USA

Charles Oman
Man Vehicle Laboratory, Massachusetts Institute of Technology, Cambridge, USA

Laure Pisella
INSERM, Université Claude Bernard, and Hospices Civils de Lyon, Bron, France

Luigi Pizzamiglio
Department of Psychology, University of Rome "La Sapienza", Italy

Delphine Preissmann
Department of Physiology and Institute of Psychology, University of Lausanne, Switzerland

John Rauschecker
Department of Physiology and Biophysics, Georgetown University School of Medicine, Washington, USA

Patrice Revol
INSERM, Université Claude Bernard, and Hospices Civils de Lyon, Bron; Institut Fédératif des Neurosciences de Lyon; Hôpital Henry Gabrielle, St Genis Laval, France

Lynn Robertson
Veterans Administration Medical Research, Martinez, CA; Department of Psychology and Helen Wills Neuroscience Institute, University of California, Berkeley, USA

Gilles Rode
INSERM, Université Claude Bernard, and Hospices Civils de Lyon, Bron; Hôpital Henry Gabrielle, St Genis Laval, France

Yves Rossetti
INSERM, Université Claude Bernard, and Hospices Civils de Lyon, Bron; Institut Fédératif des Neurosciences de Lyon; Hôpital Henry Gabrielle, St Genis Laval, France

Anne Sylvie Sacri
INSERM, Université Claude Bernard, and Hospices Civils de Lyon, Bron, France

K. Sathian
Department of Neurology, Emory University School of Medicine, Atlanta, USA

Chiara Sautter
Department of Physiology and Institute of Psychology, University of Lausanne, Switzerland

Françoise Schenk
Department of Physiology and Institute of Psychology, University of Lausanne, Switzerland

Georg Schweigart
Neurology, Neurocenter, University of Freiburg, Germany

Jennifer Shephard
Department of Psychology, Harvard University, Cambridge, USA

William Thompson
Department of Psychology, Harvard University, Cambridge, USA

Ranxiao Frances Wang
Department of Psychology, University of Illinois at Urbana-Champaign, USA

Preface

"Spatial Processing in Navigation, Imagery and Perception"

Since the decade of the brain cognitive processes have found their way to the study of brain functions and an increasing number of research studies are dealing with the aspect of spatial processing. In fact, a tremendous part of the cognitive domains studied pertain to spatial processing. However, there is also a growing tendency for diversification in relation to the subprocesses underlying spatial processing. Not only are there studies looking at the well known place cells in rats, rabbits and other animals, there is also an increasing number of studies looking at related topics in humans and monkeys such as spatial orientation, spatial construction, and spatial imagery. These studies, although diverse at first glance, have many aspects in common. We are now on the root to understand the underlying neuroanatomy and neurophysiology much better than ever before. This is made possible by the advent of novel techniques such as structural and functional in vivo anatomy, modeling, and several sophisticated behavioral research tools such as virtual reality techniques and simulators.

Spatial processing is fundamental for understanding human cognition. However, compared to other domains such as memory, language, and attention the exploration of spatial functions has been understudied in the past years. Besides the fact that the neural underpinnings of spatial processing are much more complex than they have been conceived before it has turned out that spatial functions have been shown to be involved in almost any cognitive function, even in auditory processing (e.g., music perception). Moreover, a modular organization of cognitive functions is challenged by recent findings showing that cognitive functions are nested and intertwined.

Since the processing of spatial information is so centrally involved in controlling cognitive functions it might help to understand better how basic cognitive functions operate such as language, attention, perception, movement control and mental imagery. Recent research has shown that perception and action are well linked with spatial processing. Visual representations of tools are obviously located in the parietal lobe because they are automatically linked to tool use. Not only the parietal lobe is delineated in a much more precise way, it is now evident that its connections to frontal areas play a major role in spatial processing. Spatial processing is distributed in complex cortical and sub-cortical structures. For example, it also involves sensory information of vestibular origin, the role of which has been widely neglected in previous research using cognitive tasks.

On the one hand modern techniques from the neurosciences have been shown to be the catalyst of this research, there is on the other hand a revival of behavioral approaches. It is in fact the fruitful combination of both why this exciting field has progressed so far and is still progressing for many years to come. The importance of research on spatial processing does not only concern basic researchers it is rather most important for its application in professional areas. It has gotten tremendously important to know how the human brain is accomplishing spatial tasks in real life scenarios such as driving a car, orienting oneself in large scale cities, postural control or playing various sports like baseball, soccer or tennis. Several researchers are devoted to develop strategies to cure people from various diseases or to learn more about how to counteract against the declining spatial functions with age. A promising applied area for research on spatial processing will be the plasticity and training related influences on spatial functions (e.g., stimulation of the parietal cortex can enhance spatial functions to a certain degree). Demands are substantially increasing in our culture with a steadily growing use of computer games, simulator techniques and video oriented teaching tools. Thus, a better understanding of the spatial functions is a necessary prerequisite for efficiently inserting new technologies in everyday life.

Despite the fact that the environment strongly influences spatial behavior more emphasis needs to be given to the genetic underpinnings. This track of research will benefit from tying together genetic screening and brain imaging (genetic brain imaging of spatial functions). Hopefully, this approach may be included in a future edition of this book. This book aims to provide a common platform for researchers from different fields or disciplines studying spatial functions. We were successful in having as chapter authors the most respected and internationally renowned researchers in the field.

The first chapter written by Stephen Kossslyn, Jennifer Shephard and William Thompson is entitled "*Spatial processing during mental imagery: A neurofunctional theory*". Based on the neuroscience of the human visual system the authors propose a model that consists of seven processing subsystems, each of which has a coarsely defined anatomical localization. The model describes key functions of visual perception and – most importantly – it explains how mental images can be generated from memory and how patterns in images are interpreted.

In chapter 2, Cecilia Guariglia and Luigi Pizzamiglio focus on the relation between disorders in mental representation of space and environmental navigation ("*The role of imagery in navigation: Neuropsychological evidence*"). The authors focus on the crucial role mental imagery plays in navigation as it is shown in patients with unilateral neglect who are impaired in their use of cognitive maps.

The nature of spatial images is addressed in the third chapter, entitled "*Functional equivalence of spatial images produced by perception and spatial language*". Jack Loomis, Roberta Klatzky, Marios Avraamides, Yvonne Lippa and Reginald Golledge favor the hypothesis that spatial images are based on amodal representations. They present studies showing that behavioral performance is widely independent of source modality such as vision, spatial hearing or language.

Ranxiao Frances Wang has written a chapter (4), entitled "*Spatial processing and view-dependent representations*". She discusses the relationship between different types of view-dependent representations. Based on previous studies and new data, she proposes a model that contains an egocentric spatial working memory and a representation stored in long-term memory.

Thomas Barkowsky's chapter (5) is concerned with "*Modeling mental spatial knowledge processing*" from an artificial intelligence perspective. After discussion of the existent models of intelligent spatial processing he presents a novel architecture as a framework for modeling spatial reasoning with mental models and mental images.

In chapter 6 entitled "*Optic ataxia: A gateway to the human visual action system*" Marc Himmelbach und Hans-Otto Karnath review the anatomical foundations of optic ataxia allowing the identification of brain areas necessary for the control of hand in space. In addition, they evaluate behavioral findings in patients with optic ataxia and patients with visual form agnosia. On this background, the actual validity of the "two-visual stream-model" is discussed.

Chapter 7 is authored by Bruce Bridgeman and Brian Lathrop: "*Interactions between cognitive space and motor activity*". The authors present data on the relationship between unconscious spatial processing of a spatial frame and subsequent conscious perception and action.

"*Cross-modal involvement of visual cortex in tactile perception*" is the title of chapter 8, written by K. Sathian and Simon Lacey. It is nowadays clear that visual cortical areas are not only involved in visual discrimination tasks but also then when the input modality is tactile. The authors discuss their own studies and those of other groups in the light of different perspectives such as mental imagery or cross-modal plasticity, which could account for the recruitment of visual cortex during tactile perception.

Lutz Jäncke is focusing in chapter 9 (entitled "*Neuroanatomy of the parietal cortex*") on the anatomical underpinnings of parietal functions. Specific emphasis is given to anatomical connections of the parietal lobe with frontal brain areas. In addition, a refined anatomical description of the parietal areas is given.

Lynn Robertson reviews in chapter 10 entitled "*Spatial maps, feature integration and parietal function: Implications from the study of spatial deficits*" how spatial awareness and spatial functions are intermingled with various other psychological functions. In particular she demonstrates how spatial processing is involved in binding surface features such as color and shape and how multiple spatial maps can guide attention.

Michael Corballis, Branka Milivojevic and Irina Harris describe in chapter 11 (entitled "*Pigs in space: How we recognize rotated objects*") how mental rotation functions are part of a general perception process. In particular they argue that mental rotation of objects is a fundamental strategy to recognize familiar visual objects. Their argumentation is based on a review of mental rotation research including behavioral and brain-imaging experiments.

In chapter 12 entitled "*Functional neuroanatomy of mental rotation performance*" Lutz Jäncke and Kirsten Jordan review the current knowledge about the functional neuroanatomy of mental rotation performance. In this chapter the authors also argue about the relationship between mental rotation and other spatial functions, with particular emphasis on the role of different strategies to solve mental rotation tasks and their relation to cortical activation patterns.

In chapter 13 Charles Oman ("*Spatial orientation and navigation in microgravity*") describes the spatial disorientation problems and navigation difficulties astronauts and cosmonauts experience during exposure to microgravity. He relates them to ground-based research findings on human spatial orientation and animal models on navigation.

In chapter 14 entitled "*Spatial representations in the rat: Case study or perspective on episodic memory?*" Françoise Schenk, Delphine Preissmann and Chiara Sautter focus on the rat hippocampus and its role in spatial behavior. They argue that the study of spatial memory in mammals, and more precisely in laboratory rats, sheds light on the development and evolution of other memory systems, in particular on episodic memory.

The brain needs to know head orientation relative to gravity. For this, it must parse the afferent information from the otolith signals into its gravitational and inertial components. The contribution of Bernhard Hess (chapter 15: "*Sensorimotor transformations in spatial orientation relative to gravity*") describes the computational steps necessary to resolve the ambiguous vestibular sensory information, and thus providing reliable spatial orientation and appropriate motor behavior.

The control functions of a hominoid robot capable of performing sensorimotor tasks are described by Thomas Mergner, Christoph Maurer and Georg Schweigart in their chapter (16), entitled "*Sensorimotor control of human dynamic behavior in space implemented into a hominoid robot*". Using a systems approach, their modeling is based on psychophysical research and implements an internal reconstruction of the external physics required for sensorimotor feedback control of coordinated motor behavior. The robot will be used for simulations with the aim to better understand sensorimotor deficits in neurological patients and to develop new clinical therapy designs.

Vittorio Gallese's chapter (chapter 17, entitled "*The ventro-dorsal stream: Parieto premotor neural circuits and their role in primate cognition*") is a demonstration of how primate research fosters the understanding of human brain functions. He reviews anatomical and functional findings suggesting that visual processing is carried out along three distinct streams. Two of them include the parietal lobe, and one of them includes the inferotemporal lobe. These three streams are qualified as dorso-dorsal, ventro-dorsal and ventral streams.

Fred Mast, Laura Bamert and Nathaniel Newby review the most recent research on motor imagery, entitled: "*Mind over matter? Imagined body movements and their neuronal correlates*" (chapter 18). Numerous clinical and neuroimaging studies suggest that many areas involved in the process of motor execution, planning, and preparation are also drawn upon during motor imagery. This concerns not only imagined movements of body parts but also when one imagines a movement of the whole body. The authors present new empirical evidence showing an influence of imagined whole body movements on vestibular perception.

In chapter 19 entitled "*Bottom-up effects of sensory conflict and adaptation on mental imagery: Sensorimotor grounds for high level cognition?*" Gilles Rode, Sophie Jacquin-Courtois, Patrice Revol, Laure Pisella, Anne Sylvie Sacri, Dominique Boisson and Yves Rossetti review the current status on neglect research. They emphasize new findings suggesting an important influence low-level sensorimotor transformations can have on higher cognitive levels of space representation. Thus, they explain why cognitive deficits like neglect may be positively modulated by passive physiological stimulation such as caloric vestibular stimulation or via a prism adaptation procedure.

John Rauschecker has written a chapter (20) entitled "*Cortical processing of auditory space: pathways and plasticity*" in which he describes current knowledge on anatomical and functional auditory spatial processing. He presents the concept of a "where"- and "what"- stream in the auditory systems.

George Mangun and Sean Fannon highlight in their chapter (21) entitled "*Networks for attentional control and selection in spatial vision*" the special role attentional processes play in spatial vision. In particular, they focus on the neural mechanisms underlying voluntary visual spatial attention. They propose specialized neural mechanisms for voluntary spatial attention.

Fred Mast and Lutz Jäncke

Chapter 1

SPATIAL PROCESSING DURING MENTAL IMAGERY: A NEUROFUNCTIONAL THEORY

Stephen M. Kosslyn, Jennifer M. Shephard and William L. Thompson
Department of Psychology, Harvard University, Cambridge, MA, USA

Abstract: Diverse methodologies, from animal research to neuroimaging, have begun to paint a coherent picture of the neural underpinnings of the human visual system. We outline a model that consists of seven coarsely defined processing subsystems. We begin with a discussion of the gating function of attention within a set of retinotopically mapped areas, which we call the visual buffer. This subsystem is implemented in the occipital lobe. Two major pathways lead forward, one going down to the inferior temporal lobe and the other up to the posterior parietal lobe. The functions of these systems are discussed, as are the roles of structures that integrate the two types of information and processes that use information to guide visual search. These same processes are used to generate visual mental images on the basis of stored information and to interpret patterns in images. We summarize a variety of types of evidence that support the putative role of each subsystem, as well as research investigating the anatomical localization of each subsystem.

Key words: vision; mental imagery; top-down processing; spatial processing.

SPATIAL PROCESSING DURING MENTAL IMAGERY: A NEUROFUNCTIONAL THEORY

Visual perception may be the best understood type of information processing in the brain, and thus it is fortunate that perception and mental imagery share many of the same neural systems. In this chapter we argue that visual mental imagery will become one of the best understood cognitive functions, in large part because it draws on mechanisms used in perception. Indeed, Ganis et al. (2004) estimated that the two functions rely on over 90% of the same brain areas.

Why is the study of perception such a "success story"? Consider five reasons. First, perception involves an easily observable stimulus-response relationship (even if the response is simply a report of the qualities of the observed stimulus). Because properties of the stimulus are easily controlled, it is relatively simple to measure and describe the stimulus-response relationship. Second, because several nonhuman species have visual systems similar to our own, the study of animal brains has greatly illuminated the neural mechanisms of perception. Third, the attempt to develop artificial visual systems has led to useful insights into the nature of vision per se (e.g., Marr, 1982). Fourth, methods from cognitive psychology have been used to study patients who have suffered focal brain lesions (e.g., see Farah, 1984). Finally, the advent of neuroimaging methodologies has allowed researchers to study the human brain in action. In the case of visual mental imagery, the confluence of these methods has underscored the parallels between visual imagery and visual perception, while at the same time highlighting the ways in which the two functions differ (Kosslyn et al., 2001a, 2006; Ganis et al., 2004).

The convergence of these methodologies has led to an emerging picture of many of the features of the visual system. We want to build on that understanding here, extending it to visual mental imagery processing. Not all aspects of perceptual processing are used in imagery. Rather, we can conceive of visual perception in terms of two phases. "Early" visual processing relies solely on signals from the eyes; "late" visual processing relies in part on information stored in memory. The fundamental idea underlying our theory is that visual mental imagery, which arises from stored knowledge, relies on many of the same mechanisms as late visual perception (Ganis et al., 2004; Kosslyn et al., 1997, 2001a, 2006). This chapter will provide an overview of the processing mechanisms that implement late visual perception and visual mental imagery.

SUBSYSTEMS OF LATE VISUAL PERCEPTION AND VISUAL MENTAL IMAGERY

The brain relies heavily on parallel processing to function effectively. Different areas of the brain have different processing functions, and various regions work together as systems to accomplish tasks. Studies of monkey neuroanatomy and neurophysiology have revealed at least 32 distinct visual areas in the monkey cortex, and there are probably still more to be discovered. Later visual processing and visual mental imagery can be divided into seven major components, as illustrated in Figure 1-1.

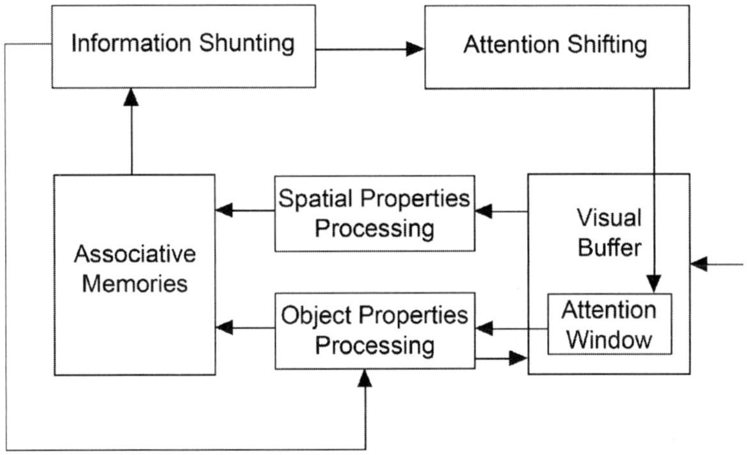

Figure 1-1. Subsystems of late visual processing.

THE VISUAL BUFFER

The neurons in the first visual area (known as V1, primary visual cortex, striate cortex, Area 17, and Area OC) are organized in a way that (roughly) preserves the spatial organization of the cells in the retina. This organizational structure is referred to as retinotopic mapping – the geometric layout of space in the real world is preserved in these first visual regions as it is on the retina. About half of the visual areas in the monkey brain have this organization. These areas may be grouped together as a single functional structure, called the *visual buffer*. In perception, such spatially organized cells detect edges and regions with like qualities (similar colors, textures, patterns, and so on; e.g., see Marr, 1982).

Perhaps the most dramatic evidence of the retinotopic structure of visual cortical regions was provided by Tootell et al. (1982). They trained a monkey to stare at the center of a bullseye-like pattern with spokes radiating from the center, consisting of staggered flashing lights. While the monkey looked at the stimulus, the investigators injected a radioactively tagged sugar (2-deoxyglucose) into its bloodstream. Tootell et al. sacrificed the animal to observe this marker of activity in the animal's brain while it looked at the pattern. The marker showed an image of the pattern essentially projected onto Area V1, the first cortical area to receive signals from the eyes. However, the image was a distortion of the original pattern, with greater cortical area devoted to the high-resolution foveal portion of the visual field.

Particularly strong evidence of retinotopic mapping in humans was initially provided by Fox et al. (1986), who used positron emission tomography (PET) to measure blood flow to Area V1 while participants looked at different types of visual patterns. Neuroimaging studies of visual imagery often document that retinotopically mapped areas are activated when one visualizes (Kosslyn and Thompson, 2003). Indeed, several studies have shown that the focus of activation in human V1 shifts when participants visualize patterns at larger sizes. In one study, for example, Klein et al. (2004) used event-related functional magnetic resonance imaging (ER-fMRI) to show that the pattern of activation in V1 mirrored the orientation of a shape. In this study, the vertical meridian of V1 was activated when participants visualized a flickering "bow-tie" pattern vertically, whereas the horizontal meridian was activated when they visualized the pattern horizontally. Moreover, when transcranial magnetic stimulation (TMS) is used to impair V1 temporarily, imagery for patterns is also impaired (Kosslyn et al., 1999).

However, not all studies have reported such effects during visual imagery. Meta-analyses indicate that activation arises in areas that implement the visual buffer when participants need to see "details" of shapes with high resolution, but not when images of spatial relations are processed (e.g., see Thompson and Kosslyn, 2000; Kosslyn and Thompson, 2003).

THE ATTENTION WINDOW

In perception, the visual buffer receives far more information than can be passed downstream for further processing. Therefore, there must be a mechanism for selecting certain information for further processing while (temporarily at least) ignoring other information. This selective aspect of processing is known as *attention*. The attention window functions the same way in both perception and imagery by selecting a pattern within the visual buffer to be processed in detail. The visual buffer's spatial organization allows the attention window to select information from contiguous locations for more detailed processing. Thus, properties of the attention window constrain the input received by each subsequent processing subsystem.

Sperling (1960) and many other researchers have reported evidence supporting our inference of an adjustable attention window. Much of this research focuses on the covert shifting of attention. For example, participants can detect a target more quickly when its location is cued just before its presentation than when there is no cue (Posner et al., 1980), presumably moving the attention window to the cued location even before they can move their eyes. Only about 30-50 milliseconds are required to shift attention covertly (without moving one's eyes).

Because the attention window functions within the visual buffer, it should be localized with the areas that implement the visual buffer. However, the size and location of the attention window may be dictated by structures elsewhere in the brain, such as the pulvinar nucleus of the thalamus. The pulvinar may gate outputs from the low-level visual areas to higher ones through its reciprocal connections to V4, the inferior temporal lobe (IT), and the posterior parietal lobe. Single cell recordings in rhesus monkeys have provided evidence for the pulvinar's role in delineating the attention window. Pulvinar neurons projecting to cortex "within the attention beam" show increased activity whereas pulvinar neurons projecting to other cortical areas show little or no response. However, the effect was found only in neurons in the dorsomedial pulvinar (projecting to dorsal areas in the posterior parietal cortex), but not in the inferior or lateral pulvinar (projecting to V1 and ventral areas) (Petersen et al., 1985).

During imagery, the attention window may underlie some types of image scanning. That is, numerous studies have shown that participants require more time to scan greater distances across an imaged object. However, the rate of scanning is the same when people scan between two locations that are "visible" initially as when they scan between a "visible" location to one that was "off screen" initially (Kosslyn, 1980). The attention window is confined to the visual buffer, and thus it cannot be shifted "off screen." Rather, the image may be shifted across the visual buffer, much as a picture is shifted across a TV screen when the camera pans across a scene. Nevertheless, the attention window may underlie scanning over images when only short distances must be traversed (Pinker, 1980).

THE VENTRAL AND DORSAL SYSTEMS

Patterns of activity in the visual buffer provide input to two major systems. The ventral system is a set of brain regions running from the occipital lobe through the inferior temporal lobe. The ventral system encodes object properties, such as shape, color, and texture. Some cells in these areas are so highly tuned that they respond to very specific stimuli, such as faces seen in profile. "Higher" cells in the system (i.e., those further down the processing pathway) have very large receptive fields and respond similarly to objects across a wide variety of locations and positions. Critically, the ventral system stores visual memories of shape and color. Some evidence suggests that such details are stored preferentially in the left cerebral hemisphere (Marsolek et al., 2002).

However, by processing information from various locations equivalently, the ventral system sacrifices information necessary for navigation and reaching. A second network of brain areas preserves this information. The dorsal system is a set of brain regions running from the occipital lobe through the parietal lobes. The dorsal system registers spatial properties of objects, such as location and size. The dorsal system also plays a key role in storing spatial information, not simply encoding it. Information is processed in the ventral and dorsal systems concurrently. These systems have sometimes been referred to, respectively, as the *what* and *where* systems (Levine, 1982; Ungerleider and Mishkin, 1982).

Levine (1982) reviews results from studies of brain-damaged patients. Damage to the occipital-temporal region leads patients to have difficulty recognizing and identifying objects, whereas damage to the occipital-parietal region results in difficulty in processing spatial information. PET and fMRI studies provide further support for the distinction between the two visual cortical pathways in humans. PET studies have shown occipital-temporal activation in recognition tasks, and occipital-parietal activation during spatial tasks (see Kosslyn et al., 2004).

Because one of the primary purposes of spatial information is to guide movements, it is not surprising that most neurons in the posterior parietal lobe are active either as a result of making a movement or in anticipation of moving. There is evidence that motor processes are also engaged during imagery, particularly during specific types of mental rotation. For instance, Kosslyn et al. (2001b) showed that motor cortex was activated when participants imagined mentally rotating figures manually but not when they imagined the figures being rotated by an external force. In addition, Wraga et al. (2003) asked one group of participants to mentally rotate Shepard and Metzler (1971) objects in two conditions and another group to mentally rotate pictures of hands in the first condition, and Shepard-Metzler objects in the second condition. When the second (i.e., object) conditions for each group were compared, motor areas were found to be activated only in the group that first imagined hand rotations, suggesting that these participants transferred a motor strategy to a non-motor rotation task. Moreover, Wraga et al. (2005) suggest that different motor areas are activated for imagined self versus object rotations.

In contrast, the ventral system uses object properties, including shape, color, and texture, to match input to information stored in visual memory. Such processing attempts to make the best match between a stored object and current input; if successful, the viewed object is recognized.

At this point, it is important to draw a distinction between *recognition* and *identification*. If visual input matches a representation in visual memory, one will know that the object is familiar, that is, will recognize it. However, identification occurs only when the input goes on to access multimodal, conceptual information about the object; one identifies an object when one can access a full range of stored explicit information about it, such as its name, its preferred environment, its sounds, smells, and so on. Identification involves knowing more about the object than can be discovered from immediate input from the senses. The ventral system does not contain the structures and mechanisms needed for identification; rather it affords only recognition, by matching input with stored representations.

Turning to imagery, a critical fact is that visual memories are stored in the temporal lobe (in the ventral system) via a population code; they are not stored as topographic images. A second crucial fact is that there are rich connections running backwards from the inferior temporal lobe to the areas that comprise the visual buffer. Thus, according to our theory, when an image is recalled the information stored in the ventral system is—at least in some conditions—"unpacked" and made explicit as a depictive image in early visual cortex (in the visual buffer). This reconstructed pattern can then be reinterpreted, it can be "inspected" much as an actual object can be inspected during perception. In both imagery and perception, input from the attention window is compared to stored visual memories, and spatial properties are registered.

ASSOCIATIVE MEMORIES

Associative memory receives input from all of the modality specific mechanisms that allow us to recognize objects. For example, we can recognize a dog by seeing it, hearing its bark, or petting its fur; and once we recognize it, using different "what" systems, we can identify it, using *long-term associative memory*. Outputs from the ventral and dorsal pathways converge in long-term associative memory. The mere fact that you can "picture" where certain items are in your kitchen cabinets is evidence that object representations are cross-indexed with spatial representations. Moreover, whenever we use landmarks to navigate, we are associating shape information with location information, and learning the layout of a building or the route to work requires storing such associations.

At the same time, the outputs from the ventral and dorsal systems converge in dorsolateral prefrontal cortex (DLPFC), in an area that appears to serve as a kind of short-term associative memory (e.g., see Wilson et al., 1993; Rao et al., 1997). Approximately half the neurons in this area conjoin shapes with specific locations, and keep this information "on line". Such representations are important for guiding eye movements.

Long-term associative memory not only cross-indexes information from the different perceptual pathways (and multiple sensory modalities), but also organizes concepts that do not arise directly from the senses (mathematical truths, meanings of abstract nouns such as "love", categorical classifications, etc.). One type of associative memory representation researchers have hypothesized is called a "structural description", which specifies how parts are arranged to form an object. Inputs from both the ventral and dorsal systems are used to build up such representations. Computer vision researchers, especially, have argued that such representations are important for object identification, in large part because such a description will apply equally well to the various configurations of a flexible shape. For example, when one sees a gymnast performing a floor routine, the same structural description of a human form will apply to all her various contortions. In order to identify an object, a structural description is constructed on the basis of information from the dorsal and ventral pathways, and then compared with stored structural descriptions.

To construct a structural description, one must be able not only to recognize all the individual parts of an object, but also be able to compute relatively abstract spatial relations among those parts. The dorsal system can compute two distinct types of spatial relations representations. One *coordinate* type preserves metric information, and is useful for guiding movement; the other type discards metric information, and instead specifies only the broad category of the spatial relation (such as "attached to," "above," "left of"). Structural descriptions apparently rely on this second, *categorical*, type of spatial relation representation. Kosslyn et al. (2005) provide evidence that these two types of processing, categorical and coordinate, may be implemented in different hemispheres. They asked participants to generate mental images of patterns that they had learned in one of two ways. In one (categorical) condition, participants memorized verbal descriptions of how an object's parts were connected; in the other (coordinate) condition, participants viewed the object's parts sequentially, in specific locations, and had to mentally attach the pieces together. Comparison of the two imagery conditions using PET showed relatively more activation in the left hemisphere during the categorical condition and relatively more activation in the right hemisphere during the coordinate condition (although a majority of brain areas were activated in common across conditions, suggesting that both types of imagery also relied on the other subsystems described here).

Long-term associative memory plays two important roles in visual imagery. First, it stores descriptions of how parts are arranged into shapes. Such descriptions are used when one needs to construct a high-resolution image. Second, long-term associative memory stores "visual codes." These codes are associated with all the other information about an object, and can be used to access specific stored memories in the ventral system. Thus, when given an object's name, the corresponding visual code is activated, which in turn can access the appropriate visual memory.

Parts of the superior, posterior temporal lobes and temporal-parietal-occipital junction area (near the junction of the angular gyrus and Brodmann's Area 19; see Kosslyn et al., 1995) may be the seat of long-term associative memory. The first sources of evidence that these areas are crucial for long-term associative memory are studies of area STP (short for superior temporal polysensory) in the monkey. This area is in the posterior superior temporal lobe of the monkey. Long-term associative memory requires input from visual, auditory, and somesthetic systems, and cells in monkey STP have such connections (from IT, superior temporal auditory cortex, and from posterior parietal cortex). Furthermore, more than half the neurons in STP respond to input from multiple modalities, and studies have shown STP to be implicated in shifting attention. STP's role in associative memory is further supported by its connections from AIT (the anterior part of IT) and Area 7a (in the parietal lobe, via the hippocampus), which may facilitate coordination of activity in the two visual pathways. Finally, STP has connections to the frontal lobes; the importance of connections between associative memory and the frontal lobes will be discussed shortly.

Although STP is a bilateral area of the monkey brain, and is not found in the human brain, it may be related to what has classically been called Wernicke's area (in the left posterior, superior temporal lobe of the human brain); this area implements processes used in language comprehension. In addition, nearby cortex in Area 19 and the angular gyrus may also play a role in implementing long-term associative memory in humans. Research with brain-damaged patients also provides support for the role of the posterior, superior temporal lobe or nearby cortex in implementing human long-term associative memory. For example, Warrington and Taylor (1978) and others have found evidence of long-term associative memory deficits in patients with left posterior lesions. For a more extensive review of the brain lesion literature with respect to visual imagery and perception, see Ganis et al. (2003).

INFORMATION SHUNTING

During perception, activated visual and spatial memories (from the ventral and dorsal systems, respectively) often are sufficient to identify an object. However, this is not always the case. For instance, if one glances briefly at an up-ended card table with folded legs, leaning against a door, there may not be a good match in visual memory. It is likely that only certain parts or characteristics could be matched. In this case more information is needed.

If the input to long-term associative memory does not result in object identification, one does not randomly look around to acquire more information. Certain properties of the object, in particular motion or intensity changes, can serve to redirect our attention. In addition, one can use the stored information that is partially activated in long-term associative memory to guide a search. We treat the activated information as a hypothesis, and use this hypothesis to guide further processing by seeking information that has direct bearing on the hypothesis. Part of this process involves priming the expected distinctive property or shape in the ventral system. Indeed, there are direct connections between the regions of the frontal lobes that are the likely site of the information shunting system and the inferior temporal lobes, allowing rapid transmission of information from stored memories, as needed (Naya et al., 2001). The use of stored information to direct further encoding is called *top-down processing*.

According to the present theory, visual mental images arise when stored visual memories, in the ventral system, are activated. At least in some cases, these memories are activated so strongly that activation propagates backwards from the temporal lobes and an image representation is formed in the areas that compose the visual buffer. At the same time, an "object map" is constructed in the dorsal system. This object map is a spatial representation that indicates where specific details and parts belong on an object (or objects belong in a scene). At least in some tasks, this spatial representation alone is sufficient and an image of high-resolution details need not be reconstructed in the visual buffer. The representations in the visual buffer are spatially indexed to the object map. Thus, scanning images "off screen" (to parts that were not initially "visible" in the image) is accomplished by activating adjacent parts of the object map, in the dorsal system, which — if the task requires interpreting high-resolution details — in turn modulates activation in the ventral system so that images of different portions of the stimulus are activated in the visual buffer.

The DLPFC plays a crucial role in hypothesis generation and testing. Patients with damage to this area often show "perseveration" — they have difficulty stopping an on-going activity and initiating a new one (which may reveal a problem in accessing stored information, both about when to stop and what to do next). In addition, PET studies have shown activation in the DLPFC when participants are asked to seek specific information in memory (e.g., Petersen et al., 1988). Kosslyn et al. (1993) and Kosslyn et al. (1997) also found left-hemisphere activation of DLPFC in an imagery task where participants were asked to decide whether an "X" mark would or would not cover a letter in a grid if the letter (which they were cued to visualize) were in fact present. This task is likely to have required categorical spatial relations processing because the participants used the grid as their guide to place the individual segments of the letters. In this context, it makes sense that left-hemisphere DLPFC would be activated (see also Laeng et al., 2003). It is also of interest that some patients with frontal lobe damage do not perform targeted eye movements, which suggests that they have a deficit in directing top-down search. In addition, PET results have provided evidence for the notion that the information shunting system and long-term associative memory work together (for example, see Kosslyn et al., 1995).

ATTENTION SHIFTING

When testing a hypothesis, one must seek additional information. When the visual system uses top-down processing during perception to seek characteristics of a hypothesized object, information about the size and location of the object must be considered. The attention-shifting system is loosely defined as the set of mechanisms responsible for computing the direction and degrees that attention must be shifted, and for actually sending the commands to make the appropriate eye, head, and attention window movements. Posner et al. (1987) decomposed this system into three subsystems; one disengages attention from its current fixation; another shifts attention; and a third engages attention at the new location.

According to our theory, this system operates similarly in perception and imagery, except that during visual mental imagery it is sufficient to move the attention window in order to shift attention (the eyes and head are not required to move). In addition, at the same time that the attention window is shifted to the location of an important part of the imaged object, the "visual code" of the object (stored in long-term associative memory, along with the locations of important parts and characteristics) is activated. This code in turn primes the part or characteristic as seen in perception – and primes it so strongly during imagery that a pattern of activation is induced in the visual buffer.

During imagery the attention shifting subsystem may help to create the object map representation. Outputs from the pulvinar may send signals to the posterior parietal cortex, setting the regions of space to be attended to. This pointing of attention may be thought of as the general layout of the image to which more complete details may later be filled in; thus, attention allocation may be considered a crucial step in the formation of a high-resolution visual image. These pointed-to positions may then be stored within posterior parietal cortex (see Sereno et al., 2001). The object map may permit the real world to be used as a type of external store (see O'Regan and Nöe, 2001). Knowing where an object is located in world space, we can return for another look. Kosslyn et al. (1993), using PET, showed that the pulvinar is more activated during an imagery task than a corresponding perception task (Experiment 1). This makes sense; attention allocation may play a primary role in the formation of the general layout of a mental image. In perception, with the object's boundaries already defined, there would be no need for the pulvinar to be activated in setting attention to the locations of space delineating the layout of the object in space.

CONCLUSIONS

To summarize: The image-formation process can be used iteratively, to construct very detailed images. To do so, all of the subsystems must work together.

For example, we've several times discussed a methodology in which participants were asked to form images of block letters within an empty grid (a task that features both spatial and depictive components and requires attentional allocation). An X was presented in one cell of the grid, and the participants were asked whether the X would cover part of the letter if the letter were in fact present. No matter where the X was presented, participants required comparable amounts of time to make this decision if they were allowed to form the image fully before the X appeared (Kosslyn et al., 1988). However, if the X was presented at the outset, the participants required different amounts of time to evaluate whether it would have fallen on the letter, depending on the location of the X. In fact, the times were predicted by the order in which the segments of the letter are typically drawn; the more segments that had to be drawn to reach the location of the X, the more time participants required to evaluate it.

This finding is as expected if the letters were visualized a segment at a time, using a description of how they are arranged to direct attention to the location of each successive segment – at which point an image of the segment would be generated. According to our theory, the description is stored in long-term associative memory, and the information shunting system looks up this description. The attention shifting subsystem would then shift attention to each successive location while the information shunting subsystem sends a visual code to the ventral system. The ventral system in turn would create a pattern of activation in the visual buffer to depict each segment. To test this theory, Kosslyn et al. (1997) asked participants to perform this task while their brains were scanned, using PET. They found activation in brain areas corresponding to each of the subsystems described above, as predicted (see Thompson and Kosslyn, 2000, Kosslyn and Thompson, 2003, for meta-analyses providing further evidence for the general theory).

Late vision and visual mental imagery are complex capabilities, and the brain has adopted a strategy of "divide and conquer" for such functions. Rather than trying to accomplish a complex process in a single step, multiple subsystems each carry out specialized aspects of the processing, and the subsystems work together to accomplish the entire task. Although we do not know the specific workings of each component subsystem, nor have the neural substrates been completely delineated, a varied set of converging evidence suggests that it is useful to organize late level vision into seven distinct subsystems. These subsystems play a role both in visual perception and visual mental imagery. If nothing else, this emerging framework allows us to focus on more precise and detailed questions than was possible previously.

ACKNOWLEDGEMENTS

This material is based partially upon work supported by the National Science Foundation under Grant No. REC-0411725 and the National Institutes of Health under Grant No. 2 R 01 MH60734. Any opinions, findings, and conclusions or recommendations expressed in this material are those of the authors and do not necessarily reflect the views of the National Science Foundation of National Institutes of Health.

REFERENCES

Farah, M. J., 1984, The neurological basis of mental imagery: A componential analysis, *Cognition* **18**:245-272.

Fox, P. T., Mintun, M. A., Raichle, M. E., Miezin, F. M., Allman, J. M., and Van Essen, D. C., 1986, Mapping human visual cortex with positron emission tomography, *Nature* **323**:806-809.

Ganis, G., Thompson, W. L., and Kosslyn, S. M., 2004, Brain areas underlying visual mental imagery and visual perception: An fMRI study, *Cognitive Brain Res* **20**:226-241.

Ganis, G., Thompson, W. L., Mast, F. W., and Kosslyn, S. M., 2003, Visual imagery in cerebral visual dysfunction, *Neurol Clin* **21**:631-646.

Klein, I., Dubois, J., Mangin, J-F., Kherif, F., Flandin, G., Poline, J-B., Denis, M., Kosslyn, S. M., and Le Bihan, D., 2004, Retinotopic organization of visual mental images as revealed by functional Magnetic Resonance Imaging, *Cognitive Brain Res* **22**:26-31.

Kosslyn, S. M., 1980, *Image and mind*, Harvard University Press, Cambridge.

Kosslyn, S. M., Alpert, N. M., and Thompson, W. L., 1995, Identifying objects at different levels of hierarchy: A positron emission tomography study, *Hum Brain Mapp* **3**:107-132.

Kosslyn, S. M., Alpert, N. M., Thompson, W. L., Maljkovic, V., Weise, S. B., Chabris, C. F., Hamilton, S. E., and Buonanno, F. S., 1993, Visual mental imagery activates topographically organized visual cortex: PET investigations, *J Cognitive Neurosci* **5**: 263-287.

Kosslyn, S. M., Cave, C. B., Provost, D. A., and von Gierke, S. M., 1988, Sequential processes in image generation, *Cognitive Psychol* **20**:319-343.

Kosslyn, S. M., Ganis, G., and Thompson, W. L., 2001a, Neural foundations of imagery, *Nat Rev Neurosci* **2**(9):635-642.

Kosslyn, S. M., Pascual-Leone, A., Felician, O., Camposano, S., Keenan, J. P., Thompson, W. L., Ganis, G., Sukel, K. E., and Allport, N. M., 1999, The role of area 17 in visual imagery: Convergent evidence from PET and rTMS, *Science* **284**: 267-170.

Kosslyn, S. M., and Thompson, W. L., 2003, When is early visual cortex activated during visual mental imagery?, *Psychol Bull* **129**:723-746.

Kosslyn, S. M., Thompson, W. L., and Alpert, N.M., 1997, Neural systems shared by visual imagery and visual perception: A positron emission tomography study, *Neuroimage* **6**: 320-334.

Kosslyn, S. M., Thompson, W. L., and Ganis, G., 2006, *The case for mental imagery*. Oxford University Press, New York.

Kosslyn, S. M., Thompson, W. L., Shephard, J. M., Ganis, G., Bell, D., Danovitch, J., Wittenberg, L. A., and Alpert, N. M., 2004, Brain rCBF and performance in visual imagery tasks: Common and distinct processes, *Eur J Cogn Psychol* **16**:696-716.

Kosslyn, S. M., Thompson, W. L., Sukel, K. E., and Alpert, N. M., 2005, Two types of image generation: Evidence from PET, *Cogn Affect Behav Neurosci* **5**:41-53.

Kosslyn, S. M., Thompson, W. L., Wraga, M., and Alpert, N. M., 2001b, Imagining rotation by endogenous and exogenous forces: Distinct neural mechanisms for different strategies, *Neuroreport* **12**:2519-2525.

Laeng, B., Chabris, C. F., and Kosslyn, S. M., 2003, Asymmetries in encoding spatial relations, in: *The Asymmetrical Brain*, R. Davidson and K. Hugdahl, eds., MIT Press, Cambridge, pp. 303-339.

Levine, D. N., 1982, Visual agnosia in monkey and man, in: *Analysis of visual behavior,* D. J. Ingle, M. A. Goodale, and R. J. W. Mansfield, eds., MIT Press, Cambridge, pp. 629-670.

Marr, D., 1982, *Vision: A computational investigation into the human representation and processing of visual information*. W. H. Freeman, New York.

Marsolek, C. J., Nicholas, C. D., and Andersen, D. R., 2002, Interhemispheric communication of abstract and specific visual-form information, *Neuropsychologia* **40**:1983-1999.

Mishkin, M., and Ungerleider, L. G., 1982, Contribution of striate inputs to the visuospatial functions of parieto-preoccipital cortex in monkeys, *Behavioural Brain Research* **6**:57-77.

Naya, Y., Yoshida, M., and Miyashita, Y., 2001, Backward spreading of memory-retrieval signal in the primate temporal cortex, *Science* **291**:661-664.

O'Regan, J. K., and Noë, A., 2001, A sensorimotor account of vision and visual consciousness, *Behav Brain Sci* **24**:939-1011.

Pinker, S., 1980, Mental imagery and the third dimension, *J Exp Psychol Gen* **109**: 354-371.

Petersen, S. E., Fox, P. T., Posner, M. I., Mintun, M., and Raichle, M. E., 1988, Positron emission tomographic studies of the cortical anatomy of single-word processing, *Nature* **331**:585-589.

Petersen, S. E., Robinson, D. L., and Keys, W., 1985, Pulvinar nuclei of the behaving rhesus monkey; visual resonses and their modulation, *J Neurophysiol* **54**: 867-886.

Posner, M. I., Snyder, C. R., and Davidson, B. J., 1980, Attention and the detection of signals, *J Exp Psychol Gen* **109**: 160-174.

Posner, M. I., Inhoff, A. W., Friedrich, F. J., and Cohen, A., 1987, Isolating attentional systems: A cognitive-anatomical analysis, *Psychobiology* **15**:107-121.

Rao, S. C., Rainer, G., and Miller, E. K., 1997, Integration of what and where in the primate prefrontal cortex, *Science* **276**: 821-824.

Sereno, M. I., Pitzalis, S. and Martinez, A., 2001, Mapping of contralateral space in retinotopic coordinates by a parietal cortical area in humans, *Science* **294**:1350-1354.

Shepard, R. N., and Metzler, J., 1971, Mental rotation of three-dimensional objects, *Science* **171**:701-703.

Sperling, G., 1960, The information available in brief visual presentation, *Psychol Monogr General and Applied* **74**:1-29.

Thompson, W. L., and Kosslyn, S. M., 2000, Neural systems activated during visual mental imagery: A review and meta-analyses, in: *Brain Mapping: The Systems*, A. W. Toga and J. C. Mazziotta, eds., San Diego, Academic Press, pp. 535-560.

Tootell, R. B., Silverman, M. S., Switkes, E., and De Valois, R. L., 1982, Deoxyglucose analysis of retinotopic organization in primate striate cortex, *Science* **218**:902-904.

Ungerleider, L. G., and Mishkin, M., 1982, Two cortical visual systems, in: *Analysis of Visual Behavior*, D. J. Ingle, M. A. Goodale, and R. J. W. Mansfield, eds., MIT Press, Cambridge, pp. 549-586.

Warrington, E. K., and Taylor, A. M., 1978, Two categorical stages of object recognition, *Perception* **7**:695-705.

Wilson, F. A. W., O'Scalaidhe, S. P., and Goldman-Rakic, P. S., 1993, Dissociation of object and spatial processing domains in primate prefrontal cortex, *Science* **260**:1955-1958.

Wraga, M., Shephard, J. M., Church, J. A., Inati, S., and Kosslyn, S. M., 2005, Imagined rotations of self versus objects: An fMRI study, *Neuropsychologia* **43**:1351-1361.

Wraga, M., Thompson, W. L., Alpert, N. M., and Kosslyn, S. M., 2003, Implicit transfer of motor strategies in mental rotation, *Brain Cognition* **52**:135-143.

Chapter 2

THE ROLE OF IMAGERY IN NAVIGATION: NEUROPSYCHOLOGICAL EVIDENCE

Cecilia Guariglia and Luigi Pizzamiglio
Dipartimento di Psicologia, Università di Roma "La Sapienza", Italy

Abstract: In this chapter a brief review of studies analyzing the relation between disorders in mental representation of space and environmental navigation is reported. Most of the studies concern the role on navigation of unilateral neglect, that is the inability to represent and to attend to the contralesional side of the space, that often follows lesion in posterior regions of the right hemisphere. Different studies demonstrate that unilateral neglect does not affect the ability to use some basic navigational processes such as path integration, but it affects the ability to develop and use cognitive maps of the environment for navigation. A case is also described of a patient who never developed navigational skills due to a congenital brain malformation. The only remarkable deficits the patients presented concerned mental imagery, supporting the hypothesis that mental imagery plays a crusial role in navigation.

Key words: unilateral neglect; representational neglect; topographical disorientation; mental imagery disorders.

Navigation in the environment requires planning a trajectory to follow and using mental maps. However, up until now the role of mental imagery disorders in determining neuropsychological deficits in navigation has not been a popular research topic. Indeed, it is quite surprising that neuropsychological deficits of visuo-spatial imagery have hardly been investigated in the study of human navigation. In this context, the study of one disorder in particular, namely, spatial neglect, seems very promising to provide a better understanding of the relationship between mental representation and navigation. Spatial neglect is a unilateral disorder of space representation affecting the contralesional side of space that is usually produced by lesions of posterior areas of the right hemisphere (Bisiach,

1999). The inability to process contralesional stimuli may involve different visuo-spatial processing, from the perception and handling of visual, auditory and tactile stimuli to the mental representation of objects and scenes.

In particular, in daily life activities or during formal neuropsychological testing right brain damaged patients affected by neglect are unable to detect stimuli on the contralesional, left side of space. Patients may fail to answer the examiner if he/she is seated on their left side and may even fail to detect the examiner's presence. They eat only from the right side of a dish, bump into left-sided obstacles when walking, read only the right columns in the newspaper and sometimes read only the rightmost letters in a word.

During formal testing, they fail to cross out left-sided stimuli such as letters, circles, lines or bells, bisect horizontal lines with consistent leftward errors and are not subject to leftward optical illusions. In some patients, neglect affects personal space; thus, these patients do not comb their hair on the left, do not shave or make up their left cheek, do not pull on their left pant leg or the left arm of shirts and jackets. In some cases, mental imagery may also be affected. For example, patients may omit the left side of drawings from memory and may fail to describe the left side of familiar places from memory such as public squares, their own office or apartment. Neuropsychological tests assess representational neglect by asking patients to compare pairs of mental images that may differ on the left or on the right side (Slit test, Bisiach et al., 1979; Ogden, 1985) or asking them to imagine two analogical clocks showing two different times on the left or on the right side in order to judge which angle formed by the clock-hands is wider (O'Clock Test, Grossi et al., 1989).

It is well known that patients affected by unilateral neglect may show topographical disorientation (De Renzi, 1982). Indeed, patients who are unable to perceive landmarks on the left, to perform in the left hemispace, to measure lengths on the left side or to compare the length of leftward and rightward stimuli may be impaired in navigating in the environment. However, it is important to note that navigational impairment may not be present in some patients with representational disorders (see, for example, Passini et al., 2000). Due to the extensive range of disorders that comprise the unilateral neglect syndrome, it is plausible to hypothesize that topographical disorientation appears only when neglect affects some specific aspects of space representation. Identifying which specific neglect disorders affect navigational skills would help us to understand navigational processes on one side and the organization of space processing on the other.

To investigate what links specific neglect deficits to specific navigational processes, in the following paragraphs we will first analyze deficits in representing familiar places, which can be considered impairments in mentally mapping the environment, and then deficits in different navigational tasks.

Patients have been reported who neglect the left side of visual images of familiar places (Bisiach and Luzzatti, 1978; Guariglia, 1993; Coslett, 1997; etc) and maps (Rode et al., Bisiach et al., 1993, etc); however, very few attempts have been made to investigate their ability to describe pathways in detail. Bisiach et al. (1993) asked two patients affected by unilateral neglect to describe well-known pathways in their hometown. In both cases, the descriptions included complex detours to avoid left turns, and when the leftward turns could not be avoided the patients failed to reach the goal and were lost.

Some reports suggest that imagery deficits refer specifically to a representational system devoted to constructing mental maps of the environment for navigation. In these cases, in fact, there is an amazing dissociation between a full representation of visual events (objects, faces, written material, etc.) and a defective representation of the left side of the environment.

Guariglia et al. (1993) reported a patient with a fronto-temporal lesion who showed severe and persistent imagery neglect when required to describe familiar public squares from memory but not when required to process mental images of an object. Ortigue et al. (2003) reported a patient with a right temporo-occipital junction lesion who was unable to describe the left side of Place Neuve in Geneva from memory or to describe the left side of an imaged map of France. However, she was able to provide a fully detailed description of the left side of the interior of her car from memory and the left side of an array of objects she had just explored on a table a few minutes before. The authors interpreted these results as demonstrating that perceptual and representational neglect are supported by independent cortical systems and that inferior temporal areas might be important for the mental representation of far space when a viewer-centered reference is imposed. Alternatively, in this case it can be hypothesized that representational neglect affects a specific representational system devoted to constructing cognitive maps of environmental space; such cognitive maps, whose frame of reference is egocentric, should be used to drive the subject's navigation.

A recent fMRI study (Committeri et al., 2005) showed the segregation of the neural substrates involved in perceiving the spatial relationship between different items and the relationship of the same items to an environmental frame of reference.

Rode et al. (2004) described a patient with representational neglect who, when asked to describe France as it would appear from Marseille, was unable to report the cities on the left. However, when asked to name as many cities as possible, the patient reported the same number of cities on the right and left sides of France. Here also, this can be interpreted as demonstrating damage to a system representing space in egocentric co-ordinates for navigation.

Further support comes from Pizzamiglio et al.'s (1996) study of a double dissociation between perceptual and representational neglect. Their results strongly support the notion that the presence of representational neglect, not that of even very severe perceptual neglect, affects the ability to construct environmental cognitive maps. In this study, the ability of a patient affected by representational, but not perceptual, neglect (MC) to explore and mentally represent a novel environment was compared to that of a patient affected by severe and persistent perceptual neglect without any sign of representational neglect (BM). Both patients were brought into a room they had never seen before, placed in the centre and asked to describe the four walls in detail. The number of elements reported on each side of the room was recorded. Soon after, the patients were brought into a different room where they spent one hour performing verbal neuropsychological tests that did not tax either memory or visuo-spatial abilities. At the end of this time, the patients were asked to imagine entering the first room, standing at the door and describing all the objects they saw; a description from memory from the opposite vantage point was also requested. The number of elements reported on each side of the mental image was recorded.

When required to visually explore and describe the room from four different vantage points, MC correctly reported all features (furniture, objects, windows, etc.) in detail. However, after the one-hour interval he was unable to describe the contralateral side of the room from memory but correctly described the ipsilesional one. This demonstrates his failure to construct and store a mental map of the environment.

On the other side, BM described only the ipsilesional side of the same room while visually inspecting it. When required to describe it from a given vantage point from memory, he correctly reported both ipsilesional and contralesional features, thus demonstrating he had processed a complete mental map of the environment. These observations suggest that representational neglect involves damage to an imagery system devoted to processing environmental information for the construction of cognitive maps for navigation; in the case of perceptual neglect without representational neglect, this system is unaffected. Therefore, the system that guides visual exploration and directs visuo-spatial manipulation does not prevent tying up the partially perceived environmental elements into a correct cognitive map by means of the system representing space for navigation.

From the above-reported observations, it can be hypothesized that two different types of space representation exist that may be selectively affected by neglect and may be classified as topographical images and non-topographical images. Topographical images are the mental representations of stimuli such as rooms, cities, public squares, etc., the subject can navigate in and they can be transformed into (or correspond to) mental maps of the environment. Non-topographical images are mental representations of stimuli such as a desktop, the interior of a car (Ortigue et al., 2003), single objects or arrays of objects that can be manipulated but can never be navigated. Topographical and non-topographical mental images may share some common mechanisms but, based on the above-described dissociations, they are essentially generated by different processes.

Additional support for the existence of separate processes subserving the generation of topographical and non-topographical visual images comes from a very recent study investigating the presence and the nature of imagery deficits in neglect (Guariglia et al., in preparation). In this study, the incidence of imagery disorders in a sample of 96 right brain damaged patients is 35.42%; 14.43% show only representational neglect in the absence of perceptual neglect. Twenty-five patients (that is, 70.59% of the patients with mental imagery impairments) show selective impairment in generating the left side of topographical mental images without any asymmetries in processing the left side of non-topographical images.

Very few studies have looked specifically for a link between disorders in representing visuo-spatial information and navigation. Bisiach and coworkers (1997) made the first attempt by submitting neglect patients and controls to an easy navigational task. Blindfolded subjects were passively moved through short paths with two or three 90° leftward or rightward turns. The subjects' task was to indicate the point of departure. Although the subjects made a broad range of errors, the neglect patients' performances did not differ from those of the controls.

Philbeck et al. (2001) obtained similar results in a group of 6 patients affected by unilateral neglect. These subjects were required to update the remembered location of a target during passive rotations of the entire body. The patients sat on a swivel chair on which a manual pointing device was mounted at the level of their median sagittal plane. The target was a flashing light located 25° or 65° to the left or to the right of the patient's initial body midline. Passive clockwise or counter-clockwise rotations ranging from 25° to 125° (25° increments) were used in the presence or absence of visual control. After the patients saw a lit target, they were passively rotated and then had to set the pointer in the centre of the previously seen target. In the absence of a visual control condition, the patients were blindfolded for both rotations and pointing.

In both visual control conditions, the neglect patients updated the target location equally well on either side of the body midline even though they generally underestimated the rotations.

At variance with these observations of intact navigational skills in neglect are Pizzamiglio et al.'s (1998) findings in a re-orientation task. Right brain damaged patients with and without neglect and healthy controls had to point to a previously seen target in a rectangular room in two different conditions. In the first condition, the walls were completely covered by homogenous curtains in order to mask any environmental cues. In this condition, the patients had to re-orient themselves by relying solely on the rectangular shape of the environment. In the second condition, a wall was covered with a red panel to introduce a salient environmental cue that could be used as reference for re-orienting. In the no visual cue condition, the normal controls and the brain damaged patients without neglect pointed with equal frequency to the corner where the target was located or to the diagonally opposite corner. In the same task, the neglect patients performed completely at random, that is, they pointed indiscriminately to all four corners of the room. These results suggest that neglect patients are completely unable to use the geometric information in the environment to guide their spatial exploration.

In the second condition, the presence of a landmark facilitated the responses of controls and non-neglect brain damaged patients who identified the correct location of the target at ceiling. The neglect patients' performances were improved but they still made a significant number of errors. Therefore, in this condition the latter group was impaired in processing geometric information as well as in integrating visual landmarks with the shape of the environment in order to reorient.

In neither condition was a difference found related to the target position (left or right of the subject's starting position). This suggests that neglect affects the possibility of representing environments in toto and not just in the contralesional hemispace.

The ability to measure the spatial linear translation of the entire body was investigated in a subsequent study (Pizzamiglio et al., 2003). Right brain damaged patients with and without neglect and matched controls without any neurological or psychiatric impairment sat on a robotized wheelchair that was linearly moved forward, leftward or rightward in a rectangular room stripped of all visual cues (landmarks). The subject's task was to reproduce the distance of the passive translation in the same or in a different direction by actively driving the robot with a joystick. The neglect patients' performances did not differ from those of the right brain damaged patients or the controls in any condition or direction. Thus, the conclusion may be drawn that the deficit in computing distances in contralesional space, which is typical of neglect patients in desktop tasks, does not involve the vestibular, motor and proprioceptive computation of spatial translations. In a second experiment, some visual cues were introduced that could be used to more accurately measure the extent of the passive displacement. In this condition, the neglect patients' performances differed significantly from those of both controls and brain-damaged patients without neglect.

Overall, the above-reported data suggest that neglect does not affect the simple computation of distances and angles in navigation when visual information is unavailable or unnecessary for the task. However, when visual information has to be taken in account to solve the navigational task, as in Pizzamiglio et al.'s (1998; 2003 - experiment 2) studies, the presence of neglect strongly affects performances by interfering with the construction of a mental map of the environment. In other words, it seems that neglect only interferes with the development of cognitive maps of the environment that tie up metric information about the shape and the dimension of environments with the presence and the relative position of landmarks.

Thus, the question is what happens in more complex navigational tasks that require the development of cognitive maps even in absence of landmarks, or when more complex navigational processes such as path integration or re-orientation have to be activated.

Guariglia et al. (2005; Guariglia et al., in preparation) developed a human analogue of the Morris water maze to test the use of different navigational processes in neglect. The first study (Guariglia et al., 2005) investigated path integration and re-orientation. To eliminate all environmental cues, the walls of a rectangular room (5 x 6m) were completely covered with homogeneous grey curtains. A photocell was mounted on the ceiling and directed toward a target location (TL) on the floor, and whenever a subject passed through the TL an acoustic signal was delivered. The subjects were brought into the centre of the room blindfolded. The curtain covering the door was closed and the blindfold was removed. The subjects' task was to explore the room to find the TL and then to memorize its location. Subsequently, they were blindfolded, disoriented and again placed in the centre of the room facing the same or a different wall. The blindfold was removed and the subjects had to reach the TL in the shortest and quickest way possible. After six trials, the blindfolded subjects were taken out of the experimental room; 30 minutes later they were brought back into the room; the blindfold was removed and they had to reach the TL. Three manipulations were introduced:

1. The subjects were placed in the centre of the room facing the same wall as in the exploratory trial,
2. The subjects were placed in the centre but facing a different wall,
3. The subjects were required to replicate the same task starting from the same position as in the exploration, but with a of 30' delay.

In the first condition, the task could be accomplished by relying on different processes; in the second and third conditions, just one process could be used successfully. In fact, in the first condition the participants could rely on both path integration and a mental representation of the environment. In the second condition, due to the change in the starting position, they could no longer use path integration and had to reorient themselves in the environment using a mental map or the geometric module. In the third condition, the subjects had to organize their navigation based on a stored map of the environment because path integration quickly deteriorates and is completely disrupted after such a long delay.

Five different groups of subjects took part in the experiment: control subjects with no history of psychiatric or neurological disease; right and left brain damaged patients without neglect; right brain damaged patients with perceptual neglect; right brain damaged patients with representational neglect. No differences in path integration were detected in the five groups of subjects. This finding confirms that unilateral neglect does not affect the ability to process idiothetic information. Instead, re-orientation was severely damaged in the representational neglect patients but not in the perceptual ones. This suggests that an impairment in representing the contralesional side of space affects the ability to construct cognitive maps based on the geometric shape of the environment. This was confirmed by the fact that representational neglect patients (but not perceptual ones) were impaired in performing the delayed reaching of the TL. In other words, the inability of representational neglect patients to construct a cognitive map of the experimental environment prevented them from storing the target position in long-term memory. The previous experiment shows that path integration is possible for perceptual and representational neglect patients, and that representational neglect patients fail to reorient themselves.

The next question is whether perceptual and representational neglect patients can integrate the relative position of various elements placed in the environment, that is, whether they can guide their navigation independently of the two above-mentioned processes.

In another very recent study (Guariglia et al., in preparation), this issue was further investigated in the above-mentioned, modified version of the human analogue of the Morris water maze. Two distinct elements (a lamp and a clothes hook), which were similar in size and general appearance but not in color and function, were brought into the room. Procedures and task were identical to those of the previous study (Guariglia et al., 2005). However, the presence of two landmarks allowed the participants to rely on the representation of the target position relative to the landmarks (that is, to use the view-dependent place recognition process), without necessarily relying on path integration or representation of the geometric information. Since perceptual neglect patients were able to orient their navigation using landmarks, their performances did not differ from those of controls and right brain damaged patients without neglect.

Instead, representational neglect patients were unable to use the landmarks. This confirms that the impairment of mental representation due to parietal lesions affects the ability to construct cognitive maps of the environment and, therefore, to efficiently store in long-term memory the location of the target relative to the configuration of the room or to the position of the landmarks.

In sum, the presence of neglect per se does not affect vestibular and proprioceptive (idiothetic) processes for the computation of linear and angular translation (Bisiach et al., 1997; Pizzamiglio et al., 2003 experiment 1). Instead the presence of representational neglect, even when perceptual neglect is absent, not only affects the possibility of memorizing environments and routes but also destroys the possibility of constructing environmental maps for navigation. This is true even in very easy tasks and in very modest and simplified experimental environments (Pizzamiglio et al., 1996, 1998; Guariglia et al., 2005; Guariglia et al., in preparation).

To our knowledge, very few attempts have been made to assess imagery components of topographical disorders and no attempts have been made to assess the possible presence of topographical and navigational deficits of patients affected by imagery disorders different from neglect.

A recent case of topographical disorientation due to a congenital brain malformation has been studied. The patient presents some specific imagery impairments directly linked to her navigational difficulties (Iaria et al., 2005). MGC is affected by a congenital cerebral malformation bilaterally involving the middle occipito-temporal regions. Despite almost complete absence of the right middle occipito-temporal cortex and the polimicrogyria of the left one, her development was quite normal (MGC successfully attended and completed high school) and her IQ within the normal range. Before coming under our observation when she was 22, MGC had never been able to learn pathways or to navigate in familiar environments by herself. An extensive neuropsychological assessment showed a moderate long-term memory impairment for visuo-spatial material and a mental rotation deficit. The assessment of MGC's navigational skills in ecological environments revealed the inability to select landmarks useful for orienting, a somewhat preserved ability to recognize familiar landmarks, but severe impairment in detecting their orientation and in using the recognized landmark to direct navigation. This impairment seems to be directly linked to her mental rotation deficit. In fact, her inability to detect whether a landmark has been reached from the left or the right or from the back corresponds well with her inability to recognize rotated stimuli on formal neuropsychological tests. Indeed, she was able to find the shortest path connecting two points on a map but was unable to follow the path even when she was allowed to rotate it. This deficit was linked to MGC's representational deficit on imaginal tasks where she was unable to imagine herself moving on the map and was unable to perform the mental transformation of her own body in mental rotation tests.

In conclusion, even though the role of mental visual imagery in navigation is not a matter of debate, thus far very few attempts have been made to assess the involvement of imagery deficits in navigation or to test visual imagery processes in topographical disorientation.

It should be noted that visual mental imagery and navigation are both complex cognitive functions in which several distinct processes and sub-processes can be recognized (see Farah, 1984, 1995; Kosslyn, 1984; Kosslyn et al., 1995; Redish and Touretzky, 1997; Wang and Spelke, 2002). Therefore, the demonstration of a generic link between visual mental imagery deficits and navigation impairments is not sufficient for understanding the nature and the relationship between these two functions. Indeed, several questions need to be answered before a model can be drawn of the interaction between mental imagery and environmental navigation.

At the moment, existing data indicate that an inability to mentally represent the contralesional side of mental images affects the construction of cognitive maps of environments and has specific consequences on some navigational processes. However, nothing is known about the effects of other imagery disorders, such as the inability to generate mental images or to perform different types of mental transformations, on specific navigational processes. Indeed, at this point we can make the following speculative hypotheses, which are as yet unsupported by neuropsychological data: 1) a deficit in generating mental images affects the ability to utilize verbal instructions for navigation; 2) mental rotation disorders affect the ability to return to the starting point in a new environment after pursuing a long path full of turns; 3) a deficit in visualizing colors or shapes affects landmark recognition. Different studies of navigational impairments suggest that specific impairments in mental imagery may affect some navigational skills. However, current studies analyzing the nature of navigational impairments include the assessment of several perceptual and memory functions (i.e., face and object recognition, verbal and non-verbal learning, etc.) but they very rarely investigate mental imagery specifically. The first question is whether only some types of visual mental images are generated defectively by patients affected by navigational disorders, more specifically whether their generation of skeletal or complex mental images of objects is intact while their generation of mental images of buildings, landmarks and views is lacking or defective. Further, the possibility that patients affected by navigational disorders may present specific impairments in specific mental transformation processes also needs to be analyzed.

In conclusion, we suggest that future studies of navigational disorders should assess representational abilities by referring to actual models of mental imagery.

REFERENCES

Bisiach, E., 1999, Unilateral neglect, in: *Handbook of clinical and experimental neuropsychology*, G. Denes, and L. Pizzamiglio, eds., Psychology Press, Hove.

Bisiach, E., Pattini, P., Rusconi, M. L., Ricci, R., and Bernardini, B., 1997, Unilateral neglect and space constancy during passive locomotion, *Cortex* **33**:313-322.

Bisiach, E., Luzzatti, C., and Perani, D., 1979, Unilateral neglect, representational schema and cousciousness, *Brain* **102**:609-618.

Bisiach, E., Brouchon, M., Poncet, M., Rusconi, M. L., 1993, Unilateral neglect in route description, *Neuropsychologia* **31**:1255-1262.

Committeri, G., Galati, G., Paradis, A. L., Pizzamiglio, L., Berthoz, A., and LeBihan, D., 2005, Reference frames for spatial cognition: different brain areas are involved in viewer-, object-, and landmark-centered judgments about object location, *J Cognitive Neurosci* **16**:1517-1535.

Coslett, H. B., 1997, Neglect in vision and visual imagery: a double dissociation, *Brain* **120**:1163–1171.

De Renzi, E., 1982, *Disorders of space exploration and cognition*, John Wiley & Sons, Norwich.

Farah, M. J., 1984, The Neurological basis of mental imagery: a componential analysis, *Cognition* **18**:245-272.

Farah, M. J., 1995, Current issues in the neuropsychology of image generation, *Neuropsychologia* **33**:1455-1471.

Grossi, D., Modafferi, A., Pelosi, L., and Trojano, L., 1989, On the different roles of the cerebral hemispheres in mental imagery: the "o'clock test" in two clinical cases, *Brain Cognition* **10**:18-27.

Guariglia, C., Padovani, A., Pantano, P., and Pizzamiglio, L., 1993, Unilateral neglect restricted to visual imagery, *Nature* **364**:235-237.

Guariglia, C., Piccardi, L., Iaria, G., Nico, D., and Pizzamiglio, L., 2005, Representational neglect and navigation in real space, *Neuropsychologia* **43**:1138-1149.

Guariglia, C., Piccardi, L., Iaria, G., and Nico, D., in preparation, Effects of perceptual and representational neglect on landmark-based navigation.

Guariglia, C., Nico, D., Iaria, G., Picardi, L., and Palermo L., in preparation, Neglecting the left side of the public square but not the left side of its clock, *incidence, characteristics and nature of representational neglect.*

Kosslyn, S. M., 1980, *Image and mind*, Harvard University Press, Cambridge.

Kosslyn, S. M., Maljkovic, V., Hamilton S. E., Horwitz, G., and Thompson, W. L., 1995, Two types of image generation: evidence for left and right hemisphere processes, *Neuropsychologia* **33**:1485-1510.

Iaria, G., Incoccia, C., Piccardi, L., Nico, D., Sabatini, U., and Guariglia C., 2005, Lack of orientation due to a congenital brain malformation. A case study, *Neurocase* **11**(6): 463-474.

Ogden, J. A., 1985, Contralesional neglect of constructed visual images in right and left brain damaged patients, *Neuropsychologia* **23**:273-277.

Ortigue, S., Viaud-Delmon, I., Michel, C. M., Blanke, O., Annoni, J. M., Pegna, A., Mayer, E., Spinelli, L., and Landis, T., 2003, Pure representational neglect for far space, *Neurology* **60**:2000-2002.

Passini, R., Rainville, C., and Habib, M., 2000, Spatio-cognitive deficits in right parietal lesion and its impact on wayfinding: A case study, *Neurocase* **6**(3):245-257.

Philbeck, J. W., Behrmann, M., and Loomis, J. M., 2001, Updating of locations during whole-body rotations in patients with hemipsatial neglect, *Cogn Affect Behav Neurosci* **1**:330-343.

Pizzamiglio, L., Guariglia, C., Nico, D., and Padovani, A., 1996, Two separate systems for space cognition, in: WJM, Levelt, ed., Advanced Psycholinguistics: A Bressanone Retrospective for Giovanni B. Flores D'Arcais, Nijmengen, NL, pp. 157-162

Pizzamiglio, L., Guariglia, C., and Cosentino, T., 1998, Evidence for separate egocentric and allocentrico space processing in neglect patients, *Cortex* **34**:719-730.

Pizzamiglio, L., Iaria, G., Berthoz, A., Galati, G., and Guariglia, C., 2003, Cortical modulation of whole body movements in brain-damaged patients, *J Clin Exp Neuropsyc* **25**:769-782.

Redish, D. A., and Touretzky, D. S., 1997, Cognitive maps beyond hippocampus, *Hippocampus* **7**:15-35.

Rode, G., Rossetti, Y., Perenin, M. T., and Boisson, D., 2004, Geographic information has to be spatialised to be neglected: a representational neglect case, *Cortex* **40**:391-397.

Wang, R. F., and Spelke, E. S., 2002, Human spatial representation: insights from animals. *Trends Cogn Sci* **6**:376-382.

Chapter 3

FUNCTIONAL EQUIVALENCE OF SPATIAL IMAGES PRODUCED BY PERCEPTION AND SPATIAL LANGUAGE

Jack M. Loomis[1], Roberta L. Klatzky[2], Marios Avraamides[3], Yvonne Lippa[4] and Reginald G. Golledge[5]

[1]Department of Psychology, University of California, Santa Barbara, USA; [2]Department of Psychology, Carnegie-Mellon University, Pittsburgh, USA; [3]Department of Psychology, University of Cyprus, Nicosia, Cyprus; [4]Department of Psychology, Miami University, Oxford, USA; [5]Department of Geography, University of California, Santa Barbara, USA

Abstract: The chapter is concerned with abstract "spatial images", which spatially coincide with visual, auditory, and haptic percepts but continue to exist after the percepts are gone. Spatial images can also be produced by language specifying environmental locations. Three experiments are reviewed which demonstrate that the spatial images produced by space perception and by language are functionally equivalent, or nearly so. The first experiment deals with the spatial updating of single images produced by language and by spatial hearing. The second experiment deals with the updating of multiple images produced by vision, spatial hearing, and language. The third experiment deals with judgments of allocentric distance and direction between spatial images corresponding to multiple target locations specified by vision and by language.

Key words: amodal; crossmodal; functional equivalence; multisensory; spatial language; spatial representations; spatial updating.

1. SPATIAL REPRESENTATIONS FROM LANGUAGE AND PERCEPTION

Vision is the primary means for gathering information about space while navigating and while learning new environments. Audition, haptics, and prorioception are important in complementing vision or substituting for vision when it is lacking. In addition to these direct perceptual inputs arising from immersion with the environment, spatial information can be obtained indirectly by means of symbolic media such as maps, diagrams, and language.

An important question is how similar are abstract spatial representations resulting from these different input modalities. Of these, language and vision differ greatly in terms of the sensory input used for creating abstract spatial representations. Language is used to describe spatial layout in a sequential fashion; whereas, vision provides experience of layout that is much more simultaneous. Thus, with visual perception one can observe more directly spatial relations such as inter-object distances and direction; whereas, with language such relations generally have to be inferred. In addition, vision provides metric information about space; whereas, language is typically used to convey spatial relations that are more qualitative and categorical. For example, the statement "the chair is on my right" leaves unspecified egocentric distance (although it is possible to infer an approximate value using knowledge about the present environment, e.g., the size of the room) as well as the spatial relations among objects. In general, language lacks the level of metric precision available with visual perception. Despite these differences between vision and language, the possibility remains that more abstract representations in memory are functionally similar.

Empirical findings in the spatial cognition literature indicate that spatial representations are indeed possible through linguistic input and that they share properties with those deriving from spatial perception. Taylor and Tversky (1992) showed that people can create very accurate spatial representations for environments conveyed linguistically. In their experiment participants read about an environment either from a route or a survey perspective. They then responded to true/false statements of two types: verbatim statements about spatial information mentioned in the text and inference statements about spatial information that was not mentioned in the text and therefore had to be inferred. Results showed that whereas verbatim statements were responded to faster than inference statements, participants answered equally accurately and rapidly to inference statements that were or were not compatible with the studied perspective. Accuracy for responding to inference statements suggests that participants during the study phase had built a spatial mental model which they used during the response phase to read out spatial information.

A number of studies using a selective-interference paradigm have shown that concurrent spatial tasks tend to hinder performance in primary tasks entailing comprehension of spatial terms. Oakhill and Johnson Laird (1984), for example, found that the concurrent execution of a visuospatial tracking task interfered with the construction of a coherent mental model from a verbal description. Furthermore, Noorzji et al. (2004) showed that a concurrent spatial tapping task lowered performance on a sentence-picture verification task for participants that were categorized as having used a visuo-spatial strategy to process the sentence. No detrimental effect was found for participants that were categorized as having used a verbal strategy instead.

The above results are compatible with the claims of the literature on mental models (e.g., Johnson-Laird, 1983) and situation models (e.g., van Dijk and Kintsch, 1983). This literature has established that when processing language, people typically go beyond representing the actual text (its surface and propositional structure) in order to create mental representations for the state of affairs that is described. Situation models are embodied in the sense that people readily adopt the perspective of the protagonist, thus experiencing the described situation as if they were themselves present in the scene. Space is therefore proposed as one of the fundamental dimensions that people represent when comprehending texts (see Zwaan and Radvansky, 1998 for a review).

While the historical view on situation models argues that these models are representations built from amodal propositions (e.g., Kintsch, 1998), a newer perspective posits that situation models are based on personal experience with the environment. With his Immersed Experiencer Framework, Zwaan (2004) argues that processing words activates experiences with their referents. This argument is supported by empirical findings showing that reading comprehension activates both perceptual and action representations (e.g., Glenberg and Kaschak, 2002; Klatzky et al., 1989; Zwaan et al., 2002; see Zwaan, 2004 for a more detailed discussion). According to Zwaan, language provides set of cues that enable readers to reinstate experiential representations previously formed through perception and action. Situation models therefore constitute perceptual simulations based on previous experience. Zwaan's theory is in line with Barsalou's (1999) perceptual symbol systems theory, according to which people routinely construct perceptual symbols based on how they interact with the world. Zwaan argues that these perceptual symbols are the building blocks of situation models. If indeed language comprehension is experienced-based, as Zwaan (2004) suggests, then the representations that stem from linguistic and perceptual input should exhibit common properties.

Other evidence supporting the idea that spatial representations from perception and language are functionally equivalent (or at least similar) comes from studies using the mental scanning paradigm. The typical finding in mental scanning experiments is that the latency to move covert attention from one location in the mental representation to another varies as a function of the distance between the two locations (e.g., Kosslyn et al., 1978). A number of studies in the mental scanning literature (e.g., Denis and Cocude, 1989; Denis and Zimmer, 1992) have examined mental scanning latencies for layouts of objects learned through language. Denis and Cocude (1989), for example, had subjects listen to a verbal description of a fictitious island and then carry out mental scanning from one island location to another. Results were similar to those obtained from control participants who learned the island by visually inspecting a map in showing the typical latency-distance correlation. Overall, the findings from mental scanning experiments with perceptual and linguistic layouts suggest that spatial representations created from verbal descriptions are functionally similar to those derived from perception.

Similar results for groups of people learning a spatial layout from a verbal description and a map are also reported by studies using other measures such as spatial priming and distance comparisons (Denis and Zimmer, 1992; Noorzij and Postma, 2005). Spatial priming refers to the finding that when items in spatial representations are accessed, they tend to increase the accessibility of other items that occupy nearby locations (McNamara, 1986). Research has shown that such priming effects are also present when spatial layouts are learned from verbal descriptions. An important finding from these studies is that the critical factor is the spatial proximity of the objects in the layout and not whether the objects were mentioned close to or far from each other in the text (Noorzij and Postma, 2005).

Consistent with the functional equivalence hypothesis, McNamara and his colleagues have proposed a general theory for how people represent spatial information in memory (McNamara, 2003; Mou et al., 2004). According to their theory, when people learn a spatial layout, they use cues from the environment (e.g., alignment of objects in the scene) or egocentric experience to determine the orientation of a set of axes intrinsic to the spatial layout. Then they use these intrinsic axes as the reference frame for representing objects and their locations in memory. Research by McNamara and colleagues has supported this theory in experiments that used visual layouts (e.g., Mou and McNamara, 2002). Recently, a study by Mou et al.

(2004) has established that intrinsic axes are used to represent spatial scenes that are learned through verbal descriptions.

Going beyond the hypothesis of functional equivalence of spatial representations from perception and language, a number of researchers have suggested that a common representational format underlies representations created from different modalities (Bryant, 1997; see also Clark, 1973, and Landau and Jackendoff, 1993). Bryant (1997), for example, proposed that humans possess a spatial representational system (SRS) that enables them to create amodal representations about space. According to Bryant (1997) linguistic and perceptual input is first analyzed by modality-specific systems, but then information is directed to the SRS, which operates to represent it in a format that is neither perceptual nor linguistic. The resulting spatial representation is therefore completely detached from the input modality. Findings compatible with Bryant's conjecture that input is first encoded by separate modality systems, but then a common representation is created, are provided by neuroimaging studies showing that (1) the same brain structures are involved when mentally scanning a spatial layout that is encoded from perceiving a map (Mellet et al., 2000) and reading a verbal description (Mellet et al., 2002) and that (2) a trace of the input modality remains present well after encoding. Both studies revealed that in addition to common activation, selective activation of areas that are commonly associated with vision and language also occurred.

2. SPATIAL IMAGES

As the foregoing indicates, the nature of abstract spatial representations derived from language is a broad and complex issue. In what follows, we focus on spatial representations of one or several distinct locations in space, representations that we call "spatial images" for reasons to be made clear. In the case of language, these spatial images are induced by simple utterances of the form "two o'clock, 10 m". Even with such simple utterances and the resulting spatial images, it is of great interest whether the images resulting from language function like those resulting from visual, auditory, and haptic perception. We review recent experiments showing a high degree of functional equivalence of spatial images derived from language and perception.

Figure 1-1 depicts spatial images corresponding either to a visual or auditory percept of a location in 3-D space or to a location specified by spatial language. In panel A, visual or auditory stimulation results in a perceived location, which may or may not coincide with the target. Coincident with the percept is the spatial image. This is a more abstract representation of location, which continues to exist even after the stimulus and resulting percept have disappeared. Spatial language, after linguistic processing, can lead to the creation of a spatial image as well (panel B). At the outset, we acknowledge the possibility that spatial images resulting from language might be functionally dissimilar from those resulting from perception, but the evidence to be presented supports functional equivalence or at least a high degree of functional similarity. Once the spatial image has been created (panels A and B), it can serve as the basis for various spatial judgments and actions when the percepts are no longer present. For example, if two spatial images corresponding to two distinct locations are simultaneously in memory, the participant can judge allocentric relations between them, such as the distance between them and the direction from one to the other. Alternatively, a spatial image can serve as the goal for action. As shown in panel C, the person can walk along different paths in space toward the location specified by the spatial image, long after the visual or auditory stimulus has been removed. When spatial updating of the spatial image is performed accurately, the spatial image remains fixed in relation to the physical environment and the person will arrive at the same location regardless of path taken. There is abundant evidence that spatial images from perception are updated accurately when people walk to targets (Loomis et al., 1998, 2002; Ooi et al., 2001; Philbeck et al., 1997). However, even when updating is accurate with respect to the percept, it need not be accurate with respect to the original stimulus location. If the percepts are reliably discrepant with respect to the targets (usually an error in distance rather than in direction, as depicted in panel A), the spatial images will inherit the same errors. Thus, the judgments and actions based on the spatial images will reflect these perceptual errors. For an extended treatment of spatial images resulting from perception, see the article by Loomis and Philbeck (in press).

The term "spatial image" was coined by Loomis et al. (2002), in connection with spatial updating of locations specified by perception and by language. Many studies had already been done on spatial updating of perceived locations prior to this (e.g., Böök and Gärling, 1981; Loarer and Savoyant, 1991; Loomis et al., 1992; Rieser, 1989); this prior research often referred to the locations being updated simply as updated locations or made vague mention of some type of internal representation of the target location. The term "spatial image" is more evocative, has a meaning not confined to the updating context, and is more congenial to thinking in terms of neural correlates. With respect to the latter, there is growing evidence that spatial updating is associated with posterior parietal cortex, which would indicate that the correlates of spatial images are located there.

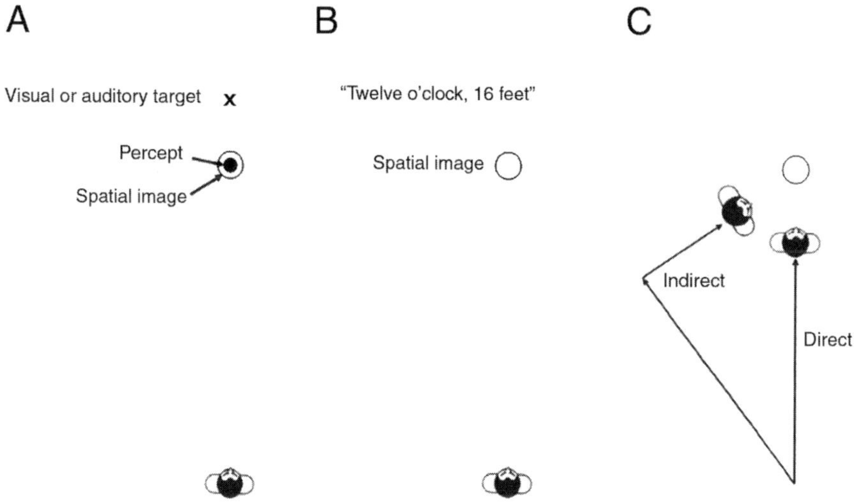

Figure 3-1. A. Formation of spatial image from perception. B. Formation of spatial image from language. C. Flexible updating over different paths with respect to the spatial image.

There is an extensive literature on visual imagery, with the most comprehensive model being that of Kosslyn (1980, 1994). The model encompasses imagery mechanisms at multiple functional and cortical levels. In particular, the term "visual buffer" is used collectively for imagery involving occipital areas including and beyond V1 (up to V4). Kosslyn and Thompson (2003) summarized evidence that imagery-induced activation occurs in early visual areas particularly when precise visual details must be retrieved and compared. The model posits that although images at this level are subjected to processes akin to early visual perception and can be formed at multiple scales, they are topographically organized, if not retinotopic. A later component of imagery is called spatiotopic, conveys the location of elements in the visual buffer, and is associated with posterior parietal cortex (PPC). We will refer to the perspective-specific, 2-D imagery associated with early visual areas as "visual", reflecting their association with primary visual areas. This form of imagery clearly contrasts with what we call the "spatial image" (although the later, spatiotopic imagery may be directly related to spatial images, when it represents space around the self).

Arguably the fundamental difference between spatial images and early visual images is what Kosslyn (1980, 1994) has called *depictiveness*. By this is meant that there is a direct correspondence between regions of the represented object and those of the image, such that inter-point distance in the object is preserved in the image. Kosslyn (1994, p. 409), states that

"A depictive representation typically corresponds to a planar projection of an object, and hence each part of the representation corresponds to a part of the object as viewed from a specific perspective", although he goes on to state that depictiveness need not be restricted to 2-D representations. In our treatment, we choose to contrast depictiveness of visual images, in the 2-D projective sense, with the non-projective nature of spatial images.

As a consequence of depictiveness, spatial images are different from the visual images that activate early visual areas in important respects. Precision in a visual image is limited by the pixel size relative to the image content, which can be changed by mental panning or zooming (Kosslyn, 1994). In contrast, spatial images presumably function for people to orient, prepare, and act in a 3-D world, spatial functions that are associated with PPC. These tasks do not require precise representation of details of the object at a spatial location. A spatial image directly conveys distance and direction in 3-D space. A 2-D image could provide pictorial cues (e.g., occlusion) to inter-object distance and bearing but would lack parallax. If the image is translated, for example, left to right in the visual buffer, it is like moving a picture back and forth in front of one's real eyes. The relative spatial ordering of depicted objects remains the same. In contrast, if one moves left to right in space while maintaining a 3-D spatial image of the environment, the 2-D projection changes (Loarer and Savoyant, 1991; Loomis et al., 1992). Objects enter and emerge from occlusion, and parallax can cause objects to shift their relative left-to-right locations.

Another important characteristic of spatial images is that they are fully externalized relative to the body, head, and hand. Once formed (regardless of input modality), the spatial image is referenced relative to the body and head and can be updated accurately with respect to the environment if body and head movements are correctly perceived. If, on the other hand, an image is tied to the retina, it will move with the eyes. If, instead, it is tied to the head, like a picture carried in front of the eyes, it will remain in the same head-centric coordinates as the person moves through space. In principle, spatial images can fully represent space in back of the individual in addition to the space in front; indeed, Horn and Loomis (2004) have shown that spatial dating of spatial images is as good in back as in front.

A particularly important aspect of spatial images, demonstrated in part by the work reported below, is that they can be produced by stimuli from all of the spatial modalities—vision, hearing, touch, and language. Once created, they can be used in various spatial tasks and seem to be treated equivalently, or nearly so, without regard to the input modality. Because of this, blind people are able to perform many of the complex spatial tasks, including wayfinding, that sighted people perform.

3. A MODEL OF SPATIAL IMAGE PROCESSING

Figure 3-2 gives a model of spatial images and their role in updating and various forms of spatial judgment. The upper part of the figure shows input modalities of spatial language, visual space perception, and auditory space perception. We assume linguistic inputs to be of the form "x degrees, y meters", specifying the egocentric coordinates of targets on the ground plane. Linguistic processing results in meaning which can, upon further processing, give rise to the putative spatial image. Visual and auditory processing result directly in spatial percepts of the visual and auditory stimuli. As mentioned above, these percepts can be spatially discrepant with respect to the stimulus locations. Post perceptual processing, referred to as "conversion" here, gives rise to spatial images that are spatially coincident with the percepts. Positional errors in the percepts are reflected in the spatial images.

When the person walks or is transported in the absence of visual and auditory position information, vestibular and proprioceptive cues, along with efference copy of commands to the musculature, cause an updating of the person's estimated position and orientation (e.g., Loomis et al., 1999; Mittelstaedt and Glasauer, 1991; Sholl, 1989). This updating, in turn, causes an updating of the egocentric coordinates of each spatial image (e.g., Böök and Gärling, 1981; Loarer and Savoyant, 1991; Loomis et al., 1992; Rieser, 1989).

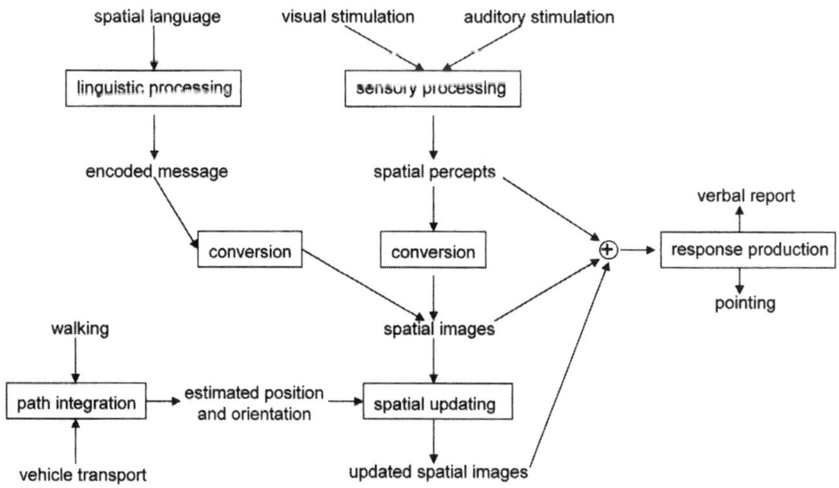

Figure 3-2. Model of spatial image processing. See text for details.

Spatial updating, however, is not an automatic process, for cognitive load and strategy can influence the updating process (Böök and Gärling, 1981; Waller et al., 2002), but it does appear to be obligatory under some conditions (Farrell and Robertson, 1998).

There are a variety of tasks that depend on processes that can operate upon either spatial percepts or spatial images when the percepts are absent. For example, a person judging the distance and direction of a stimulus can make the judgment using either the percept when the stimulus is present or using its spatial image in memory. Another example is judging the separation of two stimuli, whether using percepts or spatial images. It is assumed that spatial percepts and spatial images of sparsely situated targets (i.e., not complex layouts with many surfaces and objects) are treated equivalently by all subsequent processes provided (1) that they have the same coordinates in representational space and (2) the spatial precision of the percept and image are the same. The latter assumption is violated if the spatial image is much more diffuse than the percept. Small differences in precision would not be detectable behaviorally.

The focus of this chapter is on whether spatial images from language and spatial images from perception are functionally equivalent. Again, the basic premise is that if a spatial image from perception and a spatial image from language occupy the same location in representational space and have the same degree of precision, they will be treated equivalently by all subsequent processes that depend only on spatial images as their positional inputs. We review three studies we have done on this topic. Because we have found in our work so far that spatial precision is not of major consequence, we do not discuss it further here.

4. SPATIAL UPDATING OF SINGLE TARGETS

The first study (Loomis et al., 2002) dealt with the spatial updating of a single target (Figure 3-1 C). The target location on a given trial was conveyed either by a loudspeaker positioned there and emitting recorded speech ("Speaker 3") or by an utterance of the form "2 o'clock, 12 feet" spoken by the experimenter who was standing near the participant. In the former case, the participant would localize the sound in 3-D perceptual space and then form a spatial image of its perceived location (Figure 3-1A), and, in the latter case, form a spatial image based on the meaning of the utterance (Figure 3-1 B). For either input modality (3-D sound or spatial language), once the spatial image has been formed, behavior ought to be flexibly directed toward the spatial image while the participant moves through environment without further sensory information about the initially specified target location. To the extent that spatial images from perception and language are functionally equivalent, participants ought to show the same accuracy of updating performance, especially when care has been taken to match the locations of the spatial images of the two modalities. Before we began this research, we did not know whether spatial updating of targets specified by language was even possible.

Ten of the participants were sighted and 6 were blind from early in life. The 4 panels of Figure 3-3 depict the layouts for the experiment. The blindfolded participant stood at the origin facing in the direction corresponding to the top of the figure. The locations of the targets used in a grassy field are depicted by the X's. The nominal target distances ranged from 1.83 m to 4.27 m (6 to 16 ft) and the nominal target azimuths ranged form -90° (left of straight ahead) to 90° (right of straight ahead). On a given trial the participant heard speech from the target speaker in the field or the utterance by the experimenter. The participant then walked without vision and with sound-blocking hearing protectors to the estimated target locations. On some trials, the participant walked directly to the target and on other trials, was first led by the experimenter to the turn point 2.7 m in front and then attempted to walk unaided the rest of the way to the target. In the case of spatial language, the performance on direct walking trials is, by itself, not of great interest, because participants need not update relative to a spatial image in order to perform well. Given an utterance like "10'oclock, 8 feet", the participant need only turn to 10'oclock and walk what seems like 8 ft. However, this strategy does not work for the indirect walking trials. To perform well on these trials, the participant needs to convert the meaning of the utterance into a spatial image and then perform updating with respect to it (panels B and C of Figure 3-1).

If participants performed spatial updating perfectly, they should have walked to the same location on direct and indirect trials for the same nominal stimulus. The measure of updating performance used was the spatial separation between the terminal points of the walking trajectories on direct trials and the terminal points on indirect trials. The 4 panels of Figure 3-3 give the results in the form of centroids of terminal points, averaged over participants, for the direct and indirect paths to the different target locations. Sighted participants performed better as shown by the smaller separations between direct and indirect terminal points. However, because the separations are all quite small in comparison with the target distances and the distances walked, the data indicate spatial updating for both 3-D sound and for language, as well as for both groups of participants. The statistical analysis revealed just slightly poorer updating performance overall for language compared to auditory perception.

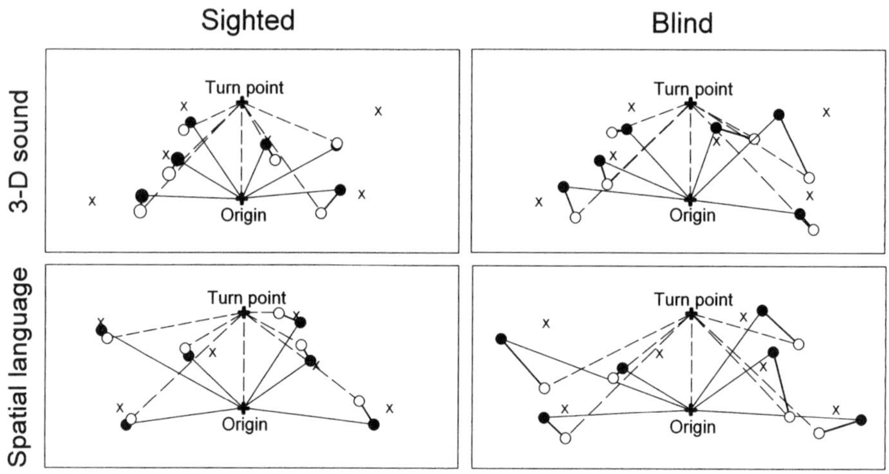

Figure 3-3. Stimulus layout and results of an experiment on spatial updating of single auditory targets ("3-D sound") and single targets specified by spatial language (Loomis et al, 2002). The experiment was conducted outdoors with both blind and blindfolded sighted participants. While standing at the origin, the participant heard a sound from a loudspeaker at one of the locations (X) or heard an utterance specifying one of the same locations (e.g., "10 o'clock, 12 feet"). The participant then attempted to walk to the location either directly or indirectly. In the latter case, the participant was guided forward 2.74 m to the turn point and then walked the rest of the way to the estimated target location. The open circles are the centroids of the indirect path stopping points, and the closed circles are the centroids of the direct path stopping points. Reprint of Figure 7 from Loomis et al. (2002). Spatial updating of locations specified by 3-D sound and spatial language. *Journal of Experimental Psychology: Learning, Memory, & Cognition,* 28, 335-345. Reprinted with permission.

These results indicate that spatial images from perception and language exhibit near functional equivalence with respect to the processes involved in spatial updating of one target at a time.

Of secondary interest is the fact that the terminal points for the 3-D sound condition were generally somewhat displaced in distance from the target locations. This is consistent with other research showing that the perceived distance of sound sources is contracted relative to the range of physical source distances, with under-perception of far distances being the norm (e.g., Loomis et al., 1999; Zahorik et al., 2005).

5. SPATIAL UPDATING OF MULTIPLE TARGETS

The second of the three studies dealt with spatial updating of multiple targets (Klatzky et al., 2003). Because the previous study involved only a single target, working memory was sufficient for retaining the target location. In this second study, participants were unable to retain up to 5 targets specified by language and auditory perception in working memory. Thus, in order to examine updating of multiple targets, we had participants spend up to 10 minutes learning the multiple target locations to criterion during a learning phase. They were subsequently tested on updating performance during a test phase. Thus, this study involved spatial images held in long term memory.

Unlike the previous study, this was done indoors where stronger reverberation cues likely resulted in the slightly more accurate auditory distance perception that was observed. There were two experiments in this study. In the first, participants used a pointer to indicate their estimates of the directions of the targets (before and after updating) and used verbal reports to indicate their estimates of the distances of the targets. In the second experiment, participants used the same procedure of direct and indirect walking to targets as in the previous study. Based on a recent analysis by Loomis and Philbeck (in press), we now believe that verbal reports of distance are systematically biased toward underestimation, producing what appears to be an updating error that is added to whatever true updating error there is. Accordingly, we focus here on the second experiment involving direct and indirect walking, which resulted in much better updating performance. Besides the use of multiple targets in an indoor setting, the experiment differed from the previous study in including a vision condition as well as including indirect walking in which participants sidestepped to the right while facing forward.

Figure 3-4 gives the spatial layouts for the experiment. The vision and language targets, indicated by the X's, were at the same nominal locations, ranging in distance from 0.91 m to 3.66 m (3 to 12 ft). Because of the tendency for under-perception of auditory distance, the auditory stimuli were delivered by loudspeakers at head level that were placed at slightly larger distances, 1.22 to 4.57 m (4 to 15 ft), in order to produce perceived distances close to those of the visual targets, which were perceived quite accurately. The turn points ("indirect waypoints") for indirect walking were either 2.5 m in front or to the right side of the origin. In the learning phase, participants in the 3-D sound condition heard spoken labels (e.g., "baby", "cat") from the positioned loudspeakers. In the language condition, participants heard the utterance of the target coordinates followed by two presentations of the label. For both 3-D sound and language conditions, synthetic speech of a male voice was used. In the vision condition, participants saw labels presented at head level. On a learning trial, the participant was exposed to each of the 5 targets and then, when prompted with a label, attempted to report the direction and distance of the target using a pointer and verbal report, respectively. The learning phase terminated when both pointing accuracy and accuracy of distance reports (assessed using rank order correlation) met strict criteria. In the updating phase, participants responded to each target label by walking either directly to the target or walking indirectly to the target after being passively guided to the turn point.

The results are given in Figure 3-4. As in the previous study, updating performance was measured by the separation between the terminal points of the direct and indirect paths. The generally small separations in all 6 panels of Figure 3-4 indicate that updating performance was good for all conditions.

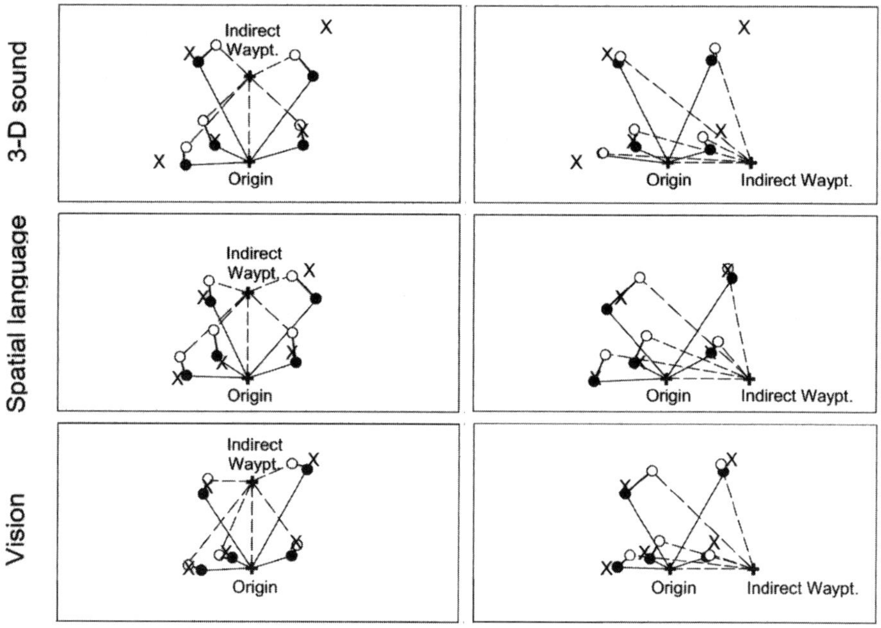

Figure 3-4. Stimulus layout and results of an experiment on spatial updating of multiple auditory targets ("3-D sound"), visual targets, and targets specified by spatial language (Klatzky et al., 2003). The experiment was conducted indoors. While standing at the origin during the learning phase, the participant memorized the locations of multiple targets (Xs) presented using vision, spatial hearing, or spatial language. Then on each trial during the test phase, the participant attempted to walk to one of the locations either directly or indirectly. In the latter case, the participant was guided forward and sideways 2.5 m to the turn point and then walked the rest of the way to the target. The open circles are the centroids of the indirect path stopping points, and the closed circles are the centroids of the direct path stopping points. Modified version of Figure 4 from Klatzky et al. (2003). Encoding, learning, and spatial updating of multiple object locations specified by 3-D sound, spatial language, and vision. *Experimental Brain Research*, 149, 48-61. Reprinted with kind permission of Springer Science and Business Media.

The statistical analysis revealed no significant differences between vision and 3-D sound, but language was just slightly worse than vision. Overall, the results support near functional equivalence of spatial images from perception and language.

6. JUDGMENT OF ALLOCENTRIC DIRECTION AND DISTANCE

The third of the three studies (Avraamides et al., 2004) dealt not with spatial updating, but with judgments of allocentric distance and direction between pairs of targets. The experiment was conducted indoors. In egocentric coordinates, the targets varied from 90° left to 90° right and from 0.91 m to 4.57 m (3 ft to 15 ft).

There were 3 experiments, in each of which there was a learning phase followed by a test phase. In the learning phase, participants memorized the locations of 4 targets presented visually or by spatial language. In the test phase, participants recalled pairs of target locations from memory and verbally estimated the distances between the two targets of each pair and estimated the inter-target directions by aligning a pointer with the direction from one of the targets in each pair to the other target. The two conditions of recall were Visual Memory (VM) and Spatial Language (SL). Experiment 1 included a third condition (Visual Perception = VP) during the test phase in which participants made allocentric judgments while being able to view the targets. Experiment 2 compared just VM and SL conditions, but differed from Experiment 1 in that during the learning phase, participants viewed only one target at a time. Experiment 3 also compared only VM and SL conditions but differed from Experiments 1 and 2 by the insertion of a spatial updating phase between the learning and test phases (i.e., the subject moved backward .9 m before making the allocentric judgments). This was done to insure that participants converted the utterances into spatial images.

The primary measures for evaluating functional equivalence of spatial images from perception and language, as well as percepts in the Visual Perception condition of Experiment 1, were pointing latency averaged over all target pairs and the product-moment correlation coefficients between input modalities computed over target pairs for (1) latency to complete the pointing response after being told the labels of each pair of targets, (2) the signed pointing error for each pair, and (3) the signed distance error. It is notable that the two conditions involving recalling from memory (VM and SL) had very different latencies in Experiments 1 and 2; the latencies for VM averaged about 5.6 sec and those for SL about 7.6 sec. When the updating phase was added in Experiment 3, insuring that participants formed spatial images prior to the test phase, pointing latencies for the two conditions were now both about 5.7 sec. We interpret this to mean that when spatial images are formed for both language and perception, they are functionally similar with respect to the processes involved in judging allocentric relations. The correlation for signed pointing error supports this interpretation. The correlation between VM and SL was statistically non-significant in Experiments 1 and 2 but increased to a highly significant value of 0.93 in Experiment 3. This high correlation of Experiment 3 can be seen in the error values of Figure 3-5. The correlation between VM and SL for signed distance error provides less convincing support. The correlation was non-significant in Experiments 2, but had similar values of 0.68 and 0.67 in Experiments 1 and 3, respectively.

Also of interest are the correlations for pointing latency (0.52), signed pointing error (0.64), and signed distance error (non-significant) between VP and VM in Experiment 1. The fact that that they were far from 1.0 means

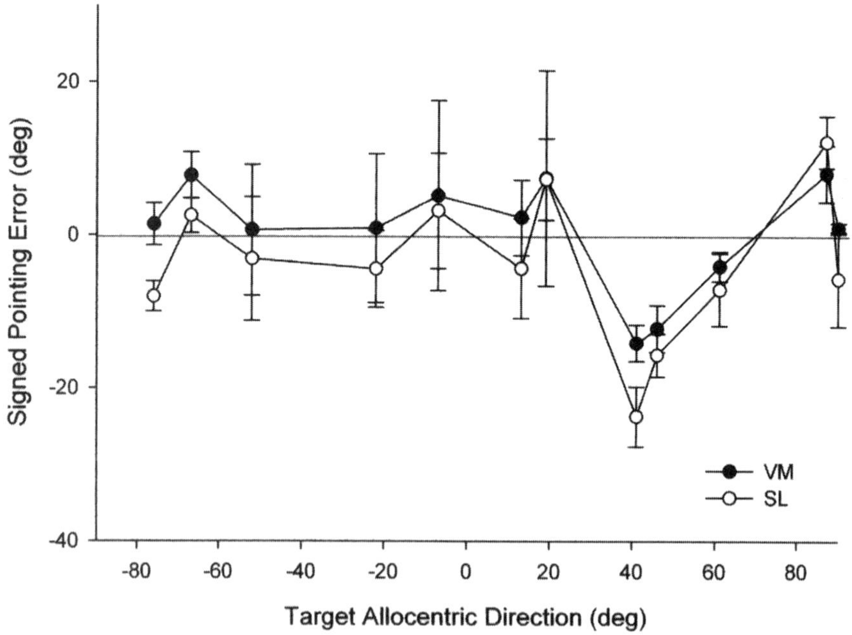

Figure 3-5. Signed pointing error as a function of the physical direction between each pair of targets. VM = visual memory; SL = spatial language. Reprint of Figure 8 from Avraamides et al. (2004). Functional equivalence of spatial representations derived from vision and language: Evidence from allocentric judgments. *Journal of Experimental Psychology: Learning, Memory, & Cognition*, 30, 801-814. Reprinted with permission.

that in this setting involving multiple targets recalled from memory, visual percepts formed from simultaneous viewing and spatial images from those visual percepts are not functionally equivalent, in contrast to the assumption made in connection with Figure 3-2. It is likely, however, that functional equivalence between percepts and spatial images still holds for spatial images of isolated targets in working memory (Loomis and Philbeck, in press).

The results of Experiment 3 of the above study (Avraamides et al., 2004) indicate that when participants are induced to create spatial images from language, there is considerable functional similarity between the spatial images from visual perception and language, especially as indicated by the signed pointing errors. However, the less impressive correlations for pointing latency and signed distance errors indicate that some functional differences remain between the spatial images from perception and language.

7. CONCLUSIONS AND FUTURE RESEARCH

Overall, the results of the two studies on spatial updating and that on judgments of allocentric relations demonstrate near functional equivalence between spatial images from perception and language. Although these experiments do not speak to whether spatial images from perception and language are functionally similar but distinct or are amodal in nature, with their properties being completely independent of the source modality. We favor the hypothesis of amodal representations. In light of the growing evidence that posterior parietal cortex (PPC) is the basis for spatial updating (Cohen and Andersen, 2002, 2004; Merriam et al., 2003; see also Creem and Proffitt, 2001 for review), it is not a far stretch to posit PPC as the substrate for spatial images. Functional MRI and transcranial magnetic stimulation (TMS) are surely going to be important tools for determining whether spatial images reside in PPC and whether they are amodal or just functionally similar but distinct representations. Future behavioral research will also be useful for addressing the question of amodality. If capacity limits in learning, storing, and acting upon spatial images are independent of the mix of perception-based and language-based images, this would be evidence of amodal representations. So would be experiments in which spatial distractors can interfere with the storage and updating of target stimuli, if the degree of interference is independent of whether targets and distractors are from the same or different modalities.

REFERENCES

Avraamides, M., Loomis, J. M., Klatzky, R. L., and Golledge, R. G., 2004, Functional equivalence of spatial representations derived from vision and language: Evidence from allocentric judgments, *J Exp Psychol Learn* **30:**801-814.

Barsalou, L. W., 1999, Perceptual symbol systems, *Behav Brain Sci* **22:**577–609.

Böök, A. and Gärling, T., 1981, Maintenance of orientation during locomotion in unfamiliar environments, *J Exp Psychol Human* **7:** 995-1006.

Bryant, D. J., 1997, Representing space in language and perception, *Mind Lang* **12:**239–264.

Clark, H. H., 1973, Space, time, semantics, and the child, in: *Cognitive development and the acquisition of language*, T. E. Moore, ed., Academic Press, New York.

Cohen, Y. E., and Andersen, R. A., 2002, A common reference frame for movement plans in the posterior parietal cortex, *Nat Rev Neurosci* **3:**553-562.

Cohen, Y. E., and Andersen, R. A., 2004, Multisensory representations of space in the posterior parietal cortex, in: *The Handbook of Multisensory Processes*, G. A. Calvert, C. Spence, and B. E. Stein, eds., MIT Press, Cambridge, MA, pp.463–479.

Creem, S. H., and Proffitt, D. R., 2001, Defining the cortical visual systems: "what", "where" and "how", *Acta Psychol* **107:**43-68.

Denis, M., and Cocude, M., 1989, Scanning visual images generated from verbal descriptions, *Eur J Cogn Psychol* **1:**293–307.

Denis, M., and Zimmer, H. D., 1992, Analog properties of cognitive maps constructed from verbal descriptions, *Psychol Res* **54:**286-298.

Farrell. M. J. and Robertson, I. H., 1997, Mental rotation and the automatic updating of body-centered spatial relationships, *J Exp Psychol Learn* **24:**227-233.

Glenberg, A. M., and Kaschak, M. P., 2002, Grounding language in action, *Psychon B Rev* **9:**558-565.

Horn, D. L. and Loomis, J. M., 2004, Spatial updating of targets in front and behind, Paidéia **14:**75-81.

Johnson-Laird, P. N.,1983, *Mental Models*, Cambridge University Press, Cambridge, UK.

Kintsch, W., 1998, *Comprehension: A Paradigm for Cognition*, Cambridge University Press, Cambridge, UK.

Klatzky, R. L., Lippa, Y., Loomis, J. M., and Golledge, R. G., 2003, Encoding, learning, and spatial updating of multiple object locations specified by 3-D sound, spatial language, and vision, *Exp Brain Res* **149:** 48-61.

Klatzky, R. L., Pellegrino, J., MeCloskey, B. P., and Doherty, S., 1989, Can you squeeze a tomato? The role of motor representations in semantic sensibility judgments, *J Mem Lang* **28:**56-77.

Kosslyn, S. M., 1980, *Image and Mind*, Harvard University Press, Cambridge, MA.

Kosslyn, S. M.,1994, *Image and Brain,* MIT Press, Boston.

Kosslyn, S. M., Ball, T. M., and Reiser, B. J., 1978, Visual images preserve metric spatial information: Evidence from studies of image scanning, *J Exp Psychol Human* **4:**47–60.

Kosslyn, S. M., and Thompson, W. L., 2003, When is early visual cortex activated during visual mental imagery? *Psychol Bull* **129:**723–746.

Landau, B., and Jackendoff, R., 1993, "What" and "where" in spatial language and spatial cognition, *Behav Brain Sci* **16:**217-265

Loarer E. and Savoyant A., 1991, Visual imagery in locomotor movement without vision, in: *Mental Images in Human Cognition*, R. H. Logie and M. Denis, eds., Elsevier, The Hague, pp. 35-46.

Loomis, J. M., Da Silva, J.A., Fujita, N., and Fukusima, S. S., 1992, Visual space perception and visually directed action, *J Exp Psychol Human* **18:**906-921.

Loomis, J. M., Klatzky, R. L., and Golledge, R. G., 1999, Auditory distance perception in real, virtual, and mixed environments, in: *Mixed Reality: Merging Real and Virtual Worlds,* Y. Ohta and H. Tamura, eds., Ohmsha, Tokyo, pp. 201-214.

Loomis, J. M., Klatzky, R. L., Golledge, R. G., and Philbeck, J. W., 1999, Human navigation by path integration, in: *Wayfinding: Cognitive Mapping and Other Spatial Processes,* R. G. Golledge, ed., Johns Hopkins, Baltimore, pp. 125-151.

Loomis, J. M., Klatzky, R. L., Philbeck, J. W., and Golledge, R. G., 1998, Assessing auditory distance perception using perceptually directed action, *Percept Psychophys* **60:**966-980.

Loomis, J. M., Lippa, Y., Klatzky, R. L., and Golledge, R. G., 2002, Spatial updating of locations specified by 3-D sound and spatial language, *J Exp Psychol Learn* **28:**335-345.

Loomis, J. M. and Philbeck, J. W., Measuring perception with spatial updating and action, in: *Embodiment, ego-space, and action,* R. L. Klatzky, M. Behrmann and B. MacWhinney, eds., Erlbaum, Mahwah, NJ, *in press.*

McNamara, T. P., 1986, Mental representations of spatial relations, *Cognitive Psychol* **18:** 87-121.

McNamara, T. P., 2003, How are the locations of objects in the environment represented in memory? In: *Spatial Cognition III: Routes and Navigation, Human Memory and Learning, Spatial Representation and Spatial Reasoning*, C. Freksa, W. Brauer, C. Habel, and K. Wender, eds., Springer-Verlag, Berlin, pp. 174-191.

Mellet, E., Bricogne, S., Crivello, F., Mazoyer, B., Denis, M., and Tzourio-Mazoyer, N., 2002, Neural basis of mental scanning of a topographic representation built from a text, *Cereb Cortex* **12:**1322-1330.

Mellet, E., Bricogne, S., Tzourio-Mazoyer, N., Ghaëm, O., Petit, L., Zago, L., Etard, O., Berthoz, A., Mazoyer, B., and Denis, M., 2002, Neural correlates of topographic mental exploration: The impact of route versus survey perspective learning, *Neuroimage* **12:** 588-600.

Merriam, E.P., Genovese, C. R., and Colby, C. L., 2003, Spatial updating in human parietal cortex, *Neuron* **39:**361-373.

Mittelstaedt, M. L., and Glasauer, S., 1991, Idiothetic navigation in gerbils and humans, *Zool Jahrb Allg Zool* **95:**427-435.

Mou, W., and McNamara, T. P., 2002, Intrinsic frames of reference in spatial memory, *J Exp Psychol Learn* **28:**162–170.

Mou, W., Zhang, K., and McNamara, T. P., 2004, Frames of reference in spatial memories acquired from language, *J Exp Psychol Learn* **30:**171–180.

Noorzji, M. L., and Postma, A., 2005, Categorical and metric distance information in mental representations derived from route and survey descriptions, *Psychol Res* **69:**221-232.

Noorzji, M. L., Van der Lubbe, R., Neggers, S. F. W., and Postma, A., 2004, Spatial tapping interferes with the processing of linguistic spatial relations, *Can J Exp Psychol* **58:** 259-271.

Oakhill, J. V., and Johnson-Laird, P. M., 1984, Representation of spatial descriptions in memory, *Curr Psychol Res Rev* **3:**52-62.

Ooi, T. L., Wu, B., and He, Z.J .J., 2001, Distance determined by the angular declination below the horizon, *Nature* **414:**197-200.

Rieser, J. J., 1989, Access to knowledge of spatial structure at novel points of observation, *J Exp Psychol Learn* **15:**1157-1165.

Sholl, M. J., 1989, The relation between horizontality and rod-and-frame and vestibular navigational performance, *J Exp Psychol Learn* **15:**110-125.

Taylor, H. A. and Tversky, B., 1992, Spatial mental models derived from survey and route descriptions, *J Mem Lang* **31:**261-282.

Van Dijk, T. A., and Kintsch, W., 1983, *Strategies of Discourse Comprehension*, Academic Press, New York.

Waller, D., Montello, D. R., Richardson, A. E., and Hegarty, M., 2002, Orientation specificity and spatial updating of memories for layouts, *J Exp Psychol Learn* **28:**1051-1063.

Zahorik, P., Brungart, D. S., and Bronkhorst, A. W., 2005, Auditory distance perception in humans: A summary of past and present research, *Acta Acust united Ac* **91:**409-420.

Zwaan, R.A., 2004, The immersed experiencer: toward an embodied theory of language comprehension, in: *The Psychology of Learning and Motivation*, B.H. Ross, ed., Vol. 44, Academic Press, New York, pp. 35-62.

Zwaan, R.A., Stanfield, R.A., and Yaxley, R.H., 2002, Language comprehenders mentally represent the shapes of objects, *Psychol Sci* **13:**168-171.

Zwaan, R. A., and Radvansky, G. A., 1998, Situation models in language comprehension and memory, *Psychol Bull* **123:**162–185.

Chapter 4

SPATIAL PROCESSING AND VIEW-DEPENDENT REPRESENTATIONS

Ranxiao Frances Wang
University of Illinois at Urbana-Champaign

Abstract: View-dependent representations have been shown in studies of navigation, object and scene recognition, and spatial reasoning. In this chapter, we discuss the relationship between different types of view-dependent representations. Based on previous studies and new data presented in this chapter, we proposed a model that contains an egocentric spatial working memory and a LTM representation of similar nature, and discussed theoretical issues that can potentially distinguish between different models of spatial representations.

Key words: spatial representation; navigation; reference frame; spatial updating; view dependent; egocentric; allocentric; spatial judgment.

1. INTRODUCTION: VIEW-DEPENDENT REPRESENTATIONS

1.1 Spatial representations and view-dependence

Representations of space can be described in various ways. One common dimension used to categorize spatial representations is view-dependence. View-dependent representations encode spatial information relative to a specific viewpoint. In contrast, view-invariant representations encode spatial information in a way so that the spatial description remains valid regardless of the viewpoint. Both types of representations have been proposed for single objects (e.g., Biederman and Gerhardstein, 1993; Edelman and Bulthoff, 1992; Tarr and Pinker, 1989), scenes (e.g., Christou and Bülthoff, 1999; Mou and McNamara, 2002; Simons and Wang, 1998), and larger scale space (e.g., Collett and Cartwright, 1983; Gallistel, 1990; Mallot and Gillner, 2000).

In principle, view-dependence includes both position and orientation dependence. Although some studies showed that spatial representations can be position-dependent (Easton and Sholl, 1995; May, 2004), the cost of mentally changing position without changing orientation (i.e., an imagined *translation*) is relatively small (e.g., Rieser, 1989). Thus, most studies have primarily focused on orientation dependence and ignored the position component.

The traditional paradigm to examine view-dependence in object and scene representations is the "angular-disparity effect." The assumption is that people should show advantage in making judgments about an object's appearance or location from the represented perspective. As the test perspective deviates farther and farther away from the represented one, performance should decrease because additional processing is needed to transform the representations in order to make the judgment. This principle applies to all types of view-dependent representations, for both position-dependence and orientation-dependence.

1.2 Three types of view-dependent representations

There are different types of view-dependent representations depending on the nature of the view that is represented. Here we discuss three types that are most important in theories of spatial representations.

1.2.1 The studied view

The most common type of view-dependent representation is the *studied-view* representation. If a viewer studies an object or a scene/place from a specific position in space, s/he can remember what the object or scene looks like from that specific vantage point, i.e., forming a view-dependent representation of the target. If the test object or scene is presented from a different vantage point than the studied view, then additional processes are needed to compare them, and thus resulting in costs in response latency and accuracy when asked to recognize or make judgments about the object or scene.

Representations of the studied view of both single objects and scenes have been shown in various studies. For example, Tarr and colleagues (e.g., Tarr and Pinker, 1989) showed that novel objects are best recognized at the views presented during the studying period, and performance deteriorates as the test view deviates from the studied views (Edelman and Bulthoff, 1992; Humphrey and Khan, 1992; Tarr et al., 1998; Ullman, 1989). Similar findings are also shown in studies of spatial layout representations. People can recognize a picture of, make a spatial judgment about, and detect changes to a scene most efficiently from the perspective they originally viewed the scene (e.g., Christou and Bülthoff, 1999; Diwadkar and McNamara, 1997; Shelton and McNamara, 1997; Simons and Wang, 1998), suggesting that people can represent a single object or a scene from the perspective they originally studied it, i.e., forming a studied-view representation of the object or the scene.

1.2.2 The canonical view

The second type of view-dependent representation is the "*canonical*" view representation. For some objects or scenes, there are certain special, privileged viewpoints that are easier to learn than others, and tend to be represented even if the object or scene has never been studied from that perspective. For example, Shelton & McNamara (2001; also see McNamara et al., 2003; Mou et al., 2004) showed that judgments of the spatial relationship among a layout of objects tend to be easier from a direction parallel to the enclosing room walls or edges of the matt on which the objects were placed. Werner and Schmidt (1999) showed that views of a city scene are better recognized from a perspective parallel to the main streets. Mou and McNamara (2002) also showed special preferred orientations defined by the structure of the object array. These special directions defined by the structure of the scene or environment are examples of canonical view representations.

1.2.3 The updated view

Both the studied view and the canonical view are "static" representations. Recent studies showed a third type of view-dependent representation, which is dynamic in nature[1]. Simons and Wang (1998; also see Amorim et al., 1997; Farrell and Robertson, 1998; Loomis et al., 1996; Presson and Montello, 1994; Rieser, 1989; Simons et al., 2002; Wang, 2003; Wang and Simons, 1999; Wang and Spelke, 2000, 2002) showed that people could recognize a scene efficiently from a novel viewpoint that had no special status, as long as they physically moved there. They proposed that an *updating process* generates new representations corresponding to what an object or scene would look like from the new vantage points using perceptual cues of ego-motion. Thus, this new view-dependent representation (updated view) is dynamically changed according to a viewer's movements to reflect the new relationship between the object/scene

1 "static" is relative to "dynamic" of the updating system and we don't mean "no-change ever". Like other type of memory, these representations are also subject to decay and distortion over time.

and the viewer. It does not require previous experience of the scene from that perspective (i.e., studying), or any special status (i.e., can be any perspective).

2. RELATIONSHIP BETWEEN VIEW-DEPENDENT REPRESENTATIONS

2.1 Studied view and canonical view representations

McNamara and colleagues conducted a series of studies comparing the effect of initial viewing direction and different types of canonical directions. For example, Shelton and McNamara (2001) showed that when people view a scene from a canonical direction following the initial viewing from a non-canonical perspective, the initial studied view may not be represented in memory. Furthermore, Mou and McNamara (2002) showed that when specifically instructed to memorize the scene from a canonical perspective while physically viewing it from a non-canonical perspective, people are capable of ignoring their true study perspective and show performance advantage of the canonical perspective rather than the actual studied perspective. These findings suggest that canonical view may have higher priority in representing a scene over the studied-view representation, and may even "replace" an existing studied-view representation under certain circumstances.

2.2 Studied view and updated view representations

Several studies have also examined the role of studied-view representation during spatial updating. For example, Mou et al. (2004) showed that after physically moving to a novel perspective relative to a scene, people can make judgments faster and more accurately from perspectives that are either close to the initial studied perspective, or close to their current perspective, suggesting that information from both the studied view and the updated view are available. Waller et al. (2002) showed that after moving to a different orientation, people showed strong advantage in making spatial judgments from their physical orientation rather than the direction they studied the scene, unless elaborate effort was made to counteract the updating process. Thus, it is not clear whether the studied-view representations are generally retained after people move to new perspectives, or only in very specific conditions. It is also not clear whether and how the studied-view and the updated-view representations interact with each other.

2.2.1 View-dependent representations after a movement

We conducted a series of studies to further examine the relationship between the studied-view and the updated-view representations. We started by examining which view(s) people show advantage at after they make a movement (turn) away from their studying perspective. Participants studied a set of objects while facing a given orientation and then turned to face a different orientation while blindfolded. Then they were asked to make judgments about object locations from either the study heading (studied view), the current heading (updated view), or other novel headings. Imagine facing X, point to Y. If people maintain a representation of the studied view, they should show advantage in pointing to objects according to the studied perspective. If they generated a new representation corresponding to their current heading, they should show advantage at the updated view. Moreover, judgments from other novel headings should be a function of their angular distance from the represented view(s) (angular disparity effect).

Eight undergraduate students were tested individually in a rectangular room (3.8m by 2.4m), as illustrated in Figure 4-1. A swivel chair was placed in the middle of the room, in which participants sat throughout the experiment. Five targets, a closet, a door, a VCR, a poster, and a computer were placed near the walls of the room.

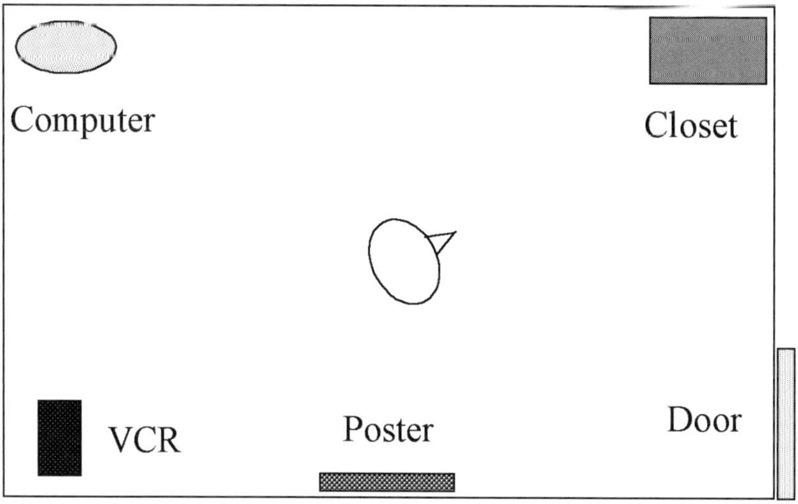

Figure 4-1. An overhead view of the rectangular testing room.

The overhead image of the room and the participants' responses were recorded with a VCR, which took input from a small video camera mounted on the ceiling just above the swivel chair. A Gateway PC4200 desktop computer randomized the order of the targets for each participant.

Participants were randomly assigned to two groups. One group studied the object array while facing the *Computer* (studied view), and was tested while physically facing the *Closet* (updated view). The other group studied the objects while facing the *Closet* (studied view), and was tested while facing the *Computer* (updated view). Thus, the studied view and updated view were counterbalanced across the groups.

Participants were blindfolded and sound-masked before they were led into the testing room and sat in the swivel chair. Then they were turned to face the pre-determined orientation according to the group they were assigned to. Then they were instructed to remain stationary, and the blindfold was removed. The experimenter then named and pointed to the five targets one by one, asking the participants to remember them. The participants were allowed to turn their head (and head only) to observe the targets, but not to move their body or feet. The participants were allowed to study the objects for as long as they desired. Then they were blindfolded and put on an earphone with constant white noise as a sound mask, and were asked to point to the targets in a random sequence. If they made a mistake (> 20º from the correct direction), the blindfold was removed and they were asked to study again. The sequence continued until they could point to all targets correctly. The learning period typically took less than 5 minutes without the need to re-learn the targets.

The directions of the targets and the pointing responses were measured from the TV monitor after the testing was completed, by superimposing a transparent radial grid on the monitor. The reaction time was measured from the ending of the target name to the completion of the pointing response, which was indicated by the stabilization of the hand. The angular error for each response was calculated as the small angle (i.e., < 180°, unsigned) between the correct direction and the pointing direction. Since the objects were 60 ± 4° away from their neighbors (123° between *Computer* & *Closet*), the AD (angular disparity) between targets were considered multiples of 60° (0°, 60°, 120°, and 180°) and combined in the analysis.

The question of interest is which view(s) do people show advantage at after they study a scene from one perspective and then physically turn to take a different perspective during testing. Performance (RT and angular error) was analyzed as a function of the angular disparity between the imagined heading and the study heading (AD_Study), and the angular disparity between the imagined heading and the current heading (AD_update), using linear regression. Both RT and angular errors showed the angular disparity effect from the updated view and increased as the imagined heading deviated from the current heading, *t*s(38) > 2.1, *ps* < .04, suggesting that participants had a representation of the updated view (see Figure 4-2, top panels). In contrast, there was no significant angular disparity effect for the studied view, *ts*(38) < 1.6, *ps* > .11. Thus, people showed little evidence that they remembered the scene from the heading they studied it after they made a small turn to face a different orientation, but at the same time showed significant advantage at making judgments from the novel, updated view. This study provides initial evidence that the studied view and the updated view representations may be independent of each other.

One concern with this design is that the AD_study and AD_update were not completely independent of each other; in fact, there is a moderate negative correlation between these two variables. Thus, it is possible that the benefit of the studied view was obscured by the strong effect of the updated view and thus difficult to detect (Mou et al, 2004). To address this issue, we selected two test perspectives that had the same AD_update (120°) but different AD_study (0° & 120°) to de-couple the two factors. Even when AD_update was held constant, there was still no significant effect of AD_study (*t*s(14) < 1.2, ps > .26), suggesting there is little evidence of the studied view representation after spatial updating occurred.

To further examine the studied view representation during the updating process, we conducted a second study that varied the AD_study and AD_update independently (c.f. Mou et al., 2004). This study was identical to the first one except for the following. Fourteen participants were randomly assigned to two groups. One group studied the targets while facing the *Computer*, and the other group studied the targets while facing the *Closet*. Then all participants turned to face the *Poster* while being tested. In both cases, participants made a 120-degree turn while blindfolded. In this design, the angular distance between the test perspective and the studied view (AD_study) and between the test perspective and people's actual perspective (AD_update) were varied independently. That is, AD_study was 60° and 180° when AD_update was 60°, and AD_study was 0° and 120° when AD_update was 120°. If participants had any memory of the studied view available after they turned, they should show an angular disparity effect of the studied view while AD_updated was held constant.

As shown in Figure 4-2, bottom panels, performance deteriorated as the test perspective deviated further from their current heading (non-significant for RT, $t(67) = 1.7$, $p = .09$; significant for error, $t\,(67) = 5.0$, $p < .001$). There was still little evidence for memory of the studied view, however. Neither RT nor angular error increased significantly as a function of angular disparity from the studied view, *t*s(67) < 1.7, *ps* > .09. Thus, there was again little evidence of the effect of the studied view after people made a small turn, even when AD_update was held constant.

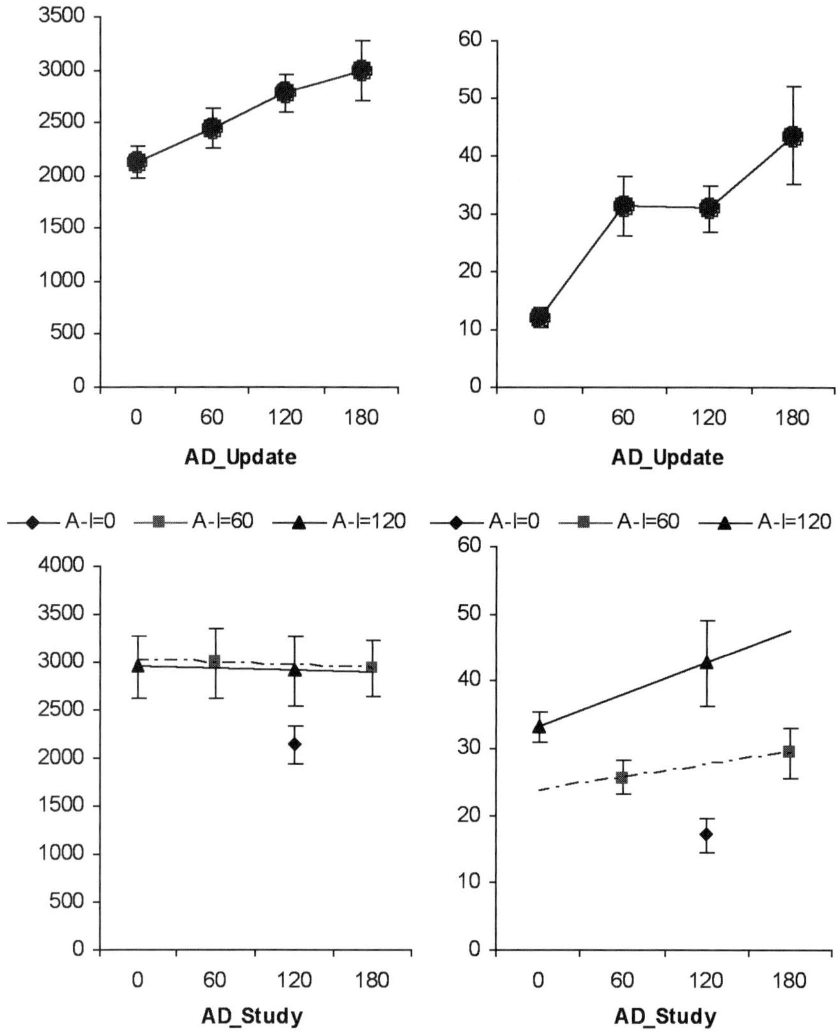

Figure 4-2. Object localization performance when people made a turn after studying targets from one perspective. The top panels showed RT (left panel) and angular errors (right panel), plotted as a function of AD_Update, The bottom panels showed performance as a function of angular disparity relative to the studied-view, grouped by AD_Update (i.e., A-I=0°, 60°, 120°).

2.2.2 View-dependent representations after disorientation

There is another possibility on the failure to show evidence of the studied view. It is possible that the studied view was never really represented under these experimental conditions. For example, because the study orientations were mis-aligned with the walls of the room, they may be very difficult to learn. Shelton and McNamara (2001) showed that when people study an array of objects from two perspectives, first from an angle mis-aligned with the room's axes and followed by one aligned with the room's axes, people ended up not remembering the scene from the initial studied perspective. Instead, they remembered the object layout from the second, canonical view. Furthermore, Mou & McNamara (2002) showed that when specifically instructed to memorize the scene from a canonical perspective while physically studying it from a non-canonical perspective, people are capable of ignoring their true study perspective and show advantage of the canonical perspective rather than the actual studied perspective. Although in the current study people never "studied" the objects from a second perspective, nor were they instructed to pay attention to the canonical view, it is still possible that people somehow never represented the initial studied view and chose to represent the canonical view instead, leading to the benefit in the updated view, which happened to coincide with the room's axis.

To further examine whether the advantage was truly a result of the representation of the updated view, or a result of the canonical view representation, we asked people to study the targets exactly the same way as before, and then disoriented them right after the studying period. Disorientation destroys the updating process by definition, because updating requires one's perception of self-movement, and cannot occur when one loses track of one's heading. Thus, the disorientation procedure should minimize the effect of updating. If people tend to represent the canonical view instead of the studied view in our environment, then they should continue to show an advantage of the canonical view but no advantage of the studied view even after disorientation. On the other hand, if the lack of studied-view effect was due to the subsequent updating process, then there should be a stronger effect of the studied view when the influence of the updating process was eliminated.

We tested ten participants who were randomly assigned to five groups, two in each group. Each group studied the objects while facing one of the target objects, i.e., each group had a different studied-view. Then immediately after the studying period, they were blindfolded and sound-masked, and were asked to turn themselves in the swivel chair for one minute, changing directions from time to time to induce disorientation. At the end of the disorientation period, they were asked to stop, facing whatever direction they happened to be facing, and sat there for 30 seconds to recover from any physical disturbance. Then they were tested in 20 trials as before, imagining facing each of the objects and pointing to each of the remaining ones. Again, this design allowed the two independent variables, AD_study and AD_canonical, be manipulated independently across subjects.

Performance (RT and error) was analyzed as a function of the angular disparity from the studied view and from the canonical view corresponding to the *Poster*. As shown in Figure 4-3, top panels, performance decreased significantly as the test perspective deviates further from the studied view, for both RT and angular errors, $ts(43) > 2.7$, $ps < .01$. In contrast, performance was not affected by the angular disparity between the test perspective and the canonical direction, $ts(43) < 1.2$, $ps > .25$.

These data suggest that people did remember the objects from the perspective they studied them, even though the room is relatively small and these studied perspectives were mis-aligned with the room's axes. These data further suggested that the lack of an effect of the studied view when people took a small turn but not disoriented was not due to their failure to represent the studied view to begin with, because the learning stage was the same in both studies, and people did show evidence of the studied-view representation if they were disoriented. Thus, it was clear that the studied view was what people initially memorized, but the subsequent updating process eliminated its effect.

There are two possible reasons that the effect of the studied view may be eliminated after updating occurred. First, the updated-view representation may replace, or wipe out the studied view representation itself, and therefore the representation is gone forever. Shelton and McNamara (2001) suggested that canonical views may replace, or override an initial "studied view" representation. Thus, it is also possible that a similar "replacement" process might occur between updated view and the studied view.

Second, the updated view may dominate, or mask the studied view and thus the studied-view representation is temporarily un-accessible when updating is in operation. To examine these possibilities, we asked people to study targets in a room from a given perspective, then make a turn while blindfolded (i.e., update), and then get disoriented for testing. If updating makes the studied-view representation temporarily un-accessible, then disorientation, which disables the updating process, should unmask the effect of the studied view. In contrast, if updating abolishes the studied-view representation once it occurs, then there shouldn't be any recovery of the studied view even after disorientation.

We tested twelve participants who were randomly assigned to two groups, one facing the *Computer* during studying and one facing the *Closet*. They were then blindfolded and turned to face the *Closet* and *Computer*, respectively, and were asked to sit there for one minute, and think about where the objects were from their current heading to ensure that updating was completed. Then they were disoriented and tested on their spatial

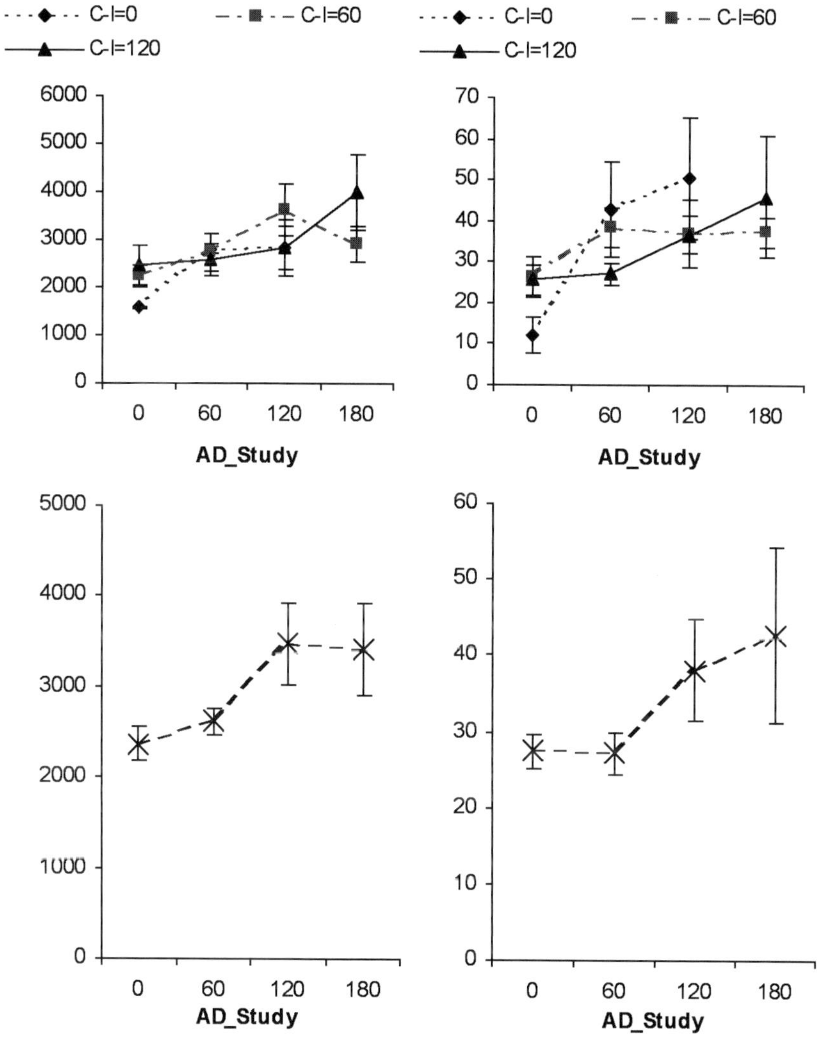

Figure 4-3. Object localization performance when people were disoriented. The top panels showed RT (left panel) and angular errors (right panel) as a function of AD_Study, grouped by angular disparity relative to the canonical direction (C-I=0°, 60°, 120°, 180°). The bottom panels showed performance as a function of AD_Study.

judgments as before. In this procedure, people should encode the studied view first, then generate an updated-view representation after they made a small turn while maintaining their sense of direction. After that, the updating process was disrupted before testing. If the updating process temporarily obscures the studied-view representation, then people should show advantage of the studied view. If the studied-view representation was replaced by the updated-view representation, then there should be no recovery of the studied view even when participants were disoriented during testing.

The results were shown in Figure 4-3, bottom panels. Performance was analyzed as a function of the angular disparity between the imagined heading and the studied view, and between the imagined heading and the updated view (i.e., the orientation they were facing right before the disorientation procedure, not their actual heading after disorientation). People's judgments of the target direction were significantly impaired when the test perspective deviated further away from the studied view (significant for RT, $t(53) = 2.1$, $p = .05$, non-significant for errors, $t(53) = 1.7$, $p = .10$)[2]. In contrast, there was no angular disparity effect relative to the updated view after people were disoriented (ts(53) < 1). A further analysis selecting test perspectives that had AD_study constant (120°) while AD_update varied (0° and 120°) again showed little evidence of the updated-view representation ($ts(22) < 1.5$, $ps > .16$).

Several conclusions can be drawn based on these findings. First, studied-view representations are preserved when people change perspectives physically. Second, the spatial updating process limits one's access to the studied view representation while it is in operation, but does not abolish the representation itself, so that the studied view can be retrieved again if people are disoriented after updating occurs. Third, the studied-view and the updated-view representations are independent of each other, updating can occur without the studied view representation. Fourth, the lack of AD effect for the updated view after people are disoriented suggested that the updated-view representation is relatively transient and is lost once the updating process is disrupted.

3. MODELS OF SPATIAL REPRESENTATION SYSTEMS

Recently McNamara and colleagues proposed a model (McNamara, 2003; also see Mou and McNamara, 2002; Mou et al., 2004) to account for new findings on the relationship between studied view, canonical view and updated view representations. McNamara (2003) wrote,

[2] Since participants were disoriented and unaware of their true physical heading, the analysis of interest was whether they retained the "updated view" representation corresponding to their heading before disorientation.

> "... the spatial structure of the environment is represented in terms of an intrinsic reference system; one defined by the layout itself (e.g., the rows and columns formed by chairs in a classroom). Intrinsic directions or axes are selected using cues, such as viewing perspective and other experiences (e.g., instructions), properties of the objects (e.g., they may be grouped together based on similarity or proximity), and the structure of the environment (e.g., geographical slant)."

According to this model, 1) the representation encodes allocentric object-to-object relationships; 2) the direction reference can be selected in various ways, either according to the viewer, the object array itself, or the environment geometry; 3) the spatial working memory system which is primarily for guiding actions updates the viewer's position and orientation relative to the same reference frame, e.g., treating the "self" as just another object, and the egocentric self-to-object relations are then computed from the object-to-object relationships in LTM. This model is similar to Sholl's model (Easton & Sholl, 1995), which also includes an allocentric LTM system and an egocentric working memory system.

We agree that there are LTM representations which are typically not updated, and working memory representations that are updated as the viewer moves. However, we disagree on the nature of the LTM and the nature of updating in the working memory system. We believe that 1) the working memory system encodes egocentric, self-to-object relationships and these relationships are updated directly based on self-motion cues; 2) the LTM system either takes an instance of the working memory representation (i.e., the studied-view), or rotated versions of this representation if rotation leads to a more symmetric, simpler form (i.e., canonical-view). In either case, the representation is of the same, or very similar, nature as the working memory representation. 3) the updating system does not typically consult the LTM representation and can operate independently. In fact, while the updating system is in operation, access to the LTM may be limited or even completely eliminated. 4) when updating is disrupted, LTM may be retrieved to re-initiate the updating process. 5) the updated representations are transient and are typically not preserved when updating is disrupted.

3.1 Two critical issues

There are several critical issues that can potentially distinguish between these models. Here we discuss two of them.

3.1.1 The nature of updating

My colleagues and I conducted a series of studies to examine the nature of spatial updating. There are three pieces of evidence supporting the hypothesis that spatial updating is an independent process that does not rely on another LTM representation. First, people's knowledge of the object-to-object relationship is disrupted by disorientation per se, above and beyond normal memory decay (Wang and Spelke, 2000), suggesting that the updating system relies on representations that are dynamic, or process-dependent, not "static" type of memory such as LTM. Second, a recent study showed that the efficiency of updating depends on the number of objects being updated (Wang et al., 2006). This effect is a direct consequence of updating self-to-object relationships, but is difficult to explain if people treat "self" as just another object and simply update its position relative to the external reference frame. Third, the studies discussed earlier in this chapter showed that updating can operate normally while people had little access to the LTM, again cast doubt on its dependence on LTM. Based on the converging evidence, we believe the updating system is an egocentric system that represents self-to-object relations directly.

3.1.2 The nature of LTM

There is little direct evidence supporting the object-to-object relationship coding, neither is there conclusive evidence against it. Mou and McNamara (2002) showed that a regular array of objects can induce view-dependent representations corresponding to its axis even if that view was never experienced. However, selection of a direction does not imply which relationships are coded. McNamara and colleagues used a judgment-of-relative-directions task to measure knowledge of the object-to-object relationship. However, relative direction judgments can also be made using representations of self-to-object relations, and there is no direct evidence that shows what the task actually addresses.

There are two important issues related to the nature of the LTM. First, the number of object-to-object relations increases exponentially as the number of objects increases. If people only represent a certain-sized subset of these relations, then the chance a specific relation is represented should decrease exponentially. Thus, performance should decrease exponentially as the number of objects increases. In contrast, the number of self-to-object relations is a linear function of the number of objects. Thus, egocentric coding predicts linear relationship between performance and setsize. Second, is the representation position-dependent or position-independent? Object-to-object relationship coding is typically considered position-independent, while self-to-object coding is position-dependent. A position-dependent representation should show cost when the perspective change involves viewer translation, while a position-independent representation should not. Answers to these questions can potentially shed light on the nature of the LTM.

3.2 Other issues

Several questions about the relationship between these view-dependent representations remained. First, when are the studied-view representations accessible during the updating process? The accessibility of the studied-view representations might be affected by their original strength (for example, familiarity, depth of encoding, etc.), and stronger representations are more likely to be accessed. The procedure of the updating process may also influence the accessibility of the studied-view representation. A third possibility is the retrieval task. It has been shown that cognitive tasks and action tasks may rely on different types of representations (Bridgeman et al., 1997; Creem and Proffitt, 1998; Goodale and Milner, 1992; Loomis et al., 1996; Wang, 2004). Thus, it is possible that when a pointing task fails to show evidence of the studied-view representation, a more cognitive task, such as verbal description task, may show it.

The second question concerns the canonical-view representation. Do canonical-view and studied-view representations belong to the same system? Shelton and McNamara (2001) showed that subsequent viewing at a canonical-view can "replace" a previous studied-view representation at an oblique perspective, and studied-view and canonical-views of multiple directions may also co-exist. However, it is not clear whether people truly possess these representations simultaneously, or only one of them at a time but different people have different ones at different time. Although this issue is not central to the nature of these representations, single-representation-at-a-time is more compatible with the egocentric model, while multiple-representation is more consistent with the allocentric model.

Finally, does disorientation truly "destroy" the updated-view representation? If so, what will happen if participants were told their true heading after they were disoriented? Can an updated-view representation ever be stored into LTM? Answers to these questions will allow us to understand fundamental properties of spatial memory and processes.

ACKNOWLEDGMENTS

Thanks to David Irwin, Dan Simons, and Jim Brockmole for discussion, and to Lauren Hennessey for collecting the data. Supported by NSF Grant BCS 03-17681 to R.F. Wang. Some of the data were presented at the 44th Annual Meeting of the Psychonomics Society, Vancouver, British Columbia, Canada, Nov. 6-9, 2003. Comments should be sent to Frances Wang (francesw@cyrus.psych.uiuc.edu).

REFERENCES

Amorim, M. A., Glasauer, S., Corpinot, K., and Berthoz, A., 1997, Updating an object's orientation and location during nonvisual navigation: A comparison between two processing modes, *Percept Psychophys* **59**:404-418.

Biederman, I. and Gerhardstein, P. C., 1993, Recognizing depth-rotated objects: Evidence and conditions for three-dimensional viewpoint invariance, *J Exp Psychol-Hum L* **19**: 1162-1182.

Bridgeman, B., Peery, S., and Anand, S., 1997, Interaction of cognitive and sensorimotor maps of visual space, *Percept Psychophys* **59**:456-469.

Christou, C. G., and Buelthoff, H. H., 1999, View dependence in scene recognition after active learning, *Mem Cognition* **27**:996-1007.

Collett, T. S. and Cartwright, B. A., 1983, Eidetic images in insects: Their role in navigation, *Trends Neurosci* **6**:101-105.

Creem, S. H., and Proffitt, D. R., 1998, Two memories for geographical slant: Separation and interdependence of action and awareness, *Psychon B Rev* **5**:22-36.

Diwadkar, V. A., and McNamara, T. P., 1997, Viewpoint dependence in scene recognition, *Psychol Sci* **8**:302-307.

Easton, R. D., and Sholl, M. J., 1995, Object-array structure, frames of reference, and retrieval of spatial knowledge, *J Exp Psychol Learn* **21**:483-500.

Edelman, S., and Bülthoff, H. H., 1992, Orientation dependence in the recognition of familiar and novel views of three-dimensional objects, *Vision Res* **32**:2385-2400.

Farrell, M. J., and Robertson, I. H., 1998, Mental rotation and the automatic updating of body-centered spatial relationships, *J of Exp Psychol Learn* **24**:227-233.

Gallistel, C. R. (1990). *The organization of learning*, MIT Press, Cambridge, MA.

Goodale, M. A., and Milner, A. D., 1992, Separate visual pathways for perception and action, *Trends Neurosci* **15**:20-25.

Humphrey, G. K., and Khan, S. C., 1992, Recognizing novel views of three-dimensional objects, *Canadian J Psych*ology **46**:170-190.

Loomis, J. M., Da Silva, J. A., Philbeck, J. W., and Fukusima, S. S., 1996, Visual perception of location and distance, *Curr Dir Psychol Sci* **5**:72-77.

May, M., 2004, Imaginal perspective switches in remembered environments: Transformation versus interference accounts, *Cognitive Psychol* **48**:163-206.

Mallot, H. A. and Gillner, S., 2000, Route navigating without place recognition: What is recognised in recognition-triggered responses?, *Perception* **29**:43-55.

McNamara, T. P., 2003, How are the locations of objects in the environment represented in memory? In C. Freksa, W. Brauer, C. Habel, and L. F. Wender (Eds.), *Spatial Cognition*

III: Routes and navigation, human memory and learning, spatial representation and spatial reasoning, Springer-Verlag, Berlin, pp. 174-191.

McNamara, T. P., Rump, B., and Werner, S., 2003, Egocentric and geocentric frames of reference in memory of large-scale space, *Psychon B Rev* **10**:589-595.

Mou, W., & McNamara, T. P., 2002, Intrinsic frames of reference in spatial memory, *J Exp Psychol Learn* **28**:162-170.

Mou, W. M., McNamara, T. P., Valiquette, C. M., and Rump, B., 2004, Allocentric and egocentric updating of spatial memories, *J Exp Psychol Learn* **30**:142-157.

Presson, C. C., and Montello, D. R., 1994, Updating after rotational and translational body movements: coordinate structure of perspective space, *Perception* **23**:1447-1455.

Rieser, J. J., 1989, Access to knowledge of spatial structure at novel points of observation, *J Exp Psychol Learn* **15**:1157-1165.

Shelton, A. L., and McNamara, T. P., 1997, Multiple views of spatial memory, *Psychon B Rev* **4**:102-106.

Shelton, A. L., and McNamara, T. P., 2001, Systems of spatial reference in human memory, *Cognitive Psychol* **43**:274-310.

Simons, D. J. and Wang, R. F., 1998, Perceiving real-world viewpoint changes, *Psychol Sci* **9**:315-320.

Simons, D. J., Wang, R. F., and Roddenberry, D., 2002, Object recognition is mediated by extra-retinal information, *Percept Psychophys* **64**:521-530.

Tarr, M. J. and Pinker, S., 1989, Mental rotation and orientation-dependence in shape recognition, *Cognitive Psychol* **21**:233-282.

Tarr, M. J., Williams, P., Hayward, W. G. and Gauthier, I., 1998, Three-dimensional object recognition is viewpoint-dependent, *Nature Neurosci* **1**:275-277.

Ullman, S., 1989, Aligning pictorial descriptions: An approach to object recognition. *Cognition* **32**:193-254.

Waller, D., Montello, D. R., Richardson, A. E. and Hegarty, M., 2002, Orientation specificity and spatial updating of memories for layouts, *J Exp Psychol Learn* **28**:1051–1063.

Wang, R. F., 2003, Spatial representations and spatial updating, in D. E. Irwin & B. H. Ross eds., *The Psychology of Learning and Motivation* 42 *Advances in Research and Theory: Cognitive Vision*, Academic Press, San Diego, CA, pp. 109-156.

Wang, R. F., 2004, Action, verbal response and spatial reasoning, *Cognition* **94**:185-192.

Wang, R. F., Crowell, J. A., Simons, D. J., Irwin, D. E., Kramer, A. F., Ambinder, M. S., Thomas, L. E., Gosney, J. L., Levinthal, B. R., and Hsieh, B. B., 2006, Spatial updating relies on an egocentric representation of space: Effects of the number of objects, *Psychon B Rev* **13**:281-286

Wang, R. F. and Simons, D. J., 1999, Active and passive scene recognition across views, *Cognition* **70**:191-210.

Wang, R. F. and Spelke, E. S. , 2000, Updating egocentric representations in human navigation, *Cognition* **77**:215-250.

Wang, R. F. and Spelke, E. S., 2002, Human Spatial Representation: Insights from Animals, *Trends Cogn Sci* **6**:376-382.

Werner, S. and Schmidt, K., 1999, Environmental reference systems for large-scale spaces, *Spatial Cognit Comput* **1**:447-473.

Chapter 5

MODELING MENTAL SPATIAL KNOWLEDGE PROCESSING
An AI Perspective

Thomas Barkowsky
Universität Bremen, Germany

Abstract: This chapter addresses mental spatial knowledge processing from an artificial intelligence (AI) point of view. It first reviews the characteristics of mental representation structures used for mental reasoning processes, motivates mental knowledge processing as a construction process, and points to the use of external media as an important factor in dealing with more complex spatial problems. A range of models of intelligent spatial information processing has been proposed both in psychology and in AI. After giving an overview of selected models, the architecture *Casimir* is presented as a framework for modeling spatial reasoning with mental models and mental images. On the basis of this architecture, the design of representation structures in working memory, the task of computationally modeling mental processing of shape information, and the issue of controlling mental resources in reasoning processes are discussed as challenging issues from an AI perspective. The chapter closes with some considerations regarding the assessment and validation of computational models of mental spatial knowledge processing.

Key words: computational modeling; spatial reasoning; spatial knowledge processing; mental imagery; spatial mental models; Casimir.

1. MENTAL SPATIAL KNOWLEDGE PROCESSING

Being an indispensable precondition of any goal-directed behavior of intelligent species in an environment, spatial knowledge processing has received intense attention in cognitive science. Spatial skills rely on mental facilities that go far beyond abstract symbol manipulation (e.g., Huttenlocher, 1968). Mental spatial reasoning capabilities are directly intertwined with perceptual and motor systems (e.g., Barsalou, 1999). Thus, research results in spatial cognition have helped to develop a radically modified perspective on intelligent information processing (cf. Wilson, 2002). The present contribution addresses mental spatial knowledge processing from an artificial intelligence (AI) point of view. Since the aim of the research reported in this chapter is the construction of a cognitively adequate computational model of mental spatial reasoning, it deals with the representation structures required for modeling mental spatial knowledge processing as well as with the processes that make use of these structures.

1.1 Analogical Representations in the Mind

Numerous research results in cognitive psychology suggest that human working memory (WM) for spatial knowledge possesses spatio-analogical characteristics. Spatio-analogical representation structures have been described as having a structural correspondence with what they stand for: entities and parts of entities in the representation together with their properties and relations correspond to objects, parts of objects, properties and relations in the represented domain (cf. Sloman, 1975). In cognitive science, two types of spatio-analogical mental representations have been identified: (visual) mental images and (spatial) mental models.

One of the first studies to detect the spatio-analogical characteristic of visual mental images was an investigation of mental rotation processes (e.g., Shepard and Metzler, 1971), showing that there must be a structural correspondence between the objects visually presented to the participants of the study and the mental representations employed by them in a mental reasoning task. Accordingly, mental images have been defined as mental re-constructions that resemble perceptions of objects or events taking place in the real world (Finke, 1989). Meanwhile, the idea of image-like mental representations has led to a vast amount of research in the field of mental imagery (e.g., Kosslyn, 1980, 1994).

However, the question whether the empirical results in mental imagery research necessarily point to truly analogical representations in the mind or whether they also could be explained through non-analogical representation structures has been an issue of ongoing discussions in the *imagery debate* (see Tye, 1991, for an overview). Nowadays, the imagery debate seems to have been decided in favor of the proponents of analogical representation structures in the mind (Kosslyn, 1994; Bartolomeo, 2002), although there are still valid arguments in favor of non-pictorial mental representation formats (Pylyshyn, 2003).

As another spatio-analogical (albeit less image-like) form of mental representation, *(spatial) mental models* have been proposed to account for the structural correspondence between the mental representation and the real-world state of affairs related to them (Johnson-Laird, 1983). Spatial mental models are conceived of as preserving the spatial relationships among represented objects, but not the more detailed visual aspects of entities in a scenario (like colors or specific shapes of entities). As such, spatial mental models can be seen as a more abstract form of mental representation of spatial knowledge.

In the human mind, the *visual* and the *spatial* form of representation (also known as *what* or *ventral* system and *where* or *dorsal* system, respectively) seem to coexist and to interact which each other (Mishkin et al., 1983; Kosslyn and Shin, 1994; Klauer and Zhao, 2004). In this way, the respective advantages of each form of representation are utilized and both forms of representation may complement each other. The use of the visual form of representation, being more detailed than the spatial one, however, may also be disadvantageous in abstract spatial reasoning tasks (Knauff and Johnson-Laird, 2002).

1.2 Spatial Knowledge Processing as Construction

The investigation of mental representation structures as mental images or mental models focuses on the form of representation in the mind. However, since any representation has to be regarded considering the corresponding processes (Palmer, 1978), also analogical representations require processes that realize their analogical facilities and exploit the representation structure in a suitable manner (Sloman, 1975).

In the attempt to find an adequate metaphor for grasping mental spatial representation, numerous alternative conceptions have been proposed, among them *cognitive maps* (Tolman, 1948), *cognitive atlases* (Kuipers, 1982), *cognitive collages* (Tversky, 1993), or *(human) geographic information systems* (Hirtle, 1998). Based on these metaphors, a vast amount of empirical experiments have been carried out, aiming at a more detailed body of knowledge about the nature of mental spatial representations. The results of these experiments revealed that mental spatial knowledge can be characterized as being distorted, incomplete, heterogeneous, and even conflicting (e.g., Tversky, 1993). A process-related aspect common to the findings gained from most studies, however, is that mental spatial representations are not available in the mind in a ready-made form, but that they are constructed on demand and depending on the given situation (Tversky, 1992).

This *constructionist* view of human mental processing of spatial knowledge applies to all mental processes and representation structures involved in dealing with spatial information:

- In long-term memory (LTM), distinct pieces of knowledge organized in semantically structured hierarchies or associative networks are retrieved for being further processed in WM (Kosslyn, 1980). Moreover, every LTM access tends to modify the memory structure used (e.g., modification of associative relations or degree of availability).
- In WM, the construction proper takes place in the form of spatial mental models or mental images as described above.
- Visual perception of a given spatial arrangement or of a spatial scene depends to a great extent on information that is already available in the mind. This information is used to generate hypotheses that are employed for making sense of the pieces of information acquired through the senses (Kosslyn and Sussman, 1995).
- When an external visuo-spatial representation is used to enhance mental reasoning (e.g. drawing a diagram on a sheet of paper, see next section), discrete spatial entities are sequentially put to the external medium to form a sensible spatial arrangement.

1.3 Coping with Mental Limitations: Using External Representations

Visual perception and mental imagery depend to a large extent on common mental subsystems (Ishai and Sagi, 1995; Kosslyn, 1994; Kosslyn and Thompson, 2003). That is, mental imagery is based on mental subsystems that are also functionally involved in scene and object recognition based on real visual input. Vice versa, mental images play a crucial role in visual recognition processes. In fact, from an evolutionary perspective, the original function of imagery was to support vision, not to serve as a mental reasoning mechanism (cf. Kosslyn and Sussman, 1995).

However, in mental reasoning contexts, this relationship between vision and imagery is also used the other way around: visual perception may facilitate reasoning. Since the capacity of WM is restricted with respect to the number of items that can be dealt with simultaneously as well as with respect to the complexity of the relationships that may hold between these entities (Miller, 1956; Cowan, 2001; Vogel et al., 2001), we often use external visual media (e.g., paper and pencil or blackboard and chalk) to extend our mental reasoning capabilities.

The advantage of this externalization of mental spatial representations is twofold: on the one hand, as just mentioned, capacity limits are overcome; on the other hand, external visuo-spatial representations extend the reasoning power compared to pure mental reasoning (Goldschmidt, 1995; Suwa et al., 1999). The reason for the latter effect is that external visual representations are much less determined regarding the identification and arrangement of entities contained in them than mental representations (cf. Chambers and Reisberg, 1992; Verstijnen et al., 1998). Consequently, the detection of new relations, features, and entities is facilitated in an external visual representation. Expert designers and architects widely make use of this effect when producing sketches during the early phases of the design process (cf. Purcell and Gero, 1998).

2. MODELS OF SPATIAL KNOWLEDGE PROCESSING

A range of diverse models of spatial knowledge processing has been proposed, both in artificial intelligence (AI) and in cognitive psychology. These models can be classified with respect to their intended cognitive adequacy: there are models that aim at explaining cognitive processes in a human's mind, and there are models – although inspired by results from psychology – that primarily aim at providing a technical solution to spatial reasoning problems. Models belonging to both classes will be briefly reviewed in the following.

2.1 Models Aiming at Cognitive Adequacy

An early comprehensive model of visual mental imagery based on empirical studies has been presented by Kosslyn (1980). This model has completely been implemented as a computer program. It employs a regular raster as spatio-analogical representation structure in which mental images are constructed, inspected, and manipulated. Like the human retina and the corresponding brain systems, this *surface representation* is characterized by a limited spatial extent and a limited resolution, which decreases towards the periphery. It roughly possesses a circular structure and the represented content fades out over time if not periodically refreshed. The information used for image construction is taken from the *deep representation* (corresponding to human long-term memory). In this representation, both literal (shape-related) and propositional information (spatial relationships) is encoded in various individual data files.

A later model of human mental imagery by Kosslyn (1994) is based on neuropsychological findings about visual perception and high-level vision processes. This functional imagery model, being conceptually compatible with the earlier model, does not come in an implemented form. Moreover, since its structure is oriented on the brain regions involved in perception and imagery, it is not organized according to representation structures and corresponding processes, but rather according to distinct functional mental subsystems. Essential subsystems in this model are the *visual buffer*, which contains the mental image, together with the *attention window* on this buffer that is used to coordinate activities on the mental image. An *associative memory* takes the role of a long-term store that provides the imagery and perception processes with the necessary information. Most subsystems dealing with mental images exist in two forms: as *categorical* systems for prototypical spatial information (e.g., to construct an image of any house) and as *coordinate* systems for specific spatial information (e.g., to evoke the mental image of the White House).

Inspired by Kosslyn's second imagery model, the computational model *MIRAGE* (Barkowsky, 2001, 2002) describes the construction of mental images for the solution of reasoning problems in geographic space. *Spatial knowledge fragments* are retrieved from a hierarchical long-term memory (LTM) component and are further processed in working memory (WM). Since the model assumes that knowledge available from LTM usually is underdetermined (i.e., it is on a coarse level and/or incomplete), pieces of knowledge retrieved from LTM are complemented by default components in a *conversion* process in WM. After this pre-processing, a visual mental image is constructed and inspected in a quasi-pictorial *visual buffer*.

A computational mental reasoning model for non-visual spatial mental models has been presented by Ragni and co-workers (2005). This *SRM* model is based on a two-dimensional array with a spatial focus that is used to locate, manipulate, and inspect objects in the array. The aim of this model is to provide a computational complexity measure corresponding to cognitive effort, which can be empirically validated. Aspects of this complexity measure are the specific difficulty of the reasoning problem dealt with as well as cognitive preferences that influence the construction of a specific mental model when there are several potential solutions to a given problem (cf. Knauff et al., 1995).

2.2 Technically Motivated Solutions

Research results in cognitive psychology related to mental imagery have inspired a number of technical reasoning systems in AI. An early application of results in vision and visual mental imagery research has been developed by Funt (1980). His diagrammatic reasoning system *WHISPER* employs a circular representation structure for performing on mechanical reasoning tasks. This *retina* represents entities perceived from a blocks world representation and carries out visuo-spatial operations like rotation, scaling, or translation. Moreover, it is used for analyzing spatial properties like symmetries or topological relations. The overall reasoning process is controlled by a propositional problem-solving system.

In a similar direction, the *DEPIC-2D* system described by Khenkhar (1991) uses a rectangular *cell matrix* to represent and reason about propositionally available spatial knowledge. On the depiction in the cell matrix, neighborhood-based processes are employed to construct spatial regions or to explore spatial relationships between represented entities.

With their *computational imagery* approach, Glasgow and Papadias (1992) present a diagrammatic reasoning system for dealing with spatial reasoning problems in geography or in chemistry. Computational imagery is inspired by Kosslyn's (1980) model and aims at applying research results in mental imagery to technical reasoning problems. The surface representation in Kosslyn's model is split into two systems in computational imagery: a positional diagrammatic representation structure for *visual* aspects of entities (e.g., shape, color, or texture) and a relational structure for *spatial* relationships between objects (cf. Logie, 1995). Moreover, as in Kosslyn's model, there is a *deep representation* for storing information necessary for the reasoning processes.

3. CASIMIR: AN ARCHITECTURE FOR MENTAL SPATIAL KNOWLEDGE PROCESSING

In the following, various aspects of our model *Casimir* (Computational Architecture, Specification and Implementation of Mental Image based Reasoning) will be presented. The aim of Casimir is to realize a cognitively adequate computational description for mental reasoning processes based on mental images and mental models. Casimir is conceptualized as a comprehensive system including mental processes in long-term memory (LTM) and in working memory (WM), as well as the interaction between mental images and external diagrammatic media. Therefore we call it a *cognitive architecture* (Newell, 1990; Byrne, 2002) rather than a *model*. Casimir is a framework for integrating partial modeling aspects rather than a specific computational model.

In comparison to established cognitive architectures (such as ACT-R, Anderson et al., 2004), Casimir's emphasis is on the retrieval of spatial facts from an extensive associative LTM, on the design of adequate structures for spatio-analogical representations in WM, and on the integration of external diagrams in the reasoning process, which so far have not been realized in other architectures (cf. Anderson and Lebiere, 2003).

The following sections give an overview of Casimir's subsystems of LTM activation, construction and exploration of WM representations, interaction with external diagrams, and memory update (see Fig. 5-1 for an overview of the system).

3.1 Long-term Memory Activation

In Casimir, spatial knowledge in LTM is represented as a graph structure. The nodes of this graph represent either spatial concepts (i.e., types of entities like 'country', 'street', or types of spatial relations like 'orientation_relation' or 'topological_relation') or specific instances of these concepts (i.e., entities like 'France', 'Broadway', or relations like 'north', 'connected'). Moreover, there are two types of edges connecting the nodes: a) edges that represent ontological relationships between nodes (i.e., 'is_a' links) and b) edges that connect instances of relations with the entities for which the respective relation holds.

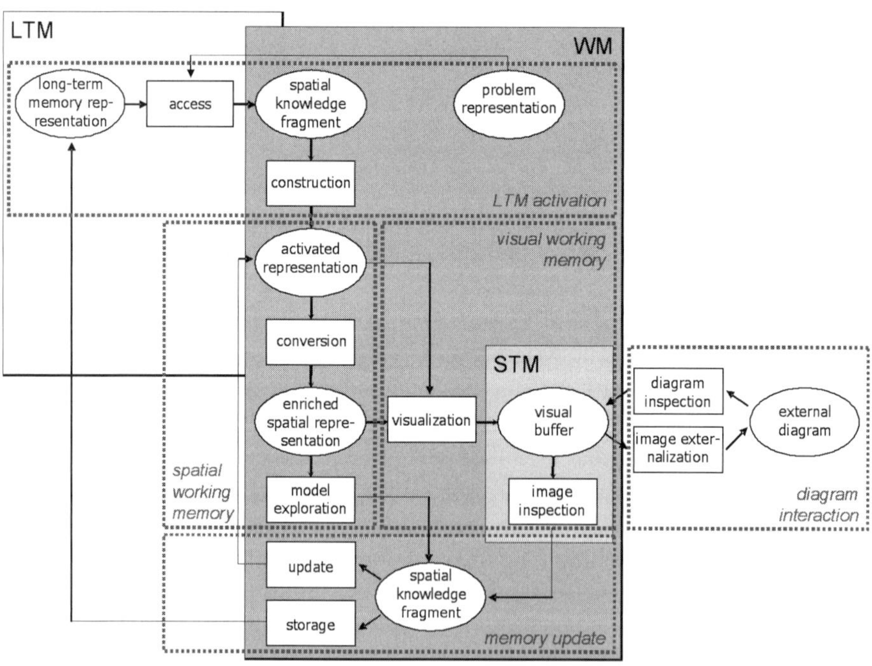

Figure 5-1. Casimir's subsystems and functional components.

For instance, if the task at hand is to find out the direction in which Paris (France) is located with respect to Rome (Italy), the three nodes 'Paris', 'Rome', and 'orientation_relation' are initially activated in LTM based on the problem representation.

As a result of the spreading activation process, a set of *spatial knowledge fragments* is identified that may be helpful for answering a specific spatial question. These spatial knowledge fragments (which are described through a propositional representation format) provide the basis for the subsequent construction processes in WM. As a first step, they build up an *activated representation* that is used for further processing steps in WM.

3.2 Working Memory Construction and Exploration

The spatial knowledge fragments contained in the activated representation are transformed into a spatio-analogical representation in WM by a *conversion process*. The resulting *enriched spatial representation* is a relational structure which is specific with respect to the type of spatial knowledge to be represented (e.g., topological or orientation knowledge) and it preserves essential properties of the represented spatial relations in the sense of mental models (Johnson-Laird, 1983). Note that there are no visual aspects (like color or specific shapes) contained in this type of representation.

As an effect of integrating the spatial knowledge fragments into the spatio-analogical representation, however, further spatial relations that do not stem from LTM are gained through the representation structure. For instance, two topological relations between entities A and B and between B and C automatically result in a topological relation between A and C being represented in the spatio-analogical structure. Thus, the enriched spatial representation can be mentally utilized by exploration processes in order to extract new spatial pieces of knowledge, which are represented as spatial knowledge fragments.

As mentioned, the enriched spatial representation is a purely spatial representation. However, when a specific reasoning task requires visual information, the enriched spatial representation can be further specified towards a visual representation in WM. This *visualization* results in a *visual mental image* in a visual buffer representation. This visual mental image either can be inspected by *image inspection* processes to extract new spatial or visual relations, or it can be externalized towards an external diagrammatic representation.

3.3 Interaction with External Diagrams

The *image externalization* process transfers the visual mental image to an external diagrammatic medium. By this, the mental capacity for mental images is extended, since the external diagram can be used as an intermediate store for images or parts of images (for example for exploring a number of alternative partial solutions during a reasoning process). Every detail of a mental image that is externalized can be reloaded at a later stage of a reasoning process.

However, the externalization of a visual mental image also may directly enhance the reasoning process. Since the visual mental image is constructed from distinct entities represented in WM, and since the spatial relations that hold between those entities are determined by the underlying spatial WM structure, it is often difficult to identify new entities (e.g. parts of entities that have not been identified before) or new relations (e.g., between spatially distant entities) in a visual mental image. In this case, externalizing a visual mental image is helpful, since visual perception of an external diagram allows aggregations or subdivisions of represented entities in a much more flexible manner than the inspection of a mental image (cf. Chambers and Reisberg, 1992).

3.4 Memory Update

Spatial knowledge fragments gained by model exploration or image inspection processes in WM can be used to update memory content, both of LTM and WM. For memory update in LTM, the spatial knowledge fragment obtained from the mental reasoning process is integrated in the LTM structure, such that it can be retrieved in later reasoning contexts if required.

However, a spatial knowledge fragment gained from model exploration or image inspection can also directly be fed into the reasoning process going on in WM. That is, it is integrated in the activated representation in WM which forms the basis for the construction of a spatial mental representation or a mental image. Through this mechanism, reasoning processes that are performed in subsequent steps can be realized.

4. CHALLENGES FROM AN ARTIFICIAL INTELLIGENCE POINT OF VIEW

Casimir is designed as a framework for modeling mental spatial knowledge processing. To turn the system into a fully specified architecture, however, the subsystems described above have to be further detailed regarding representation structures and processes. In the following, I will focus on three core questions that illustrate this process of detailing Casimir towards a specific model.

4.1 Design of Representation Structures for Working Memory Representations

As described above, spatio-analogical representations in working memory (WM) can be either spatial or visual, where the spatial representation is assumed to be constructed first during a reasoning task and may be further processed towards a visual mental image. From the perspective of cognitive economy (Collins and Quillian, 1969), however, it is sensible to assume that a mental representation is only made as specific as necessary with respect to the task to be performed on. Thus, a visual mental image is only formed when truly visual information has to be dealt with, and similarly the spatial representation is only made as specific as necessary.

From a theoretical point of view, types of spatial information can be ordered with respect to their structural complexity, starting from simple (one-dimensional) ordering aspects (e.g., A is left of B), over more complex types like topological information (e.g., A is overlapping B) or information about orientation of entities with respect to each other (e.g., A is west of B), up to fully specified metrical representations of spatial locations and detailed shapes (cf. Cohn, 1997; Vieu, 1997). With respect to spatial representations in WM, the question arises how the different types of spatial information are dealt with, i.e., whether there is one kind of representation structure that suits all types of knowledge (which would not be economic when only simple types of knowledge are processed) or whether there are diverse structures for the different types of spatial knowledge.

In the latter case, there must be mechanisms for integrating different types of knowledge (for instance when ordering knowledge has to be combined with orientation knowledge during a reasoning process). So either there is an integrating structure that subsumes the different specialized structures (a kind of meta-structure) or different types of representation structures are amalgamated to obtain more complex structures, which would result in a concept of *scalable representation structures* in the mind that are adapted to the types of knowledge to be dealt with.

This consideration can also be extended towards the relation between visual and spatial representations in the mind. Visual and spatial aspects of mental representation interact with each other and complement each other. But are they really two distinct representations (as for instance conceived in the computational imagery approach by Glasgow and Papadias, 1992) or is the visual representation rather an extension of spatial representations, meaning that a given spatial representation is enhanced by visual aspects if required by a reasoning task?

From a neuropsychological point of view, both interpretations are sensible. Specific brain areas can be assigned specific roles within a reasoning task which speaks in favor of distinct representation structures. However, since the subsystems involved in a mental reasoning task closely interact with each other, also the conception of integrated, scalable representation structures forms a sensible basis for the design of a computational model.

4.2 Modeling Mental Processing of Shape Information

An interesting case from a computational modeling point of view is the mental processing of shape information, since virtually all mental subsystems interact with each other in spatial mental reasoning with shapes. Thus, mental processing of shapes can be regarded as a valuable scenario, especially since it combines spatial and visual aspects of mental representation.

For the construction of shapes in (visual) WM, knowledge fragments that describe pieces of shape information are retrieved from long-term memory (LTM). Since these shape fragments are used to evoke specific shapes in a mental image, they can be characterized as *visual*. On the other hand, pieces of shape information often are combined to form more complex mental images. For this purpose, information about how partial shapes are related to each other in a complex scenario is needed. Obviously, the relational organization of partial shape aspects is a *spatial* knowledge aspect.

On the one hand, specific shapes are represented in WM in the form of visual mental images; on the other hand, shape properties widely interact with purely spatial aspects. For instance, whether a specific spatial relation (say, a topological relation like *A is contained in B*) holds or does not hold, directly depends on the entities' shapes. Thus, the question of how spatial and visual mental representations interact in the human mind is a central concern for mental shape processing.

When mental shape processing is regarded with respect to the interaction between mental imagery and visual perception, the relation between visual and spatial knowledge aspects become even more obvious. In image externalization, shape aspects (i.e., pieces of shape information and the spatial relationships among them) are processed through image externalization with subsequent visual perception. By this mechanism, new shape aspects can be identified and re-internalized from the external diagram, or previously not represented spatial relations may be found between parts of shapes. In object and scene recognition, the spatial relation between shape features is used to control visual attention processes (e.g., saccadic eye movements) and to verify hypotheses about object or scene characteristics (Schill et al., 2001).

4.3 Controlling Mental Resources

As illustrated in the case of mental processing of shape information, all mental subsystems highly interact in spatial mental reasoning tasks. Retrieval processes in LTM, construction processes in WM, the exploration of constructed mental representations, as well as the interaction with external diagrammatic media have to be coordinated. Therefore, control processes required to direct this interaction are a central concern in the design of computational models of mental reasoning.

Most existing conceptions of mental visuo-spatial reasoning postulate distinct mechanisms for controlling attention and the distribution of available mental resources. These mechanisms are given by explicit structural representations that guide the construction of mental representations (e.g., in the discursive representation of Kosslyn's model of 1980), by causal knowledge that is used to direct attention and reasoning processes (e.g., in mental animation, Hegarty, 1992), or by representations of sequences of eye movements corresponding to an object or a scene (e.g., sensorimotor representations, Schill et al., 2001).

However, research results in cognitive psychology and neuroscience indicate that there seems to be no central instance (like a specific representation or subsystem) that controls the overall reasoning process (e.g., Monsell and Driver, 2000; Nobre et al., 2004). Rather, cognitive control seems to be a distributed, emergent phenomenon that evolves from the interaction between all subsystems contributing to the mental reasoning process (cf. Hommel et al., 2004). From a computer science point of view, it has been assumed that corresponding analogical properties of the mental spatial representation structures involved, together with spatial processing mechanisms produce emergent control characteristic in, as well as across the diverse mental subsystems (Engel et al., 2005). Under a more general cognitive modeling perspective, cognitive control processes can be conceived of as being associative. That is, mental stimuli induce further mental activity in a given context which results in an autonomous flow of control (Schultheis, 2005).

5. ASSESSING AND VALIDATING MODELS OF MENTAL SPATIAL KNOWLEDGE PROCESSING

This chapter presented the cognitive architecture Casimir as a framework for building specific models of complex spatial reasoning processes that go on in the human mind. As it is the case in any cognitive architecture, representation structures and processes operating on these structures need to be fully specified before they can be realized in an implemented computational model. This specification task is done on the basis of research results from behavioral investigations and data gathered in the neurosciences.

However, these results have to undergo an interpretation process during the computational modeling task, since there is no strict relationship between a given observable phenomenon and a specific realization of that phenomenon in a computational model (cf. Anderson, 1978). Moreover, usually not all issues to be included in a comprehensive computational architecture are supported by results from empirical studies. As a consequence, design decisions have to be taken on the basis of sensible assumptions for the sake of the completeness of the model.

So, given the aforementioned restrictions on the modeling process, how can the validity of a model be assessed and how can we make sure that we are able to design an adequate computational model?

- From an empirical point of view, a central criterion for a model's validity is its predictive power. That is, behavior that can be observed in the model while performing on a given task (measured, for instance, in terms of reaction times or error rates) should be in line with human subjects performing on the same task.
- Since any computational model is designed on the basis of a set of empirical studies, the results of these studies have to be reflected in the model. Moreover, the design decisions taken in the model have to be continuously checked against empirical findings that may be gained after

the initial model has been completed. As soon as a new result contradicts any conception in the existing model, it will have to be revised.

- As mentioned above, comprehensive computational models always contain aspects that are not based on experimental findings, but that have been based on plausible assumptions or the model designer's intuition. To close these gaps in the overall conception of the computational model, further empirical investigations have to be carried out to make a given model more and more adequate (or to scientifically prove its adequacy).
- From a more application-oriented perspective, any computational model should be tested in real-world contexts, be it in the form of an implementation in an autonomous agent, or be it in an interactive human-computer system. In such application scenarios, the computational model can prove its capabilities off the usual lab context. Its usability in real-world problem contexts may serve as a touchstone for the modeling decisions taken.
- Finally, from a computer science point of view, any framework for constructing specific computational model should be constructed on the basis of modular components that easily can be modified and exchanged to provide sufficient flexibility for adaptations and extensions that may become necessary during the modeling process. If an existing model gets more and more complex, a modular system structure helps to keep the model comprehensible as well as maintainable.

6. ACKNOWLEDGMENTS

I would like to thank my colleagues Christian Freksa, Sven Bertel, Holger Schultheis, Inessa Seifert, and Dominik Engel for countless discussions, productive criticism, and their individual contributions to the research presented in this chapter. We are jointly developing Casimir in the project R1-[ImageSpace] of the Transregional Collaborative Research Center SFB/TR 8 Spatial Cognition: Reasoning, Action, Interaction. Funding by the German Research Foundation (DFG) is gratefully acknowledged.

7. REFERENCES

Anderson, J. R., 1978, Arguments concerning representations for mental imagery, *Psychological Rev* **85**(4):249-277.

Anderson, J. R., Bothell, D., Byrne, M. D., Douglass, S., Lebiere, C., and Qin, Y., 2004, An integrated theory of the mind, *Psychological Rev* **111**(4):1036-1060.

Anderson, J. R., and Lebiere, C., 2003, The Newell test for a theory of cognition, *Behav Brain Sci* **26**:587-640.

Barkowsky, T., 2001, Mental processing of geographic knowledge, in: *Spatial Information Theory - Foundations of Geographic Information Science*, D. R. Montello, ed., Springer, Berlin, pp. 371-386.

Barkowsky, T., 2002, *Mental representation and processing of geographic knowledge - A computational approach*, Springer, Berlin.

Barsalou, L. W., 1999, Perceptual symbol systems, *Behav Brain Sci* **22**(4):577-660.

Bartolomeo, P., 2002, The relationship between visual perception and visual mental imagery: A reappraisal of the neuropsychological evidence, *Cortex* **38**:357-378.

Byrne, M. D., 2002, Cognitive architecture, in: *Handbook of human-computer interaction*, J. Jacko and A. Searse, eds., Lawrence Erlbaum, Hillsdale, NJ, pp. 1-57.

Chambers, D., and Reisberg, D., 1992, What an image depicts depends on what an image means, *Cognitive Psychol* **24**:145-174.

Cohn, A. G., 1997, Qualitative spatial representation and reasoning techniques, in: *KI-97: Advances in Artificial Intelligence*, G. Brewka, C. Habel, and B. Nebel, eds., Springer, Berlin, pp. 1-30.

Collins, A. M., and Quillian, M. R., 1969, Retrieval time from semantic memory, *J Verb Learn Verb Behav* **8**:240-247.

Cowan, N., 2001, The magical number 4 in short-term memory: A reconsideration of mental storage capacity, *Behav Brain Sci* **24**(1):87-114.

Engel, D., Bertel, S., and Barkowsky, T., 2005, Spatial principles in control of focus in reasoning with mental representations, images, and diagrams, in: *Spatial Cognition IV Reasoning, Action, Interaction. International Conference Spatial Cognition 2004,* C. Freksa, M. Knauff, B. Krieg-Brückner, B. Nebel, and T. Barkowsky, eds., Springer, Berlin, pp. 187-209.

Finke, R., 1989, *Principles of mental imagery*, MIT-Press, Cambridge, MA.

Funt, B., 1980, Problem-solving with diagrammatic representations, *Artif Intell* **13**:201-230.

Glasgow, J., and Papadias, D., 1992, Computational imagery, *Cognitive Sci* **16**:355-394.

Goldschmidt, G., 1995, The designer as a team of one, *Design Studies* **16**:189-209.

Hegarty, M., 1992, Mental animation: inferring motion from static displays of mechanical systems, *J Exp Psychol Learn Mem Cogn* **18**(5):1084-1102.

Hirtle, S. C., 1998, The cognitive atlas: using GIS as a metaphor for memory, in: *Spatial and temporal reasoning in geographic information systems*, M. Egenhofer and R. Golledge, eds., Oxford University Press, Oxford, pp. 267-276.

Hommel, B., Daum, I., and Kluwe, R. H., 2004, Exorcizing the homunculus, phase two: Editors' introduction, *Acta Psychol* **115**:99-104.

Huttenlocher, J., 1968, Constructing spatial images: a strategy in reasoning, *Psychol Rev* **75**(6):550-560.

Ishai, A., and Sagi, D., 1995, Common mechanisms of visual imagery and perception. *Science* **268**:1772-1774.

Johnson-Laird, P. N., 1983, *Mental models*, Harvard University Press, Cambridge, MA.

Khenkhar, M., 1991, Object-oriented representation of depictions on the basis of cell matrices, in: *Text understanding in LILOG*, O. Herzog and C. -R. Rollinger, eds., Springer, Berlin, pp. 645-656.

Klauer, K. C., and Zhao, Z., 2004, Double dissociations in visual and spatial short-term memory, *J Exp Psychol Gen* **133**:(3):355-381.

Knauff, M., and Johnson-Laird, P.N., 2002, Visual imagery can impede reasoning, *Mem Cognition* **30**:363-371.

Knauff, M., Rauh, R., and Schlieder, C., 1995, Preferred mental models in qualitative spatial reasoning: A cognitive assessment of Allen's calculus, *Proceedings of the Seventeenth Annual Conference of the Cognitive Science Society*, Lawrence Erlbaum, Mahwah, NJ, pp. 200-205.

Kosslyn, S. M., 1980, *Image and mind*, Harvard University Press, Cambridge, MA.

Kosslyn, S. M., 1994, *Image and brain - The resolution of the imagery debate*, MA: MIT Press, Cambridge.

Kosslyn, S. M., and Shin, L. M., 1994, Visual mental images in the brain: Current issues, in: *The neuropsychology of high-level vision*, M. J. Farah and G. Ratcliff, eds., Lawrence Erlbaum, Hillsdale, NJ, pp. 269-296.

Kosslyn, S. M., and Sussman, A. L., 1995, Roles of imagery in perception: Or, there is no such thing as immaculate perception, in: *The cognitive neurosciences*, M. S. Gazzaniga, ed., MIT Press, Cambridge, MA, pp. 1035-1042.

Kosslyn, S. M., and Thompson, W. M., 2003, When is early visual cortex activated during visual mental imagery?, *Psychol Bull* **129**:(5)723-746.

Kuipers, B., 1982, The 'map in the head' metaphor, *Environ Behav* **14**(2):202-220.

Logie, R. H., 1995, *Visuo-spatial working memory*, Lawrence Erlbaum, Hillsdale, NJ.

Miller, G. A., 1956, The magical number seven, plus or minus two: Some limits on our capacity for information processing, *Psychological Rev* **63**:81-97.

Mishkin, J., Ungerleider, L. G., and Macko, K. A., 1983, Object vision and spatial vision: two cortical pathways, *Trends Neurosci* **6**:414-417.

Monsell, S., and Driver, J., 2000, Banishing the control homunculus, in: *Control of cognitive processes, Vol. XVIII of Attention and Performance*, S. Monsell and J. Driver, eds., MIT Press, Cambridge, MA, pp. 3-32.

Newell, A., 1990, *Unified theories of cognition*, Harvard University Press, Cambridge, MA.

Nobre, C. A., Coull, T. J., Maquet, P., Frith, D. C., Vandenberghe, R., and Mesulam, M. M., 2004, Orienting attention to locations in perceptual versus mental representations, *J Cognitive Neurosci* **16**:(3):363-373.

Palmer, S. E., 1978, Fundamental aspects of cognitive representation, in: *Cognition and categorization*, E. Rosch and B. B. Lloyd, eds., Lawrence Erlbaum, Hillsdale, NJ, pp. 259-303.

Purcell, A. T., and Gero, J. S., 1998, Drawings and the design process, *Design Studies* **19**:389-430.

Pylyshyn, Z. W., 2003, Return of the mental image: Are there really pictures in the brain?, *Trends Cogn Sci* **7**:113-118.

Ragni, M., Knauff, M., and Nebel, B., 2005, A computational model for spatial reasoning with mental models, in: *Proceedings of the 27th annual cognitive science conference*, B. G. Bara, L. Barsalou, and M. Bucciarelli, eds., Lawrence Erlbaum, Mahwah, NJ, pp. 1797-1802.

Schill, K., Umkehrer, E., Beinlich, S., Krieger, G., and Zetzsche, C., 2001, Scene analysis with saccadic eye movements: top-down and bottom-up modeling, *J Electron Imaging, Special issue on Human Vision and Electronic Imaging* **10**(1):152-160.

Schultheis, H., 2005, Distribution and association: Modeling two fundamental principles in cognitive control, in: *Proceedings of KogWis05. The German Cognitive Science Conference 2005*, K. Opwis and I. -K. Penner, eds., Schwabe, Basel, pp. 177-182.

Shepard, R. N., and Metzler, J., 1971, Mental rotation of three-dimensional objects, *Science* **171**:701-703.

Sloman, A., 1975, Afterthoughts on analogical representations, *Proc. Theoretical Issues in Natural Language Processing (TINLAP-1)*, Association for Computational Linguistics, Cambridge, MA, pp. 164-168.

Suwa, M. Gero, J., and Purcell, T., 1999, Unexpected discoveries: How designers discover hidden features in sketches, in: J. S. Gero and B. Tversky, eds., *Visual and spatial reasoning in design*, Key Centre of Design Computing and Cognition, University of Sydney, pp. 145-162.

Tolman, E. C., 1948, Cognitive maps in rats and men, *Psychological Rev* **55**(4):189-208.

Tversky, B., 1992, Distortions in cognitive maps, *Geoforum* **23**(2):131-138.

Tversky, B., 1993, Cognitive maps, cognitive collages, and spatial mental models, in: *Spatial information theory*, A. Frank and I. Campari, eds., Springer, Berlin, pp. 14-24.

Tye, M., 1991, *The imagery debate*, MIT Press, Cambridge.

Verstijnen, I. M., van Leeuwen, C., Goldschmidt, G., Hamel, R., and Hennessey, J. M., 1998, Creative discovery in imagery and perception: Combining is relatively easy, restructuring takes a sketch, *Acta Psychologica* **99**:177-200.

Vieu, L., 1997, Spatial representation and reasoning in artificial intelligence, in: *Spatial and temporal reasoning*, O. Stock, ed., Kluwer Academic Publishers, Dordrecht, pp. 5-41.

Vogel, E. K., Woodman, G. F., and Luck, S. J., 2001, Storage of features, conjunctions, and objects in visual working memory, *J Exp Psychol Human* **27**(1):92-114.

Wilson, M., 2002, Six views of embodied cognition, *Psychon B Rev* **9**:625-636.

Chapter 6

OPTIC ATAXIA: A GATEWAY TO THE HUMAN VISUAL ACTION SYSTEM

Marc Himmelbach and Hans-Otto Karnath
Section Neuropsychology, Center for Neurology, Hertie-Institute for Clinical Brain Research, University of Tübingen, Tübingen, Germany

Abstract: Stroke patients with optic ataxia have an outstanding inability to perform spatially accurate movements to visual targets located in their peripheral visual field. Neuropsychological investigations of such patients contributed essentially to the two visual stream hypothesis, which presumes dissociated action- and perception-related processing of visual information in the human brain. Here we review the anatomical foundations of optic ataxia that have been elucidated in detail quite recently and allow for the identification of brain areas that are necessary for the control of hand in space. We further evaluate the behavioral findings from crucial experimental paradigms in patients with optic ataxia, in comparison to results from patients with visual form agnosia, a disorder characterized by severely impaired visual perception without deficits of action control. On this background, the actual validity of the two visual streams model is discussed facing the (I) perceptual functions of the dorsal posterior parietal cortex, (II) sustained activation of these areas supporting the retention of spatial information, and (III) the anatomical dissociation between a foveal and an extrafoveal action system.

Key words: optic ataxia; dorsal stream; visuomotor action; grasping; reaching; neuropsychology; parietal cortex.

At the beginning of the 20th century Rudolf Bálint (1909) reported a neurological case demonstrating various deficits associated with the processing of visual stimuli. One of the most prominent characteristics of the patient was his striking inability to move his right hand to visual targets. While reaching for objects in his environment, he 'mis-reached' them; the hand deviated grossly from the target. Bálint (1909) ruled out several perceptual deficits and primary motor deficits that could have contributed to this inability. He left us with an impressively concise characterization of the patient's behavior and pathology:

"... we tested his visual fields using a perimeter; it was found that the fields were normal both for objects and for colors. [...] Stereoscopic vision was tested in the usual way by asking the patient to say which of two objects was closer to him, which one was higher, etc.; he made hardly any errors. [...] He recognized objects or images immediately. When describing the patient's general condition I mentioned that the muscular power of the upper and lower extremities was fully retained and that, for the most part, the patient executed elementary movements correctly. [...] A substantial abnormality became visible, however, in the movements of his right hand. He himself reported that while lighting a cigarette he often lit the middle and not the end. [...] Thus when asked to grasp a presented object with his right hand, he would miss it regularly and would find it only when his hand knocked against it. [...] all the movements performed deficiently with the right hand were executed perfectly or with very little error with the left hand." (Translation by Harvey, 1995).

Rudolf Bálint deduced from his thorough observations that the pathology was caused neither by a pure motor nor by a perceptual deficit alone, but rather represents a deficient sensorimotor coordination in the visual domain. Thus, he coined the term "optic ataxia" for the observed deficit. A similar observation, almost unknown until recently has also been made by the japanese ophthalmologist Tatsuji Inouye (1909). Better known were the following reports by Gordon Holmes (1918) and Holmes and Horrax (1919). They reported seven patients suffering from a disability to localize objects or visual stimuli in their surroundings. In addition to the symptoms described by Rudolf Bálint, six out of seven patients demonstrated a striking inability to verbally report the absolute or relative position of certain objects, a severe deficit in spatial orientation, as well as characteristic eye movement deficits (Holmes, 1918; Holmes and Horrax, 1919). Incorporating the reports of Bálint and Holmes, the Bálint-Holmes syndrome today is defined to consist of four cardinal symptoms (cf. Karnath, 2003; Rafal, 1997): (I) optic ataxia, (II) disturbed organization of eye movements, (III) impaired spatial orientation, and (IV) simultanagnosia (the inability to perceive more than one object at a time (Farah, 1990)). However, as in Bálint's groundbreaking observation, several patients have been reported in the following decades whose misreaching behavior could be clearly dissociated from pure perceptual impairments. Garcin and colleagues (1967) were the first submitting a convincing demonstration of isolated optic ataxia excluding perceptual, oculomotor, and visual attention deficits.

The most detailed report on a group of 10 unilateral patients suffering from isolated optic ataxia was published by Perenin and Vighetto (1988).

Summarizing the previous findings they concluded that the report of "most [...] bilateral syndromes are more reminiscent of the 'visual disorientation' of Holmes" (p. 644). However, like Garcin et al. (1967) they pointed out, that their own patients did not reveal impairments of primary vision or spatial perception that could lead to the dramatic visuomotor impairments. Moreover, they emphasized the value of hand-specific visuomotor discoordination in bilateral cases as it was already observed by Rudolf Bálint (1909) and others (Guard et al., 1984; Hecaen and de Ajuriaguerra, 1954). Such hand-specific deficits cannot be attributed to a general visual impairment alone. The latter would equally affect both hands. Investigating a group of patients with unilateral lesions of the posterior parietal cortex (PPC), Perenin and Vighetto (1988) extended the existing knowledge by demonstrating a striking difference in the consequences of lesions to the left or right hemisphere. Patients with lesions of the right hemisphere showed significant misreaching to targets in the left visual field with both hands while movements to right-sided targets remained largely unaffected ("field effect"). Lesions of the left hemisphere led to an additional "hand effect". The authors observed that only movements with the contralesional right hand to targets in the contralesional visual hemifield deviated grossly from the target position. In contrast, movements with the ipsilesional left hand to targets on either side and with the right hand to left-sided targets were precise. This pattern of lateralization has been supported by recent single-case studies of patients with unilateral optic ataxia (Himmelbach and Karnath, 2005; Khan et al., 2005; Revol et al., 2003).

In search of the typical lesion location provoking this visuomotor disturbance, modern imaging techniques such as computerized tomography (CT) and magnetic resonance imaging (MRI) have been used. Single case studies of patients with optic ataxia have shown lesions typically including the superior parietal lobule (SPL) (Auerbach and Alexander, 1981; Buxbaum and Coslett, 1998; Ferro, 1984). However, lesions of individual patients rarely are restricted to a well specified anatomical site but include various regions related and not related to the disorder. Thus, only the anatomical evaluation of a larger group can reveal the crucial lesion site. Such an analysis first has been carried out by Perenin and Vighetto (1988). The anatomical evaluation of their 10 patients with unilateral left- or right-sided lesions revealed an overlap of lesion location that was symmetrical in both hemispheres. It included the intraparietal sulcus (IPS) and either the upper part of the inferior parietal lobule (IPL) or – more often – the medial or the ventral part of the SPL. Still, the paper-and-pencil techniques available at those times held some uncertainties and, in addition, did not allow for a direct visual comparison between the pattern of lesion location in patients with vs. without optic ataxia. The necessity of such contrasts for valid anatomical conclusions in modern imaging studies was demonstrated by Rorden and Karnath (2004). A recent study therefore re-investigated the typical lesion location in a group of 16 unilateral stroke patients with optic ataxia, collected over a time period of 15 years, and compared them with 36 stroke patients without that disorder using digitized brain templates and standardized rendering algorithms for 3D visualization of the subtraction analysis (Karnath and Perenin, 2005). The authors found no evidence for the previous assumption that the disruption of visually guided reaching in humans is associated with a lesion centering on the SPL on the convexity. In both left and right hemispheres, they rather found optic ataxia associated with a lesion overlap that affected the lateral cortical convexity at the parieto-occipital junction (POJ), i.e. the junction between the IPL and superior occipital cortex in both hemispheres including – in the left hemisphere even more posteriorly – also the junction between the superior occipital cortex and the SPL (Fig. 6-1). Via the underlying parietal white matter the lesion overlap extended in both hemispheres to the medial cortical aspect where it affected the precuneus close to the parieto-occipital sulcus (Fig. 6-2).

Converging evidence is reported from an intriguing event-related functional magnetic resonance imaging (fMRI) study conducted with healthy subjects (Prado et al., 2005). The authors measured the brain activity when participants reached either towards a target represented on the fovea or towards an extrafoveal target. The analysis of the correlated BOLD effects revealed increased signals bilaterally at the POJ depending on the retinal position of the visible target (Fig. 6-2). Their results fit surprisingly well with the above mentioned finding of a reaching deficit for targets in the peripheral visual field typically following a damage to precisely this region (Karnath and Perenin, 2005). A third piece of evidence has been presented by van Donkelaar and Adams (2005) who applied transcranial magnetic stimulation (TMS) at the PPC while the subjects were pointing to peripheral targets. Without any interference, normal subjects tend to overshoot eccentric targets in their peripheral visual field (Bock, 1986; Bock, 1993). The application of interfering TMS pulses led to a bias of pointing movements of the contralateral arm towards the position of visual fixation (van Donkelaar and Adams, 2005). This effect obviously mimics the pathological movement bias in patients with optic ataxia (Carey et al., 1997; Jackson et al., 2005; Milner et al., 1999, 2003; Ratcliff and Davies-Jones, 1972). Altogether, these investigations – using three different methods of functional mapping (stroke lesions, fMRI, TMS) – support the assumption of a circumscribed region in the posterior parieto-occipital cortex specifically dedicated to the visual control of hand movements to extrafoveal targets, while movements to foveated targets seem to recruit a (cortical) network not including this area.

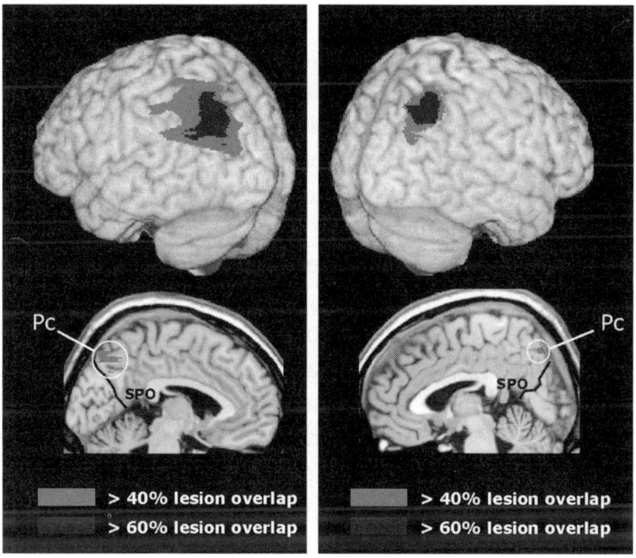

Figure 6-1. Lateral and medial surface views of the center of lesion overlap in patients with unilateral optic ataxia. The percentage of lesion overlap in these patients has been calculated after the subtraction of control subjects with unilateral lesions but without optic ataxia. SPO: Sulcus parieto-occipitalis; Pc: Precuneus. (adapted from Karnath and Perenin, 2005).

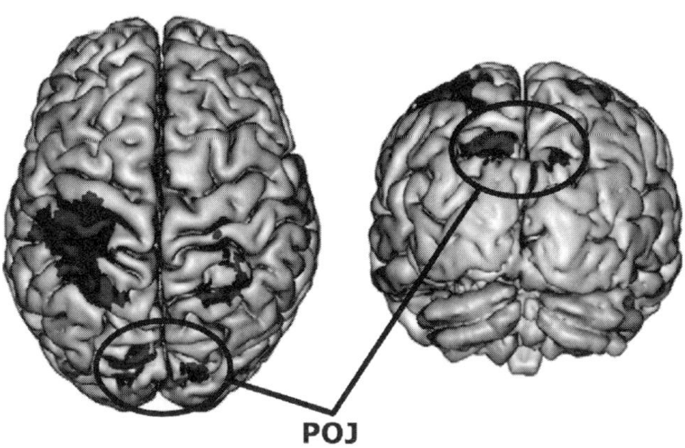

Figure 6-2. Significant fMRI activations at the bilateral parieto-occipital junction (POJ) due to the execution of pointing movements to targets in the visual periphery (adapted from Prado et al., 2005).

1. THE TWO VISUAL STREAMS HYPOTHESIS

The numerous reports about patients with visuomotor disorders have contributed substantially to the currently dominant idea of a dichotomous organization of the visual system. Based on a body of already existing evidence from behavioral and anatomical studies in animals and on their own experiments in monkeys, Ungerleider and Mishkin (1982) suggested the existence of a ventral occipito-temporal "what" pathway and of a dorsal occipito-parietal "where" pathway. They decomposed the visual system into a spatial processing system on the one hand and an identification system on the other hand. However, the aforementioned observations of neurological patients suffering from optic ataxia falsified their conclusions. These observations represent behavioral dissociations *within* the supposed "where" processing and could not be fitted easily to the original suggestions of Ungerleider and Mishkin (1982). Therefore, Milner and Goodale (1995) modified the model. Referring to the same anatomical structures, they suggested a dissociation between action- and perception-related visual processing. Such a differentiation seemingly fits to the observations in patients suffering from optic ataxia but also to the behavior of patients with damage to the ventral occipito-temporal pathway. As we said before, some of these patients demonstrated both motor as well as perceptual deficits with respect to spatial characteristics (Holmes, 1918; Holmes and Horrax, 1919). Such a general disorder of spatial information processing would be in agreement with the suggested distinction proposed by Ungerleider and Mishkin (1982). But decisively, several patients revealed exclusive spatial disorders of only goal-directed movements while their perceptual estimation of absolute and relative spatial distances was accurate (Garcin et al., 1967; Perenin and Vighetto, 1988). Furthermore, these findings of action-specific impairments in patients with uni- or bilateral lesions of the POJ were complemented by patients with the reverse dissociation of disorders, namely intact spatial action processing with concomitant severe impairments of spatial perception when lesions were located more ventrally in the occipito-temporal cortex. Damage to occipito-temporal areas of the human brain typically leads to apperceptive visual agnosia (e.g.: Farah, 1990). A well known patient suffering from such disorder, D.F., demonstrated well preserved reaching and grasping behavior while she revealed a striking disability to report the identity, size, and/or orientation of different objects (Goodale et al., 1991; Milner et al., 1991). For instance, she was able to smoothly move her hand through an oblong slot. In contrast, the adjustment of a second slot with respect to the first (a perceptual task) was imprecise. The opposite pattern is observed in patients with optic ataxia. While the same motor task cannot be executed adequately, they can easily match different line orientations (Perenin and Vighetto, 1988).

The very same dissociation between both groups of patients - optic ataxia vs. visual form agnosia patients - holds for grasping movements. Whereas a patient with optic ataxia could easily estimate the size of different objects, her grip size was just weakly correlated with the objects' size during actual grasping movements (Jeannerod, Decety, and Michel, 1994). The contrary behavior was found in a patient with visual form agnosia. She revealed a weak correlation between the actual target size and her estimation of it, while she adjusted her grip size during actual grasping movements adequately to the different objects (Goodale et al., 1991). Just recently, it has been shown that the spared visuomotor abilities of this patient are mediated (amongst other structures) by parietal areas, i.e. by areas believed to be part of the dorsal stream (Culham et al., 2003; James et al., 2003). A small number of further patients has subsequently been reported with varying degrees of behavioral impairments and dissociations (e.g.: Ferreira et al., 1998; Hildebrandt et al., 2004; Le et al., 2002; Marotta et al., 1997).

2. GRASPING VISUAL ILLUSIONS: COMPLEMENTARY EVIDENCE FOR TWO VISUAL STREAMS?

Further evidence for a dissociated processing of visual information derives from healthy human subjects. Several studies revealed a significant impact of visual illusions on perceptual estimates of size and/or location while grasping and pointing movements were unaffected (Aglioti et al., 1995; Bridgeman et al., 2000; Danckert et al., 2002; Haffenden et al., 2001; Meegan et al., 2004). However, despite numerous findings in favor of such a dissociation, subsequent studies revealed a more inconsistent view. Some authors did not find a comparable dissociation or found at least a somewhat smaller but nevertheless significant effect on actions (Daprati and Gentilucci, 1997; Elliott and Lee, 1995; Franz, 2003; Franz et al., 2000, 2003; Gentilucci et al., 1996; Pavani et al., 1999). Interestingly, the measure typically used to demonstrate effects on grasping movements in these studies has been the maximum grip aperture (MGA). Undoubtedly, the MGA appears to be a straightforward measure of grasping performance. However, in search of potential effects of illusions (i.e. perceptual context cues) on reaching and grasping, other parameters of grasping might be taken into account such as velocity and force. Some authors indeed revealed effects of visual illusions on these kinematic parameters (Brenner and Smeets, 1996;

Jackson and Shaw, 2000; van Donkelaar, 1999). Actually, the discussion is quite controversial. Evidence in favor of as well as against an influence of visual illusions on grasping has been reported (for review: Bruno, 2001; Carey, 2001; Franz, 2001; Goodale and Westwood, 2004; Milner and Dyde, 2003; Plodowski and Jackson, 2001).

One of the most valuable contributions to the debate proposes the use of different spatial attributes of a certain object during estimation and grasping tasks. Following this line of evidence, estimation relies more on size and extend information, whereas grasping is guided by discrete target positions for each finger at the respective object. It seems as if visual illusions typically exert a different influence on these different spatial attributes. Thus, the observed behavioral dissociations would not represent a divergence between perception and action but between different spatial properties used for the execution of the respective tasks (Smeets et al., 2002). Milner and Dyde (2003) on the other hand have suggested a differentiation between illusions which affect different levels of visual processing. They found a differential impact of the rod-and-frame illusion and the simultaneous-tilt illusions (Dyde and Milner, 2002). Whereas the first is assumed to be based on contextual information, the latter might to be due to local interactions within the visual field mediated by inhibitory connections in V1 (Milner and Dyde, 2003). Just recently, the work of Bruce Bridgeman and Paul Dassonville added further controversial evidence to this field of research. They explored the impact of the Roelofs effect on goal-directed hand movements and perceptual estimations of stimulus positions (Bridgeman et al., 2000; Bridgeman et al., 1997; Dassonville and Bala, 2004; Dassonville et al., 2004). The observed dissociation of the Roelofs effect on pointing and estimation has previously been assumed to be in line with the two visual streams theory (Bridgeman et al., 1997). But in their most recent work both authors interpret their findings in a very different way. The dissociation between action and perception found for this illusory change of target position might be indirectly mediated by an underlying common process involved in action control and perception. From their latest results they conclude that a shift of the subjective body midline within one and the same egocentric spatial frame induced by the Roelofs effect exerts a different impact on the accuracy of perceptual estimations and immediate goal-directed movements (Dassonville and Bala, 2004; Dassonville et al., 2004). However, while their data suggest a simple common mechanism to explain different outcomes for motor control and perception, it does not rule out dual visual processing per se. If we assume, in agreement with Milner and Goodale (1995), that the proposed midline shift is mediated by and affecting only the ventral (cognitive or perceptual) visual stream, the results of the Dassonville and Bridgeman groups fit nicely to the two visual streams theory. Furthermore, up to now it remains unclear whether their results can be generalized to other visual illusions as well.

Interestingly, it is also unclear how the performance of patients suffering from optic ataxia or visual form agnosia is affected by visual illusions. As far as we know, illusions such as e.g. the Müller-Lyer or the Ebbinghaus illusion have not been investigated in patients with these disorders. A recent study reported that patients with visual agnosia were not prone to a size-weight illusion (larger objects are felt to be lighter in comparison to smaller objects of the same physical weight). When executing the same task without visual information, i.e. when retrieving the size of the objects from kinesthetic input only, the patients showed the same illusion effect than controls (Dijkerman et al., 2004). However, although this experiment revealed a clear dissociation between visual and kinesthetic processing, it is unclear whether these patients would incorporate the illusory visual information into motor behavior or not.

3. DELAYED MOVEMENTS: TIMING MAKES THE DIFFERENCE

Several studies disclosed a behavioral dissociation between movements to visible targets and movements to remembered positions in neurological patients. Goodale and co-workers (1994) observed that a patient with visual agnosia was unable to adjust grip aperture properly when the target object was removed before movement onset. In striking contrast, patients with optic ataxia improve their performance considerably after a delay of a few seconds following the presentation of a target while their immediate action to visible targets is severely distorted (Himmelbach and Karnath, 2005; Milner et al., 2001; Milner et al., 2003; Milner et al., 1999; Revol et al., 2003). Thus, it has been assumed that the dorsal visual pathway - which is intact in patients with visual form agnosia - is dedicated to a fast processing of visual information. On the other hand, an intact ventral pathway - found in patients with optic ataxia - seems to be devoted to a longer lasting processing of visual information and its output does not seem to be immediately available for movement control. These conclusions are supported by the increased effects of visual illusions on actions after interfering time delays between stimulus presentation and movement onset in healthy subjects (Bridgeman et al., 2000; Bridgeman et al., 1997; Elliott and Lee, 1995; Gentilucci et al., 2001; Gentilucci et al., 1996; Hu and Goodale, 2000; Meegan et al., 2004; Rival et al., 2003). It has been questioned whether this increase relies on the elapsed time after the target presentation or simply on the disappearance of the visual stimulus before the execution of a movement. The latter presumption has been favored by Westwood and Goodale (2003) who reported significant illusory effect on the peak grip aperture if the target was occluded right after the start signal for movement onset. They found no additional effects of a prolonged delay between the occlusion of the objects and the start signal. This finding led the authors to assume a sudden switch between dorsal and ventral control of visuomotor performance. They concluded that the ventral visual system provides the decisive spatial information which is necessary to control hand movements as soon as no immediate visual information about target size and position is available (Goodale et al., 2004).

However, other behavioral experiments in healthy subjects yielded different results. The impact of a visual illusion on pointing accuracy increased significantly with longer time delays (Bridgeman et al., 2000; Meegan et al., 2004). So, even if there would be a dramatic shift between "two distinct modes of control" (Goodale et al., 2004) - in anatomical terms: between the dorsal and the ventral pathway - there still seems to be an additional progressive change depending on the time delay between target presentation and movement onset. Such progressive improvement of pointing accuracy has also been found in two patients with optic ataxia (Himmelbach and Karnath, 2005). Both patients demonstrated a gradual decrease of absolute pointing errors over a range of delay times from 0 to 10 seconds preceding movement onset (Fig. 6-3). In agreement with the dependence of the effect of visual illusions on the interfering time delay in healthy subjects (Bridgeman et al., 2000; Meegan et al., 2004), this gradual decrease in optic ataxia patients argues against a sudden shift between anatomically separated systems. Rather, it points to either a gradually decreasing dorsal processing of visual information in gradually delayed movements or to a gradually increasing contribution from alternative systems (which might be, e.g., the occipito-temporal stream). This concept of a gradual change - instead of a sudden switch - in the functional anatomy of movement-related information processing, is corroborated by the observation of a reverse behavioral pattern in healthy subjects, i.e. a gradual decrease of movement accuracy depending on the length of a pre-response delay (Bradshaw and Watt, 2002).

Neuroimaging studies in healthy human subjects have suggested that areas of the dorsal posterior parietal cortex are critically involved in delayed movement tasks (Connolly et al., 2003; Culham, 2004; Lacquaniti et al., 1997) and in visuo-spatial memory-tasks (for review: Owen, 2004). Most interestingly, sustained activity of the medial superior parietal cortex has been demonstrated during a delay of 9 seconds between target presentation and movement execution using an event-related fMRI paradigm (Connolly et al., 2003).

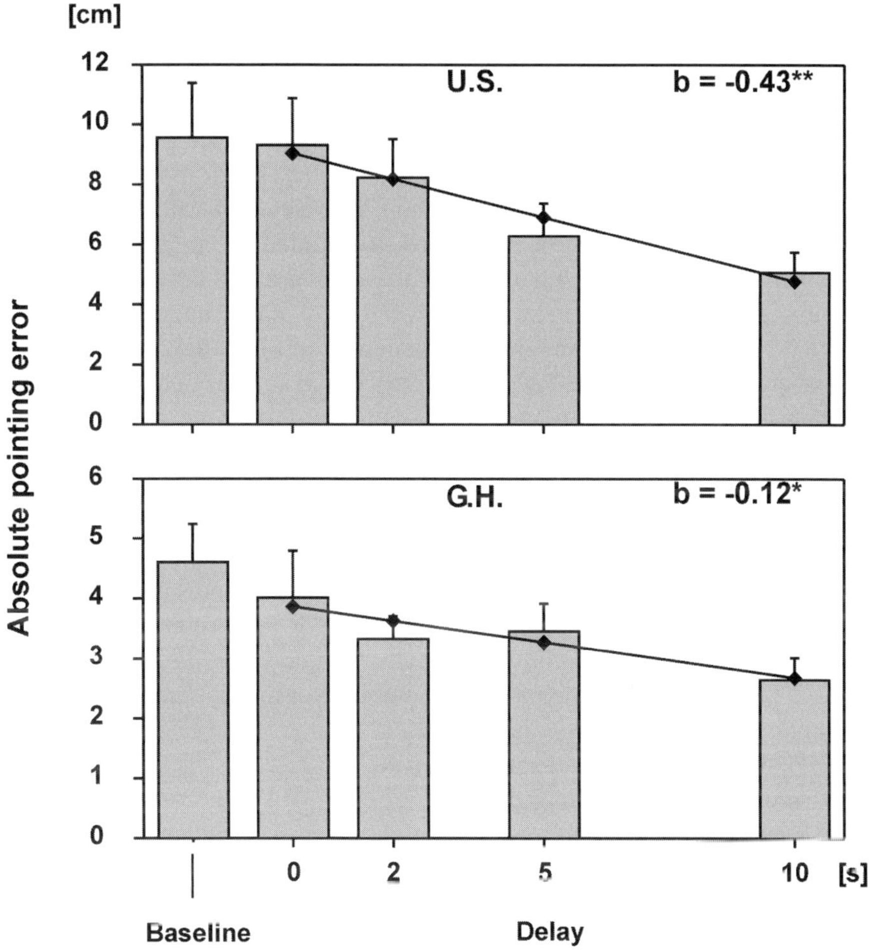

Figure 6-3. Absolute pointing errors of gradually delayed movements in two patients with optic ataxia. Movements have been performed either to a visible target (baseline) or to remembered target position after a gradually increasing delay of 0 to 10 seconds. Linear regressions over delay time have been calculated revealing negative regression coefficients (b) significantly smaller than zero (* $p < 0.05$ one-tailed, ** $p < 0.01$ one-tailed). Results show that the pointing error of optic ataxia patients decreases with an increase of the period between target offset and movement onset (adapted from Himmelbach and Karnath, 2005).

The authors convincingly distinguished between non-spatial preparation of a movement - induced by a cue lacking spatial information - and the retention of previously provided spatial information. Thus, it seems as if the prolonged superior parietal activation represents either the retention of a specific target position or hand trajectory. These findings clearly endorse a critical involvement of the dorsal stream in the maintenance of spatial information, which, in some cases, is dedicated to the execution of goal-directed movements later on. Correspondingly, single-cell recordings in monkeys revealed maintenance-related activity of neurons in dorsal posterior parietal areas in visuo-spatial memory tasks employing eye movements (Chafee and Goldman-Rakic, 1998; Gnadt and Andersen, 1988; Snyder et al., 1997) and, more specifically, in delayed hand movement tasks (Murata et al., 1996; Quintana and Fuster, 1999; Snyder et al., 1997). Quite recently, Tsutsui and co-workers (2003) demonstrated successfully sustained activity of IPS neurons correlated with the retention of surface orientation over a period of 2.3 seconds. Altogether, these data suggest a possible involvement of dorsal stream areas in the retention of spatial information in general and of visuospatial information dedicated to the guidance of hand movements in particular.

Such correlative data gathered in neuroimaging investigations of healthy humans and single cell recordings in monkeys are corroborated by brain interference methods. Inducing a transient inactivation of the PPC in rhesus monkeys interfered with the retention of spatial information - the required response direction - only. In contrast, inactivation of the prefrontal cortex in the same experiment interfered with delayed performance after the presentation of spatial and non-spatial cues (Quintana and Fuster, 1993). A recent TMS study in healthy human subjects using a memory guided pointing task complements these findings (Smyrnis et al., 2003). The application of a single TMS pulse over the PPC as early as 300 ms after target presentation had a significant effect on the accuracy of hand movements executed 3000 ms after target offset. Similar findings for memory guided saccades further support the assumption of a crucial involvement of the SPL and IPS in memory guided actions (Muri et al., 1996, 2000; Oyachi and Ohtsuka, 1995). Altogether, these findings indicate a crucial involvement of the superior PPC in the early spatial encoding of either a target position or a movement trajectory, which is required to be executed later on.

4. INTERACTING STREAMS

Although Milner and Goodale (1995) already explicitly stated the necessity of functional interactions between the two visual streams, the vast majority of subsequent contributions to the field emphasized the apparent distinction between different ways of visual processing. In fact, the experiments on delayed movement execution reviewed in the preceding chapter support a close interaction of processing systems instead of functionally distinct pathways. A further argument against separate processing systems with respect to different time constants of information processing - i.e. immediate vs. delayed onset of movements - is provided by the recent report of a patient with visual agnosia (S.B.) showing accurate delayed grasping movements (Dijkerman et al., 2004). Although there are substantial differences between this and the previously examined case D.F., this observation obviously questions those conclusions which essentially were based on the behavior of only one patient with visual agnosia (patient D.F.; Milner and Goodale, 1995).

The handling of everyday objects instead of geometric, meaningless items by patients suffering from visual form agnosia and patients with optic ataxia provides us with further evidence for the (necessary) interaction between object recognition and action control. The well known patient D.F. seems to be unable to grasp everyday objects appropriately according to their typical use (Carey et al., 1996). While she grasped these objects skillfully to pick them up, she did not appear to take account of the way these objects were supposed to be used afterwards as healthy subjects would do (e.g. grasping a hammer at its head instead of grasping it at its handle). Moreover, in this series of experiments, a general difficulty in grasping complex objects was observed in D.F. Grasping rectangular objects as well as irregularly shaped objects did not pose a problem to her as long as they provided a clear main axis which she could aim for. But if the objects lacked such an outstanding principal axis, she showed a considerable number of trials with inadequate grip posture (Carey et al., 1996). These specific impairments of grasping everyday objects according to their specific use and of grasping irregular objects with multiple spatial axes might be due to a general lack of allocentric analysis or encoding of complex object properties for appropriate visuomotor guidance (Dijkerman et al., 1998; McIntosh et al., 2004). Interestingly, the reverse behavioral dissociation was observed in a patient with optic ataxia. While being unable to adjust her grip size to cylinders of various diameters, her behavior improved considerably if she was asked to grasp familiar cylinder-shaped objects (Jeannerod et al., 1994). Obviously, tasks which require a high-level object recognition and/or an allocentric encoding of object features pose a problem to a subject with a damaged ventral system while it seems to open alternative routes of information processing for a patient with a damaged dorsal system. However, such conclusions should be drawn with great caution since these results have been shown in only one patient with visual form agnosia (patient D.F.; Carey et al., 1996; Dijkerman et al., 1998; McIntosh et al., 2004) and in only one patient with optic ataxia (patient A.T.; Jeannerod et al., 1994).

The need for an interaction between the dissociated ways of processing is apparent and, as stated before, has been already considered in the original concept of the two visual streams. It has been suggested that the ventral stream acts as an identifier which 'flags' certain goals or objects for upcoming actions (Goodale and Milner, 2004; Milner and Goodale, 1995). The incorporation of such information could be performed via the inferior parietal lobule and superior temporal areas that receive projections from the occipito-parietal as well as from the occipito-temporal pathways. Alternatively, back-projections to early visual areas could label certain objects or features, which then form the basis for information processing leading to the execution of appropriate actions. Such back-propagation has been shown in a combined electrophysiological and functional magnetic resonance study of visual spatial attention (Noesselt et al., 2002). As Milner and Goodale (1995) suggested, mechanisms of selective attention might be the mediating process between anatomically dissociated streams of processing.

The ill-defined term 'flagging' might be synonymous with the known encoding of the behavioral relevance or saliency of objects and features (Assad, 2003). In a recently reported experiment Toth and Assad (2002) demonstrated the unexpected coding of color by neurons in the lateral intraparietal area (LIP) following associative training. Two identical saccade targets on the right and left side were presented simultaneously. The investigated monkeys had to execute a saccade either to the left or to the right target based on information provided by a visual cue, which had been presented before the saccade targets. The direction of the required saccade was either indicated by the position or by the color of the cue. During a delay following cue presentation neurons within LIP revealed changes of the spike rate in correlation with the presented color. Remarkably, they did so only if color was the informative dimension during the respective trial, i.e. such encoding of color was only observed if this attribute of the cue was of behavioral relevance. This neuronal behavior fits to the requirements of "a local selective transfer of information between brain areas" (Assad, 2003) which seems to be close to the 'flagging' concept of Milner and Goodale (1995).

5. PERSPECTIVES

The existing literature on visuomotor control processes seems to clearly indicate that immediate, goal-directed visuomotor responses to point-like targets or towards objects which provide an unambiguous request for action (e.g. catching a falling cup) essentially rely on the so-called dorsal stream of visual processing. This kind of action is largely independent of detailed analyses of non-spatial aspects of the target object. However, although typically taken together as "*the* dorsal stream", there is nothing like a monolithic action system comprising the numerous functionally different areas of the PPC. For example, most of the visuomotor studies in patients with optic ataxia dealt with deficits of movements to peripheral visual targets only. An anatomical differentiation between a foveal and an extrafoveal action system has been explicitly considered quite recently (Milner et al., 2003). Most recent experimental findings suggest that this behavioral distinction of visuomotor processing is indeed reflected at the cortical level (Karnath and Perenin, 2005; Prado et al., 2005; van Donkelaar and Adams, 2005). Moreover, the dorsal parietal areas apparently do not simply represent exclusively an "action system". The clear dissociation between a "where" and a "what" stream as suggested by Ungerleider and Mishkin (1982) was falsified after the demonstration of the remarkable behavioral dissociations between action and perception in patients with optic ataxia and visual form agnosia. However, numerous reports starting with Holmes and Horrax (Holmes, 1918; Holmes and Horrax, 1919) up to recent investigations such as the one by Phan and colleagues (2000) demonstrated perceptual deficits in patients with exclusive posterior parietal brain damage. Further, some of the areas in the dorsal PPC involved in the perception of spatial relations, also are involved in memorizing spatial information.

We already emphasized the importance of functional interactions between the dorsal and ventral streams. Largely in agreement with the Milner and Goodale (1995) model, it is possible that, beyond immediate action control, the abovementioned PPC functions in spatial cognition are not mediated by the dorsal stream per se, but by systems that depend on ventral stream inputs. This hypothesis remains to be substantiated by according experiments in healthy humans and brain damaged patients. Alternatively, it could become necessary to revise the "perception vs. action model" in the version outlined by Milner and Goodale (1995) by incorporating (I) perceptual functions of the dorsal PPC, (II) sustained activation of these areas supporting the retention of spatial information, and (III) the anatomical dissociation between a foveal and an extrafoveal action system.

Further, we should be aware that much of our current knowledge about the anatomo-functional relationship in visuomotor control processes derives from few patients with (often non-acute) brain damage. Beyond the legitimate enthusiasm to observe and conclude from the behavior of stroke patients suffering from specific visuomotor disturbances after stroke, we must keep in mind that not all of this behavior necessarily reflects a pure consequence of a disturbed brain system. The reason is that many of the patients studied in the experiments reviewed above already suffered from chronic impairments at the time of the respective examination. Thus, (partly) conscious strategies might have been adopted by these patients to execute certain perceptual and visuomotor tasks. For example, Goodale and Milner (2004) anecdotally reported such strategies adopted by their patient D.F. to accomplish perceptual tasks. During a line copying task she apparently used a motor imagery strategy, tracing the line in her mind only (Dijkerman and Milner, 1997). Likewise, optic ataxia patients with chronic brain lesions might make use of spared abilities to guide their movements under conditions, which provide them with enough time. Milner and co-workers (2001) revealed a significant improvement of such a chronic patient's performance if the object to be grasped was shown to her in advance. Under this 'preview' condition, her movements seemed to rely partly on memorized spatial information instead of the actually available sight of the object. In contrast, healthy subjects simply ignored previewed object information (Milner et al., 2001). Thus, it is evident that some of these strategies have been deliberately adopted by the patients while other strategies might unfold unconsciously. These observations clearly show that contextual information and high level representations of action (including explicit knowledge of preserved abilities) are involved in action control.

Marc Jeannerod and Pierre Jacob recently broadened our view on the dualism of visual action-control and of visual perception emphasizing the involvement of such high level representations of action (Jeannerod and Jacob, 2005). They pointed out that quite simple visuomotor transformations involved in reaching and grasping movements to point-like targets or geometric objects require no or only little conceptual information concerning the goals, the environmental conditions, and the consequences of these actions. However, a comprehensive theory of action control needs to incorporate such high-level information into visuomotor representations as it crucially affects the actually required kinematics of an intended or demanded action. Future studies in patients with optic ataxia should focus on such contextual influence. Moreover, past contributions to the field focused almost exclusively on the antagonism between the behavioral consequences of optic ataxia and visual form agnosia. Investigating these patients' reactions to manipulations of contextual information and comparing them with the behavioral changes observed in patients with other higher order motor deficits (apraxia) and recognition impairments (associative and apperceptive agnosia) might reveal the anatomical substrates which are necessary to incorporate semantic information in action control. We think that uncovering these integration processes would constitute the next step on our way of understanding action control in a natural environment.

ACKNOWLEDGEMENTS

The work was supported by a grant from the Deutsche Forschungsgemeinschaft (SFB 550-A4). We are grateful to A. David Milner for stimulating discussion of a previous version of the manuscript.

REFERENCES

Aglioti, S., Desouza, J. F., and Goodale, M. A., 1995, Size-contrast illusions deceive the eye but not the hand, *Curr Biol* **5:**679-685.

Assad, J. A., 2003, Neural coding of behavioral relevance in parietal cortex, *Curr Opin Neurobiol* **13:**194-197.

Auerbach, S. H., and Alexander, M. P., 1981, Pure agraphia and unilateral optic ataxia associated with a left superior parietal lobule lesion, *J Neurol Neurosurg Psychiatry* **44:**430-432.

Balint, R., 1909, Seelenlähmung des "Schauens", optische Ataxie, räumliche Störung der Aufmerksamkeit, *Monatsschrift für Psychiatrie und Neurologie* **25:**51-81.

Bock, O. 1986, Contribution of retinal versus extraretinal signals towards visual localization in goal-directed movements, *Exp Brain Res* **64:**476-482.

Bock, O., 1993, Localization of objects in the peripheral visual field, *Behav Brain Res* **56:** 77-84.

Bradshaw, M. F., and Watt, S., 2002, A dissociation of perception and action in normal human observers: the effect of temporal delay, *Neuropsychologia* **40:**1766-1778.

Brenner, E., and Smeets, J. B., 1996, Size illusion influences how we lift but not how we grasp an object, *Exp Brain Res* **111:**473-476.

Bridgeman, B., Gemmer, A., Forsman, T., and Huemer, V., 2000, Processing spatial information in the sensorimotor branch of the visual system, *Vision Res* **40:**3539-3552.

Bridgeman, B., Peery, S., and Anand, S., 1997, Interaction of cognitive and sensorimotor maps of visual space, *Percept Psychophys* **59:**456-469

Bruno, N., 2001, When does action resist visual illusions? *Trends Cogn Sci* **5:**379-382.

Buxbaum, L. J., and Coslett, H. B., 1998, Spatio-motor representations in reaching: Evidence for subtypes of optic ataxia, *Cogn Neuropsychol* **15:**279-312.

Carey, D. P., 2001, Do action systems resist visual illusions? *Trends Cogn Sci* **5:**109-113.

Carey, D. P., Coleman, R. J., and Della, S. S., 1997, Magnetic misreaching, *Cortex* **33:** 639-652.

Carey, D. P., Harvey, M., and Milner, A. D., 1996, Visuomotor sensitivity for shape and orientation in a patient with visual form agnosia, *Neuropsychologia* **34:**329-337.

Chafee, M. V., and Goldman-Rakic, P. S., 1998, Matching Patterns of Activity in Primate Prefrontal Area 8a and Parietal Area 7ip Neurons During a Spatial Working Memory Task, *J Neurophysiol* **79:**2919-2940.

Connolly, J. D., Andersen, R. A., and Goodale, M. A., 2003, fMRI evidence for a 'parietal reach region' in the human brain, *Exp Brain Res* **153:**140-145.

Culham, J., 2004, Human Brain Imaging Reveals a Parietal Area Specialized for Grasping, in: *Functional Neuroimaging of Visual Cognition: Attention and Performance XX*, N. Kanwisher and J. Duncan, eds., Oxford University Press, Oxford, pp. 417-438.

Culham, J. C., Danckert, S. L., Desouza, J. F., Gati, J. S., Menon, R. S., and Goodale, M. A., 2003, Visually guided grasping produces fMRI activation in dorsal but not ventral stream brain areas, *Exp Brain Res* **153:**180-189.

Danckert, J. A., Sharif, N., Haffenden, A. M., Schiff, K. C., and Goodale, M. A., 2002, A temporal analysis of grasping in the Ebbinghaus illusion: planning versus online control, *Exp Brain Res* **144:**275-280.

Daprati, E., and Gentilucci, M., 1997, Grasping an illusion, *Neuropsychologia* **35:**1577-1582.

Dassonville, P., and Bala, J. K., 2004, Perception, action, and Roelofs effect: a mere illusion of dissociation, *PLoS Biol* **2:** e364.

Dassonville, P., Bridgeman, B., Kaur, B. J., Thiem, P., and Sampanes, A., 2004, The induced Roelofs effect: two visual systems or the shift of a single reference frame? *Vision Res* **44:**603-611.

Dijkerman, H. C., Lê, S., Démonet, J.-F., and Milner, A. D., 2004, Visuomotor performance in a patient with visual agnosia due to an early lesion, *Brain Res Cogn Brain Res* **20:** 12-25.

Dijkerman, H. C. and Milner, A. D., 1997, Copying without perceiving: motor imagery in visual form agnosia, *Neuroreport* **8:**729-732.

Dijkerman, H. C., Milner, A. D., and Carey, D. P., 1998, Grasping spatial relationships: failure to demonstrate allocentric visual coding in a patient with visual form agnosia, *Conscious Cogn* **7:**424-437.

Dyde, R. T. and Milner, A. D., 2002, Two illusions of perceived orientation: one fools all of the people some of the time; the other fools all of the people all of the time, *Exp Brain Res* **144:**518-527.

Elliott, D. and Lee, T. D., 1995, The role of target information on manual-aiming bias. *Psychol Res* **58:**2-9.

Farah, M. J., 1990, *Visual Agnosia*, MIT Press, Cambridge, MA.

Ferreira, C. T., Ceccaldi, M., Giusiano, B., and Poncet, M., 1998, Separate visual pathways for perception of actions and objects: evidence from a case of apperceptive agnosia, *J Neurol Neurosurg Psychiatry* **65:**382-385.

Ferro, J. M., 1984, Transient inaccuracy in reaching caused by a posterior parietal lobe lesion. *J Neurol Neurosurg Psychiatry* **47:**1016-1019.

Franz, V. H., 2001, Action does not resist visual illusions, *Trends Cogn Sci* **5:**457-459.

Franz, V. H., 2003, Planning versus online control: dynamic illusion effects in grasping? *Spat Vis* **16:**211-223.

Franz, V. H., Bülthoff, H. H., and Fahle, M., 2003, Grasp effects of the Ebbinghaus illusion: obstacle avoidance is not the explanation, *Exp Brain Res* **149:**470-477.

Franz, V. H., Gegenfurtner, K. R., Bulthoff, H. H., and Fahle, M., 2000, Grasping visual illusions: no evidence for a dissociation between perception and action, *Psychol Sci* **11:** 20-25.

Garcin, R., Rondot, P., and de Recondo, J., 1967, Optic ataxia localized in 2 left homonymous visual hemifields (clinical study with film presentation). *Rev Neurol (Paris)* **116:**707-714.

Gentilucci, M., Benuzzi, F., Bertolani, L., and Gangitano, M., 2001, Visual illusions and the control of children arm movements, *Neuropsychologia* **39:**132-139.

Gentilucci, M., Chieffi, S., Deprati, E., Saetti, M. C., and Toni, I., 1996, Visual illusion and action, *Neuropsychologia* **34:**369-376.

Gnadt, J. W. and Andersen, R. A., 1988, Memory related motor planning activity in posterior parietal cortex of macaque, *Exp Brain Res* **70:**216-220.

Goodale, M. A., Jakobson, L. S., and Keillor, J. M., 1994, Differences in the visual control of pantomimed and natural grasping movements, *Neuropsychologia* **32:**1159-1178.

Goodale, M. A. and Milner, A. D., 2004, *Sight Unseen*, Oxford University Press, Oxford.

Goodale, M. A., Milner, A. D., Jakobson, L. S., and Carey, D. P., 1991, A neurological dissociation between perceiving objects and grasping them, *Nature* **349:**154-156.

Goodale, M. A. and Westwood, D. A., 2004, An evolving view of duplex vision: separate but interacting cortical pathways for perception and action, *Curr Opin Neurobiol* **14:**203-211.

Goodale, M. A., Westwood, D. A., and Milner, A. D. (2004), Two distinct modes of control for object-directed action, *Prog Brain Res* **144:**131-144.

Guard, O., Perenin, M. T., Vighetto, A., Giroud, M., Tommasi, M., and Dumas, R., 1984, Bilateral parietal syndrome approximating a Balint syndrome, *Rev Neurol (Paris)* **140:**358-367.

Haffenden, A. M., Schiff, K. C., and Goodale, M. A., 2001, The dissociation between perception and action in the Ebbinghaus illusion: nonillusory effects of pictorial cues on grasp, *Curr Biol* **11:**177-181.

Harvey, M., 1995, Translation of 'Psychic paralysis of gaze, optic ataxia, and spatial disorder of attention' by Rudolph Balint, *Cogn Neuropsychol* **12:**261-282.

Hecaen, H. and de Ajuriaguerra, J., 1954, Balint's syndrome (psychic paralysis of visual fixation) and its minor forms, *Brain* **77:**373-400.

Hildebrandt, H., Schutze, C., Ebke, M., and Spang, K., 2004, Differential impact of parvocellular and magnocellular pathways on visual impairment in apperceptive agnosia? *Neurocase* **10:**207-214.

Himmelbach, M. and Karnath, H.-O., 2005, Dorsal and ventral stream interaction: Contributions from optic ataxia, *J Cogn Neurosci* **17:**632-640.

Holmes, G., 1918, Disturbances of visual orientation, *Brit J Ophtal* **2:**449-468.

Holmes, G. and Horrax, G., 1919, Disturbances of spatial orientation and visual attention with loss of stereoscopic vision, *Arch Neurol Psychiatry* **1:**383-407.

Hu, Y. and Goodale, M. A., 2000, Grasping after a delay shifts size-scaling from absolute to relative metrics, *J Cogn Neurosci* **12:**856-868.

Inouye, T., 1909, *Die Sehstörungen bei Schußverletzungen der kortikalen Sehsphäre*, Engelmann, Leipzig.

Jackson, S. R., Newport, R., Mort, D., and Husain, M., 2005, Where the eye looks, the hand follows; limb-dependent magnetic misreaching in optic ataxia, *Curr Biol* **15:**42-46.

Jackson, S. R. and Shaw, A., 2000, The Ponzo illusion affects grip-force but not grip-aperture scaling during prehension movements, *J Exp Psychol Hum Percept Perform* **26:**418-423.

James, T. W., Culham, J., Humphrey, G. K., Milner, A. D., and Goodale, M. A., 2003, Ventral occipital lesions impair object recognition but not object-directed grasping: an fMRI study, *Brain* **126:**2463-2475.

Jeannerod, M., Decety, J., and Michel, F., 1994, Impairment of grasping movements following a bilateral posterior parietal lesion, *Neuropsychologia* **32:**369-380.

Jeannerod, M. and Jacob, P., 2005, Visual cognition: a new look at the two-visual systems model, *Neuropsychologia* **43:**301-312.

Karnath, H.-O., 2003, Bálint-Holmes Syndrom, in: *Neuropsychologie*, H.-O. Karnath and H.-P. Thier, eds., Springer Verlag, Heidelberg, pp. 231-242.

Karnath, H.-O. and Perenin, M. T., 2005, Cortical control of visually guided reaching - evidence from 16 patients with optic ataxia, *Cereb Cortex* **15:**1561-1569.

Khan, A. Z., Pisella, L., Vighetto, A., Cotton, F., Luaute, J., Boisson, D. et al., 2005, Optic ataxia errors depend on remapped, not viewed, target location, *Nat Neurosci* **8:**418-420.

Lacquaniti, F., Perani, D., Guigon, E., Bettinardi, V., Carrozzo, M., Grassi, F. et al., 1997, Visuomotor transformations for reaching to memorized targets: a PET study, *Neuroimage* **5:**129-146.

Le, S., Cardebat, D., Boulanouar, K., Henaff, M. A., Michel, F., Milner, D. et al., 2002, Seeing, since childhood, without ventral stream: a behavioural study, *Brain* **125:**58-74.

Marotta, J. J., Behrmann, M., and Goodale, M. A., 1997, The removal of binocular cues disrupts the calibration of grasping in patients with visual form agnosia, *Exp Brain Res* **116:**113-121.

McIntosh, R. D., Dijkerman, H. C., Mon-Williams, M., and Milner, A. D., 2004, Grasping what is graspable: evidence from visual form agnosia, *Cortex* **40:**695-702.

Meegan, D. V., Glazebrook, C. M., Dhillon, V. P., Tremblay, L., Welsh, T. N., and Elliott, D., 2004, The Muller-Lyer illusion affects the planning and control of manual aiming movements, *Exp Brain Res* **155:**37-47.

Milner, A. D., Dijkerman, H. C., McIntosh, R. D., Rossetti, Y., and Pisella, L., 2003, Delayed reaching and grasping in patients with optic ataxia, *Prog Brain Res* **142:**225-242.

Milner, A. D., Dijkerman, H. C., Pisella, L., McIntosh, R. D., Tilikete, C., Vighetto, A., and Rossetti, Y., 2001, Grasping the past. Delay can improve visuomotor performance, *Curr Biol* **11:**1896-1901.

Milner, A. D. and Goodale, M. A., 1995, *The Visual Brain in Action*, Oxford University Press, Oxford.

Milner, A. D., Paulignan, Y., Dijkerman, H. C., Michel, F., and Jeannerod, M., 1999, A paradoxical improvement of misreaching in optic ataxia: new evidence for two separate neural systems for visual localization, *Proc R Soc Lond B Biol Sci* **266:** 2225-2229.

Milner, A. D., Perrett, D. I., Johnston, R. S., Benson, P. J., Jordan, T. R., Heeley, D. W., Bettucci, D., Mortara, F., Mutani, R., Terazii, E. et al., 1991, Perception and action in 'visual form agnosia', *Brain* **114:**405-428.

Milner, D. and Dyde, R., 2003, Why do some perceptual illusions affect visually guided action, when others don't? *Trends Cogn Sci* **7:**10-11.

Murata, A., Gallese, V., Kaseda, M., and Sakata, H., 1996, Parietal neurons related to memory-guided hand manipulation, *J Neurophysiol* **75:**2180-2186.

Muri, R. M., Gaymard, B., Rivaud, S., Vermersch, A., Hess, C. W., and Pierrot-Deseilligny, C., 2000, Hemispheric asymmetry in cortical control of memory-guided saccades. A transcranial magnetic stimulation study, *Neuropsychologia* **38:**1105-1111.

Muri, R. M., Vermersch, A. I., Rivaud, S., Gaymard, B., and Pierrot-Deseilligny, C., 1996, Effects of single-pulse transcranial magnetic stimulation over the prefrontal and posterior parietal cortices during memory-guided saccades in humans, *J Neurophysiol* **76:** 2102-2106.

Noesselt, T., Hillyard, S. A., Woldorff, M. G., Schoenfeld, A., Hagner, T., Jancke, L. et al., 2002, Delayed striate cortical activation during spatial attention, *Neuron* **35:**575-587.

Owen, A. M., 2004, Working Memory: Imaging the Magic Number Four, *Curr Biol* **14:** R573-R574.

Oyachi, H. and Ohtsuka, K., 1995, Transcranial magnetic stimulation of the posterior parietal cortex degrades accuracy of memory-guided saccades in humans, *Invest Ophtalmol Vis Sci* **36:**1441-1449.

Pavani, F., Boscagli, I., Benvenuti, F., Rabuffetti, M., and Farne, A., 1999, Are perception and action affected differently by the Titchener circles illusion? *Exp Brain Res* **127:** 95-101.

Perenin, M. T. and Vighetto, A., 1988, Optic ataxia: a specific disruption in visuomotor mechanisms, *Brain* **111:**643-674.

Phan, M. L., Schendel, K. L., Recanzone, G. H., and Robertson, L. C., 2000, Auditory and visual spatial localization deficits following bilateral parietal lobe lesions in a patient with Balint's syndrome, *J Cogn Neurosci* **12:**583-600.

Plodowski, A. and Jackson, S. R., 2001, Vision: getting to grips with the Ebbinghaus illusion. *Curr Biol* **11:**R304-R306.

Prado, J., Clavagnier, S., Otzenberger, H., Scheiber, C., Kennedy, H., and Perenin, M. T., 2005, Two cortical systems for reaching in central and peripheral vision - Evidence from an fMRI study in humans, *Neuron* **48:**849-858.

Quintana, J. and Fuster, J. M., 1993, Spatial and temporal factors in the role of prefrontal and parietal cortex in visuomotor integration, *Cereb Cortex* **3:**122-132.

Quintana, J. and Fuster, J. M., 1999, From perception to action: temporal integrative functions of prefrontal and parietal neurons, *Cereb Cortex* **9:**213-221.

Rafal, R. D., 1997, Balint syndrome, in: *Behavioral Neurology and Neuropsychology,* T.E. Feinberg and M. J. Farah, eds., McGraw-Hill, New York, pp. 337-356.

Ratcliff, G. and Davies-Jones, G. A., 1972, Defective visual localization in focal brain wounds, *Brain* **95:**49-60.

Revol, P., Rossetti, Y., Vighetto, A., Rode, G., Boisson, D., and Pisella, L., 2003, Pointing errors in immediate and delayed conditions in unilateral optic ataxia, *Spat Vis* **16:**347-364.

Rival, C., Olivier, I., Ceyte, H., and Ferrel, C., 2003, Age-related differences in a delayed pointing of a Muller-Lyer illusion, *Exp Brain Res* **153:**378-381.

Rorden, C. and Karnath, H.-O., 2004, Using human brain lesion to infer function - a relic from a past era in the fMRI age? *Nat Rev Neurosci* **5:**813-819.

Smeets, J. B., Brenner, E., de Grave, D. D., and Cuijpers, R. H., 2002, Illusions in action: consequences of inconsistent processing of spatial attributes, *Exp Brain Res* **147:**135-144.

Smyrnis, N., Theleritis, C., Evdokimidis, I., Muri, R. M., and Karandreas, N., 2003, Single-pulse transcranial magnetic stimulation of parietal and prefrontal areas in a memory delay arm pointing task, *J Neurophysiol* **89:**3344-3350.

Snyder, L. H., Batista, A. P., and Andersen, R. A., 1997, Coding of intention in the posterior parietal cortex, *Nature* **386:**167-170.

Toth, L. J. and Assad, J. A., 2002, Dynamic coding of behaviourally relevant stimuli in parietal cortex, *Nature* **415:**165-168.

Tsutsui, K., Jiang, M., Sakata, H., and Taira, M., 2003, Short-term memory and perceptual decision for three-dimensional visual features in the caudal intraparietal sulcus (Area CIP), *J Neurosci,* **23:**5486-5495.

Ungerleider, L. G. and Mishkin, M., 1982, Two cortical visual systems, in: *Analysis of Visual Behavior,* D.J. Ingle, M. A. Goodale, and R. J. W. Mansfield, eds., MIT Press, Cambridge, MA, pp. 549-586.

van Donkelaar, P., 1999, Pointing movements are affected by size-contrast illusions, *Exp Brain Res* **125:**517-520.

van Donkelaar, P. and Adams, J., 2005, Gaze-dependent deviation in pointing induced by transcranial magnetic stimulation over the human posterior parietal cortex, *J Mot Behav* **37:**157-163.

Westwood, D. A. and Goodale, M. A., 2003, Perceptual illusion and the real-time control of action, *Spat Vis* **16:**243-254.

Chapter 7

INTERACTIONS BETWEEN COGNITIVE SPACE AND MOTOR ACTIVITY[1]

Bruce Bridgeman and Brian Lathrop
Department of Psychology, Social Sciences 2, University of California, Santa Cruz, CA, USA

Abstract: Extensive signal processing occurs in sensory systems before perception of object positions. Normally this processing is cognitively opaque, inaccessible to experience or behavioral experiment. Several experimental techniques, however, allow analysis of relationships between unconscious processing and subsequent conscious perception and action. In the induced Roelofs effect, a visual target's perceived position is biased by a large static frame that is offset from the center, an effect that appears in perceptual judgment but not in immediate motor activity. Delayed judgment and delayed pointing both show the effect. All four of these results are due to the frame capturing the straight-ahead, a bias that disappears after stimulus offset. The subject, however, is unaware of the bias, believing the straight-ahead (which affects orientation judgments) to remain accurate. Thus an unconscious bias changes conscious behavior. In a further experiment, inattentional blindness prevents perception of the frame. Nonetheless, the induced Roelofs effect appears. This phenomenon requires two dissociable and sequential unconscious steps, processing the frame and biasing the straight-ahead, before conscious responses are altered.

Key words: consciousness; inattentional blindness; Roelofs effect; straight-ahead; unconscious processing; vision.

Orientation in space by vision is a problem that we share with all other mobile animals, and thus we expect mechanisms that achieve this orientation to be effective after a long period of biological evolution. Though the visual properties of the world appear to us instantaneously and effortlessly, a vast amount of neural processing must occur before a person perceives even the

3 A preliminary abstract of the inattentional blindness studies described here has appeared in the proceedings of the Psychonomic Society 2003 annual meeting.

simplest aspects of the environment. This processing, critical to a principled understanding of perception and of visually guided action, has been studied in several ways.

Physiologically, we know a lot about receptive fields of single neurons in the visual system, and something about how these neurons are connected. We know which of dozens of cortical areas are active during perceptual discriminations and actions involving various parameters of visual stimulation (Van Essen et al., 1992). A functional analysis of sequential stages of processing at the level of behavior, however, is just beginning. In this paper we describe recent experiments pointing to a series of reorganizations of visual material that occur before perception or action take place.

The analysis concerns spatial orientation, measuring both the perceived and the motorically indicated angles of targets relative to the body. While translations are specific to the distances of targets, angles are preserved at any distance, so the results generalize to any distance from the observer. In a natural environment, of course, it is the observer who moves in an environment that has features at every distance.

Another domain in which unconscious functional precursors of conscious action have been investigated is in language, where priming at various levels, including lexical and semantic, is well documented (Merikle et al., 1995). The primes in this domain, however, occur at the same level of stimulation as the responses, the level of linguistic material interpreted as such. Our spatially oriented sensory analysis will instead treat coding of one parameter that affects coding of another parameter, sometimes in a different modality. Finally, the second parameter affects perception and action.

One way to dissect visual processes at a functional level is by analyzing illusions. In analogy to destructive testing by engineers, psychophysical destructive testing requires pushing perceptual processes until they begin to break down, giving us perceptions that no longer match a measurable reality. It is in where and how perception breaks down that we learn about its structure.

1. ONE-STEP UNCONSCIOUS INFLUENCE: THE INDUCED ROELOFS EFFECT

An illusion that has proved particularly effective in uncovering stages of spatial processing in the visual domain is the Roelofs effect, a static illusion of visual position. In the Roelofs situation a large rectangular frame is presented to an observer off-center in an otherwise empty visual field, so that one edge is directly in front of the observer. In the absence of other visual stimulation, that edge will appear to be offset in the direction opposite the rest of the frame (Roelofs, 1935). If the right edge of the rectangle is in front of the observer, for example, the whole rectangle will appear further to the right than it really is.

Recently, the Roelofs effect has been modified to facilitate investigations of spatial processing (Bridgeman, 1991). First, the frame need not have one edge centered; illusions occur whenever the frame is presented asymmetrically in the visual field. Second, a small target within an offset rectangular frame is mislocalized in the direction opposite the frame's offset. Thus, the misperception of frame position induces an illusion of target position; this newly described 'induced Roelofs effect' is a static perceptual mislocalization.

1.1 Methods

Extensive work with the induced Roelofs effect has differentiated its properties when measured with two kinds of response (Bridgeman et al., 1997; Bridgeman et al., 2000). A cognitive or symbolic response was defined as a verbal forced choice or a button press, to indicate one of several predefined target positions. The spatial relationship between the angle of the target relative to the self on one hand, and the spatial location of the response on the other, is arbitrary. A motor act, in contrast, is defined as a response such as jabbing the target open-loop (without sight of the hand), where there is a 1:1 isomorphic relationship between target angle and motor aim. Subjects either make a decision about the angle of the target (a cognitive response) or jab it with a finger (a motor response). The responses can occur either immediately or after a variable delay. The target and the eccentric inducing frame always appear simultaneously, remain visible for one sec, and disappear simultaneously, to prevent any apparent motion from affecting the results.

1.2 Properties of the induced Roelofs effect

Using these two measures, we have obtained the following four results: 1.) A large and consistent induced Roelofs effect is seen for cognitive measures at 0 delay, but 2.) no induced Roelofs effect is seen for sensorimotor measures. 3.) With increase in response delay the effect appears in the sensorimotor measure, while 4.) the cognitive measure does not change. The motor induced Roelofs effect appears gradually after disappearance of the stimulus array. At 1 sec delay it is not significant; it begins to emerge as a small but statistically significant effect at 2 sec (Figure 7-1), and by 5 sec it is as large as the cognitive effect.

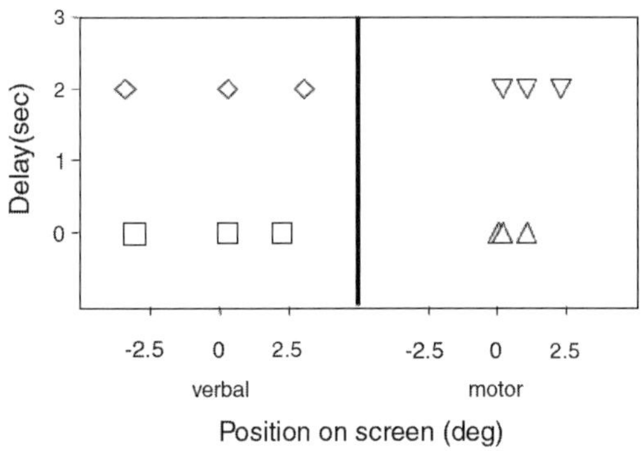

Figure 7-1. Induced Roelofs effects measured with cognitive choice and with motor action, either immediately after stimulus offset or after a 2-sec delay. The cognitive (verbal) measure is a 5-alternative forced choice for locations 2 deg apart. The motor measure is a jab at the target with the right forefinger, without vision of the hand (open-loop). Each graph shows a small representation of the screen area, with average cognitive or motor indications depicted at the corresponding delay. The three symbols shown for each of the four conditions represent average responses when the frame is in three different positions. Replotted from data of Bridgeman et al. (2000).

The result has been replicated many times, most recently in a joint effort of two laboratories (Dassonville et al., 2003), one measuring the motor response as a jab with the finger (Bridgeman) and the other measuring it as a gaze angle (Dassonville). Despite the difference in motor response measures, results collected independently in the two laboratories were closely comparable.

1.3 Interpreting the induced Roelofs effect

The difficult theoretical issue is why the induced Roelofs effect should be absent in immediate pointing but present in the other conditions with identical stimulus arrays. The original explanation was that a sensorimotor representation of space, unavailable to perception, maintained a veridical map of space even when perception was biased (Bridgeman, 1991). Recent data call this interpretation into question, however: pointing is biased toward visible non-target landmarks even for immediate motor responses (Diedrichsen et al., 2004), contradicting the landmark-independent result found when pointing to induced Roelofs targets. How can this discrepancy be resolved?

The explanation for these results begins with another measure, an effect of the Roelofs frame on the subjective straight ahead direction. To measure this, a subject simply points or gazes straight ahead either in darkness or in the presence of the visual inducing frame. In darkness, subjects do quite well at this task, pointing close to their geometric centerline, and they find it easy, but it turns out that in trials with an inducing frame, the off-center context biases the straight-ahead direction.

Subjects are just as confident in their estimates as before, but they fail to indicate the veridical forward direction. The bias, though, is in the direction opposite the induced Roelofs effect – the straight ahead deviates in the *same* direction as the lateral offset of the inducing frame. Further, the magnitude of the straight-ahead deviation in each subject correlates with the amplitude of that subject's induced Roelofs effect. Correlation coefficients were r=0.92 in the Bridgeman data and r=0.96 in the Dassonville data (Dassonville et al., 2003).

It is important that the straight ahead is not a visual variable, nor even a sensory event at all. It is an opinion, a judgment of the orientation of the body measured by the orientation of a part of the body (Harris, 1974). Thus it is qualitatively different from the sensory-based judgments and behaviors that it influences. It affects all measures that are calibrated to bodily orientation, in any modality. Indeed, in complete darkness, without any sensory indications of direction at all, subjects show a consistent opinion about their straight ahead, along what Mittelstaedt (1999) defines as an idiotropic vector, generated from within the subject. It can be influenced by external stimuli, but is distinct from them.

For cognitive measures, the straight ahead indirectly affects judgments of position. In a two-stage process, the Roelofs inducing frame pulls the straight ahead in the direction of the frame; then, the subject perceives the visual target's direction relative to the deviated straight ahead. If the inducing frame is on the left, for instance, the subjective straight ahead is pulled to the left, and as a result a target that is objectively straight ahead will be perceived to the right of straight ahead. The subject concludes that the target is to his right because the bias in the straight ahead is unconscious: he still believes that his straight ahead is accurate. The bias of the straight ahead, however, must logically precede the resulting misjudgment of the target position.

This is the sense in which an unconscious effect, the deviation of the subjective straight ahead direction, becomes a precursor to a visual perception, the conscious experience of a target's location in space. Dependence of visual perception on the unconscious, internally stored value of the straight ahead is revealed in the induced Roelofs effect.

It remains to explain why the induced Roelofs effect is always seen in cognitive measures such as verbal report, but is absent in immediate motor responses. Immediate motor action escapes the induced Roelofs effect because subjects calibrate their pointing activity toward a visible target in a biased frame of reference, based on the biased straight ahead. If a Roelofs inducing frame appears on the subject's left, for instance, the straight ahead will be captured by the frame and will also deviate to the left. In pointing, the subject perceives the target to be further to the right than it really is, and initiates a movement toward the right. But the movement is calibrated to the internally registered straight ahead, which is deviated to the left. The two distortions (perception to the right, straight ahead to the left) cancel, resulting in accurate behavior. After a delay, the straight ahead returns to its veridical location, and the subject mispoints.

The delayed cognitive estimate of position retains an induced Roelofs effect because it is based on remembered position of the comparison array. This position does not change with delay after stimulus offset. A subject who perceives a target to be located at position 4 out of a set of 5 previously defined positions, for example, will verbally indicate '4' no matter how long the delay in the response. The memory for the position is a verbal memory, not a spatial value.

1.4 Discussion

From an evolutionary or functional point of view, then, we conclude that the motor control system is organized in such a way as to maintain accurate behavior even in the presence of distortions in the straight-ahead direction. This is where geometric accuracy matters in interactions with the outside world; perception is free to wander around, driven by asymmetries in the visual environment, but action must remain on target. We can assume that small induced Roelofs effects are occurring constantly in the natural visual world, where stimulation is seldom completely symmetrical.

The function of perception, in this view, is not to localize targets accurately with respect to the self, but to inform about what is in the world, where objects and surfaces are located relative to one another, and only roughly to localize the self in the visual world. Sensorimotor coordination is more accurate, because it is calibrated with unconsciously registered information.

2. TWO-STEP UNCONSCIOUS INFLUENCE: INATTENTIONAL BLINDNESS

At this point we have established that a visually guided motor response (pointing at a small target) is affected by a precursor coded amodally and unconsciously (the straight ahead). We now go one step further to ask whether the effect of asymmetric visual stimuli on motor behavior will remain when the perception of the Roelofs inducing frame itself is prevented. For the purposes of this paper, perception will be defined as conscious, tested by the possibility of verbal report. If the presence of an object can be noted verbally, then it was perceived, however imperfect the perception might be. If it cannot, then any signal processing that occurs is not perception.

Our strategy is to present a large Roelofs inducing frame, present at high contrast and for a reasonable length of time, but to prevent its perception by diverting attention elsewhere with the phenomenon of inattentional blindness. Mack and Rock (1998) popularized inattentional blindness as a method for investigating the role of conscious awareness in perception. In their experiments, subjects were asked to report the presence of an unexpected stimulus. They often failed to report the appearance of many properties of an unexpected stimulus, even if the eyes were directly fixated on the location of its appearance. We review their methods here because similar methods will be employed in the following experiments.

Mack and Rock (1998) had subjects view a fixation target for 1500 ms, followed by distractor cross for 200 ms, which was then followed by a mask. Subjects judged whether the vertical or horizontal segment of the cross was longer. The two segments differed only slightly in length, making the task difficult. The cross was centered at fixation or parafoveally, within a few degrees of fixation. Typically there were three or four of these non-critical trials before a critical trial. On the critical trial, an unexpected 'critical stimulus' was presented along with the cross in one of the four quadrants created by its intersecting horizontal and vertical segments. Immediately following the critical trial, subjects were asked, "Did you see anything on the screen on this trial that had not been there on previous trials?" If they reported seeing something else, they attempted to identify it with a recall or recognition test. In addition to this critical trial, subjects subsequently underwent divided- and full-attention (critical) trials. There was nothing different between these trials and the inattention trial, except for the knowledge that something in addition to the cross might appear in the display. That is, subjects now had a reason to expect that something in addition to the cross might appear.

2.1 Method

Our design was similar, except that subjects had two tasks. One was to judge the relative lengths of the arms of a small cross, like the Mack and Rock task; the other was to judge the position of the cross in a 5-alternative forced choice like the choice we had previously offered in Roelofs-effect experiments. Before the experimental trials began, subjects were shown positions numbered 1-5, at -8, -4, 0, 4, or 8 degrees from the participant's midline, and asked to use these positions as choices for the target position. Subjects completed 25 training trials in order to learn the five possible locations of target (distractor cross) appearance. On each trial the target appeared at one of the five positions, in random order. During the experiment we used only positions 2, 3 and 4, anticipating that subjects might mislocalize the targets. Our distractor stimulus, a 21 deg wide x 9 deg high Roelofs inducing frame that surrounded the target, appeared on critical trials 5, 10, and 12. Its center was offset 4 deg to the right of the subject's center line (Fig. 7-2). These were categorized as our inattention, divided-attention, and full-attention trials, respectively.

During critical trials, both the cross and frame simultaneously appeared for 100 ms, with the cross at position 2, followed by a mask consisting of a grid of vertical and horizontal lines. The design requires a large number of subjects, because most of the interesting information comes from a single trial, the first appearance of the unexpected frame.

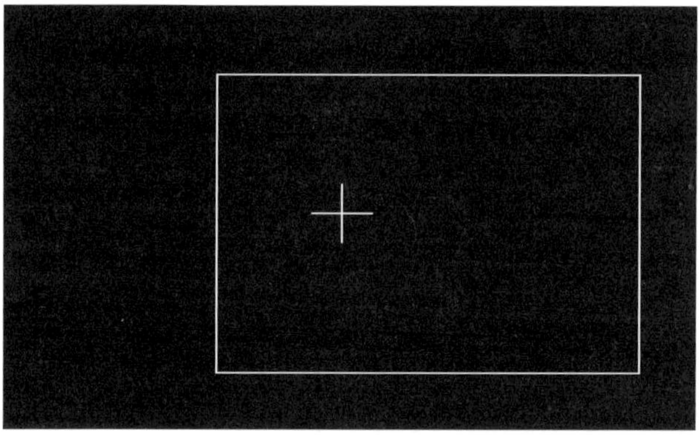

Figure 7-2. Stimulus array during each critical trial; a large laterally displaced rectangular frame is presented simultaneously with a cross.

This experiment investigates two questions: (1) Can a large illusion-inducing stimulus presented unexpectedly go unreported due to inattentional blindness? and (2) If that stimulus goes unreported, will it still impact the perceived location of the target cross?[1]

2.2 Results

More than half of 57 subjects (54%) failed to notice the frame on the inattention trial when asked about it immediately after the trial. The answer to the first question was a definite 'yes'.

To examine the second question, we looked at the localizations of the target in the inattention trials where a frame was presented but not reported. The result was a systematic distortion of the perceived target location in the direction opposite the bias of the frame. The key comparison is with 2 other trials that also had the target at position 2, but had no frame. When the frame was present, the target was estimated to be 3.5 deg to the left of its true location; with the frame absent, the same target position was estimated to be 2.4 deg to the right (Figure 7-3). Thus the presence of the frame induced a Roelofs effect of 5.9 deg.

This analysis suggests that when the target was surrounded by a frame offset to the right, participants tended to perceive the target as being more to the left. In fact, a Scheffe test showed that the difference between positions 4 with the frame and 2 without the frame was not significant ($p > .05$), showing that participants tended to perceive targets at position 2 more to the right. The deviation of position estimates toward the center in the no-frame condition is a common characteristic of cognitive measures of position, and has been observed several times previously (Bridgeman, 1991; Bridgeman et al., 1997; 2000).

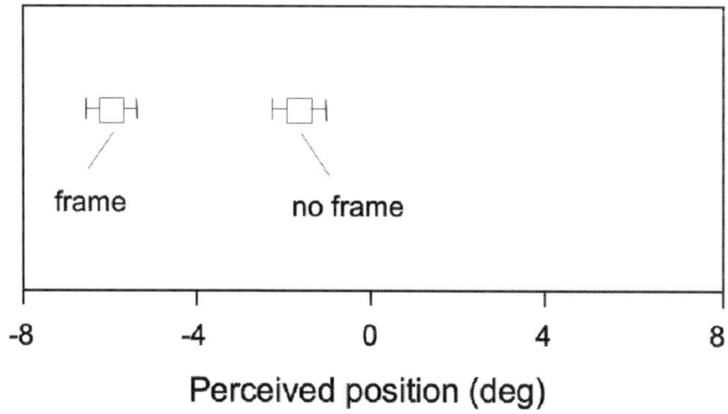

Figure 7-3. Frame vs. No Frame. Perceived position of a target located 4 deg to the left of the centerline, at a post-stimulus delay of 1 sec. Presence of a Roelofs inducing frame offset to the right causes the target to be localized about 5.9 deg to the left of its perceived position without the frame, on average.

A one-way repeated-measures ANOVA was used to evaluate differences in bias due to target location. This factor consisted of Positions 2, 3, 4, and Position 2 with the frame. Results showed significant differences in judgments across this factor [F (3, 56) = 17.15, $p < .01$]. A post-hoc Scheffe test showed that mean biases for the inattention, divided-, and full-attention trials were significantly different from the control trials ($p <.01$).

More surprising was a comparison of the subjects who detected the presence of the frame on the inattention trial and those who did not. The two groups of subjects did not significantly differ in their estimates of target position. A between-subjects t-test showed no difference in performance change between the two groups [t (55) = .826, $p > .05$]. These results suggest that the frame did not impact performance on the localization task differently for those who did and did not report the frame.

This analysis suggests that when the target was surrounded by a frame offset to the right, participants tended to perceive the target as being more to the left. Further, this effect did not depend on one being able to report the appearance of the frame, as there was no classification by trial type interaction. So the answer to the second question is also 'yes'.

2.3 Discussion

In order for the systematic mislocalization of the target to occur in the inattention trials, two distinct sequential steps of unconscious processing must occur before the conscious decision is made about which of the positions was presented. First, the unconsciously processed Roelofs inducing frame must bias the subject's subjective straight ahead in the direction of the frame. Second, the resulting unconscious bias must change the subject's perception of the target's position. Logically, the frame's position must be processed in the brain before that position can affect straight ahead, and the straight ahead must be processed before the target can be mislocalized relative to it.

Thus we can identify two sequential, unconscious processing steps that must precede the conscious decision, each coded in a different way. The Roelofs frame is a visual stimulus, processed but not perceived; the subjective straight ahead is not a sensory coding but an opinion about the orientation of the body, influenced by sensory events but separate from them. Its offset is not perceived. It affects the localization of perceptions in any modality, and slowly changes its biased value in the absence of any spatial stimulation at all (a characteristic shown in the first experiment described here, where the induced Roelofs effect reappeared in a motor measure after a delay). This result adds to recent work showing that subjects can sense whether a change has taken place in a visual scene without being able to consciously identify the change (Rensink, 2004).

REFERENCES

Bridgeman, B., 1991, Complementary cognitive and motor image processing, in: *Presbyopia research: From molecular biology to visual adaptation*, G. Obrecht and L. Stark, eds.,. Plenum Press, New York, pp. 189–198.

Bridgeman, B., Gemmer, A., Forsman, T., and Huemer V., 2000, Processing spatial information in the sensorimotor branch of the visual system, *Vision Res* **40:**3539–3552.

Bridgeman, B., Peery, S., and Anand, S., 1997, Interaction of cognitive and sensorimotor maps of visual space. *Percept Psychophys* **59:**456–469.

Dassonville, P., Bridgeman, B., Bala, J., Thiem, P., and Sampanes, A.., 2003, The induced induced Roelofs effect: two visual systems or the shift of a single reference frame? *Vision Res* **44:**603-611.

Diedrichsen, J., Werner, S., Schmidt, T., and Trommershäuser, J., 2004, Immediate spatial distortions of pointing movements induced by visual landmarks. *Percept Psychophys* **66**:89-103.

Harris, C.S., 1974, Beware of the straight-ahead shift - nonperceptual change in experiments on adaptation to displaced vision. *Perception* **3:**461-476.

Mack, A. & Rock, I., 1998. *Inattentional blindness*. MIT Press/Bradford Books series in cognitive psychology, Cambridge, MA.

Merikle, P.M., Joordens, S., and Stolz, J. A., 1995, Measuring the relative magnitude of unconscious influences. *Conscious Cogn* **4**:422-439.

Mittelstaedt, H., 1999, The role of the otoliths in perception of the vertical and in path integration, in: Otolith function in spatial orientation and movement, B. Cohen and B. Hess, eds., *Ann NY Acad Sci* **871**:334-344.

Rensink, R., 2004, Visual sensing without seeing. *Psychol Sci* **15**:27-32.

Roelofs, C., 1935, Optische Lokalisation. *Archiv für Augenheilkunde* **109**:395–415.

Van Essen, D. C, Anderson, C. H, and Fellman, D. J., 1992, Information processing in the primate visual system: An integrated systems perspective. *Science* **255**:419-423.

Chapter 8

CROSS-MODAL INVOLVEMENT OF VISUAL CORTEX IN TACTILE PERCEPTION

K. Sathian[1] and Simon Lacey[2]

[1]*Department of Neurology, Emory University School of Medicine, Atlanta, GA, USA;* [2]*School of Human Sciences, Southampton Solent University, Southampton, UK*

Abstract: It is now accepted that visual cortical areas that are specialized for processing particular aspects of vision are also involved in the corresponding tactile tasks. However, the reasons for such cross-modal recruitment of visual cortex remain unclear. Visual imagery may be a partial explanation, as may the idea that both visual and tactile inputs can access multisensory representations. Studies of connectivity between somatosensory and visual regions could aid insight into these issues. Cross-modal plasticity offers another perspective. The blind show greater recruitment of visual cortical areas in various non-visual tasks. This has been most clearly demonstrated in tasks involving language, but may also apply to tactile perception. There has been recent interest in the effects of short-term visual deprivation, which appears to result in considerable changes in visual cortical activity. This suggests that cross-modal plasticity might not require forming new connections, but instead might be grafted onto existing connectivity between modality-specific areas. This chapter reviews work from many groups on visual cortical involvement in tactile perception in both normally sighted and visually deprived humans, and considers their implications.

Key words: multisensory; plasticity; touch; visual deprivation; visual imagery.

1. INTRODUCTION

Recent research has established firmly that areas of the cerebral cortex that are traditionally regarded as exclusively visual are closely involved in normal tactile perception. These studies have revealed that various visual cortical areas are active during a number of tactile tasks in the normally sighted, in a manner that is quite task-specific. There is also some evidence that the active visual areas are functionally involved in tactile processing, since their transient or permanent dysfunction can impair tactile performance. Why does such cross-modal cortical recruitment occur? Although the answer is still not clear, there have been some attempts to address this question. One obvious idea is that visual cortical recruitment during tactile tasks reflects visual imagery: lack of familiarity with the evoked tactile representations might favor cross-modaltranslation into a visual representation, which could make for more efficient processing on account of its greater familiarity. Another possibility is that "visual" cortical areas are not exclusively visual, but rather, house multisensory representations that are accessible to input from different sensory modalities. Some might call such representations "amodal", but here the term "multisensory" is preferred, since the modality tags of the relevant inputs could well be maintained (Sathian, 2004). As will emerge later in this chapter, these two views are by no means mutually exclusive, and in fact may simply be two faces of the same crystal – which may perhaps turn out to be quite multifaceted. A different perspective on visual cortical function is offered by studies of visual deprivation, both short-term and long-term. Such studies have shown that many areas of visual cortex are involved in a variety of non-visual tasks, but dissecting the contribution of perceptual from linguistic and other cognitive factors has not been easy. In this chapter, we review the literature on visual cortical involvement in tactile perception in normally sighted humans and the visually deprived, and consider the evidence for the theoretical interpretations that have been advanced.

2. VISUAL CORTICAL INVOLVEMENT IN TACTILE PERCEPTION IN THE SIGHTED

2.1 Tactile perception of two-dimensional patterns

The first demonstration that visual cortex is normally recruited in tactile tasks came from a positron emission tomographic (PET) study (Sathian et al., 1997), using the grating orientation discrimination task that was originally introduced for psychophysical assessment of tactile spatial acuity (van Boven and Johnson, 1994a, 1994b). In this task, plastic gratings bearing alternating ridges and grooves of equal width are applied manually to the immobilized fingerpad, and subjects discriminate whether the gratings are oriented along or across the finger. In the course of a study using this task to investigate variations in spatial acuity between fingers (Sathian and Zangaladze, 1996), it was serendipitously discovered that visual imagery was being employed by subjects making the tactile discriminations. Prior interest in this possibility, and knowledge of findings by many others that visual imagery is associated with activity in visual cortical areas (e.g., Kosslyn et al., 1995, 1999), led to the question of whether tactile discrimination of grating orientation would evoke visual cortical activity. To address this, a PET study (Sathian et al., 1997) was designed employing this task as the experimental task. The control task used the same stimuli, but required subjects to report whether the grooves were wide or narrow (spacing task). The orientation task can be considered a macrospatial task, and the spacing task, microspatial (groove width was ≤3 mm). In general, vision seems to be better than touch for perceiving macrospatial features, the reverse being true for microspatial features (Heller, 1989). Further, macrospatial tactile tasks are preferentially associated with visual processing (Klatzky et al., 1987). Hence, a contrast between a macrospatial and a microspatial tactile task can be expected to show areas engaged in cross-modal visual processing.

Stimuli in this PET study were applied to the right index fingerpad. Stimulus parameters in both tasks were adjusted in individual subjects to yield approximately equal performance in the supra-threshold range. This design controlled for basic somatosensory processing as well as cognitive load. A contrast between the two tasks revealed a single focus specifically activated during processing of grating orientation, in a region of left extrastriate visual cortex, close to the parieto-occipital fissure (Sathian et al., 1997). Since this focus is also active during visual discrimination of grating orientation (Sergent et al., 1992) and spatial mental imagery (Mellet et al., 1996), it was considered to mediate spatial processes common to tactile and visual orientation discrimination (Sathian et al., 1997). Its location near the parieto-occipital fissure suggests homology with an area in the macaque parieto-occipital fissure known as V6 or PO, where a high fraction of the neuronal population appears to be orientation-selective (Galletti et al., 1991).

In order to exclude the possibility that the parieto-occipital cortical activation observed in this first PET study was merely an epiphenomenon, transcranial magnetic stimulation (TMS) was used. When applied focally over occipital cortex, TMS can briefly disrupt visual perception (Amassian et al., 1989; Epstein et al., 1996; Epstein and Zangaladze, 1996). The goal of the TMS study was to test whether TMS over parieto-occipital cortex interferes with tactile discrimination of grating orientation (Zangaladze et al., 1999). Gratings were applied electromechanically to the immobilized right index fingerpad. As in the preceding PET study, subjects discriminated either grating orientation or spacing. Single-pulse TMS, at a delay of 180 ms following the onset of the tactile stimulus, significantly disrupted performance in the orientation task, when applied directly over or near the left parieto-occipital locus found in the earlier PET study. Crucially, the effect was specific for the orientation task, since TMS over parieto-occipital cortical sites did not affect discrimination of grating spacing. In contrast to this specific effect over parieto-occipital cortex, TMS over primary somatosensory cortex at a 30 ms delay exerted a non-specific effect, impairing both orientation and spacing discrimination. This was only to be expected, since interfering with somatosensory cortical function should impair processing of all tactile input. Consistent with these TMS effects on performance, subjects reported feeling the gratings but being unsure about their orientation (but not spacing) during parieto-occipital TMS, whereas they had difficulty even feeling the gratings during TMS over primary somatosensory cortex. Thus, this study (Zangaladze et al., 1999) showed convincingly that extrastriate visual cortical activity during tactile perception is not just an epiphenomenon, but is necessary for optimal tactile performance.

Analogous findings have recently been reported in a TMS study from another group (Merabet et al., 2004). Here, repetitive TMS (rTMS) was applied in 10 min trains at 1 Hz. This is known to reduce excitability of the cortical target zone for a period that outlasts the TMS train, facilitating testing "off-line" in the absence of potential contamination from extraneous effects of TMS, such as muscle twitching. Stimuli were dot-patterns varying in inter-dot distance, which subjects explored actively with the index fingerpad. The study took advantage of the observation that subjective magnitude estimates of perceived inter-dot distance increase linearly with physical inter-dot distance (up to 8 mm), whereas those of perceived roughness scale non-monotonically, peaking at intermediate values (around 3 mm). The main finding was that rTMS over primary somatosensory cortex impaired judgments of roughness, but not inter-dot distance, whereas rTMS over medial occipital cortex, affected distance but not roughness judgments. The effects on distance judgments were smaller than those on roughness judgements, and were greatest at the largest inter-dot distances tested (~8 mm). The rTMS results were backed by converging evidence from study of a congenitally blind patient who suffered a stroke that damaged occipital cortex bilaterally. This patient had normal roughness judgments but was impaired on judging inter-dot spacing (Merabet et al., 2004). These findings support the idea that macrospatial tactile processing has a greater tendency to involve visual processing.

In another PET study (Prather et al., 2004), a global form condition, in which subjects were asked to distinguish between the upside-down letters T and V, was contrasted with an orientation condition, in which subjects reported whether a bar was oriented along the long axis of the fingerpad or at a 45° angle to it. This contrast revealed a right occipito-temporal focus in the region known as the lateral occipital complex (LOC), which is a visual object-selective region (Malach et al., 1995) that is probably homologous with macaque inferotemporal cortex (Grill-Spector et al., 1998). In a related fMRI study (Stoesz et al., 2003) focussing on the macrospatial-microspatial dichotomy, the same global form discrimination task was used as a macrospatial example, detection of a 3-4 mm gap in a bar being the microspatial example. A contrast between the tasks showed that LOC activity was greater in the form than the gap task, in agreement with the other studies (reviewed above) indicating that macrospatial tactile tasks are more likely to evoke visual cortical activity during tactile tasks.

2.2 Haptic perception of three-dimensional form

Visual cortical areas are also active during haptic object identification using active manual exploration, as shown by functional magnetic resonance imaging (fMRI) studies from many groups (Amedi et al., 2001, 2002; Deibert et al., 1999; James et al., 2002; Reed et al., 2004; Stoeckel et al., 2003; Zhang et al., 2004). The LOC has attracted particular attention in this context. A sub-region of the LOC demonstrates selectivity for objects compared to textures for both haptic and visual stimuli (Amedi et al., 2001, 2002). Multisensory object-selective activity in the LOC is stronger for graspable visual objects compared to other visual stimuli, and does not appear to extend to the auditory modality (Amedi et al., 2002). Interestingly, both visual and haptic shape perception may engage the same neural representation, as suggested by multiple lines of evidence: cross-modal priming (visuo-haptic) effects observed psychophysically (Easton et al., 1997a,b; Reales and Ballesteros, 1999) as well as in fMRI studies (Amedi et al., 2001; James et al., 2002); category-specific representations that overlap between visual and haptic modalities (Pietrini et al., 2004); and the case report of a patient with a lesion of the left occipito-temporal cortex that probably included the LOC, who had visual as well as tactile agnosia (a specific inability to recognize objects), although basic somatic sensation (and somatosensory cortex) was intact (Feinberg et al., 1986).

A special case of form perception is perception of faces. Although normally sighted humans perform face recognition using vision almost exclusively, haptic identification of faces can be surprisingly accurate, and cross-modal transfer occurs between visual and haptic identification (Kilgour and Lederman, 2002). Moreover, study of a prosopagnosic patient (i.e. one who could not recognize faces) revealed that the patient was unable to recognize faces not only visually but also haptically (Kilgour et al., 2004). While these behavioral studies suggest that visual and haptic recognition of faces tap into a common neural substrate, fMRI studies so far do not bear this out: face recognition via the haptic modality seems to involve the left fusiform gyrus, in contrast to the right fusiform gyrus for the visual modality (Kilgour et al., 2005), and face-selective voxels in ventral and inferior temporal cortex are largely non-overlapping between the two modalities (Pietrini et al., 2004). Thus, further studies are needed to resolve the nature of bimodal representation of faces, and how it differs from representation of objects.

2.3 Tactile perception of motion

Cross-modal recruitment of visual cortical areas is not limited to tactile form tasks. Tactile motion stimuli, even in the absence of any task, recruit the human MT complex (Blake et al., 2004; Hagen et al., 2002), which is a well-known area sensitive to visual motion and thought to be the human homolog of the macaque middle temporal visual area (MT/V5). Moreover, during bisensory (visual and tactile) evaluation of the direction of motion of a rotating globe, the tactually perceived direction of motion can influence the visually perceived direction (Blake et al., 2004; James and Blake, 2004), suggesting that both modalities engage a common representation, as in the case of haptic object shape reviewed earlier.

3. DOES VISUAL IMAGERY MEDIATE CROSS-MODAL INVOLVEMENT OF VISUAL CORTEX IN TOUCH?

It should be clear by now that visual cortical processing is routinely involved in normal tactile perception in the sighted, especially during macrospatial tasks, and, further, that such processing is task-specific: extrastriate visual cortical areas specialized for particular visual tasks are involved when the same tasks are performed in the tactile modality. Thus, it appears that these "visual" areas are not devoted solely to vision. One obvious possible reason for cross-modal visual cortical recruitment is the use of visual imagery, perhaps necessitated by relative unfamiliarity with the tactile stimuli or tasks. Such cross-modal translation has been proposed to be a general phenomenon, particularly when processing complex information (Freides, 1974). We consider the visual imagery explanation to be plausible because subjects consistently report mental visualization of tactile stimuli during macrospatial tasks such as discrimination of grating orientation or tactile form, which are associated with visual cortical recruitment, but not during microspatial tasks such as discrimination of grating spacing or gap detection, which do not tend to involve visual cortical activity (Sathian et al.,

1997; Stoesz et al., 2003; Zangaladze et al., 1999). Further, an fMRI study (Zhang et al., 2004) found that inter-individual variations in the magnitude of haptic shape-selective activity in the *right* LOC (ipsilateral to the hand used for haptic perception) were strongly predicted by a multiple regression on two visual imagery scores, one indexing the vividness of visual imagery in everyday situations (using the Vividness of Visual Imagery Questionnaire, VVIQ (Marks, 1973)) and the other indexing the vividness of visual imagery employed during haptic shape perception. However, activation strengths in the *left* LOC were uncorrelated with visual imagery ratings, suggesting that other factors besides visual imagery could contribute to cross-modal visual cortical recruitment (Zhang et al., 2004).

Others have argued that LOC activity during haptic perception is not due to visual imagery, based on the finding that visual imagery evoked only one-fifth of the activity in the LOC that haptic object identification did (Amedi et al., 2001). There are certainly other possible explanations for visual cortical recruitment during tactile perception. The simplest possibility is that there are direct, bottom-up somatosensory inputs to the visual cortical areas implicated in tactile perception. In contrast, a visual imagery account would predict top-down connections into visual cortical areas. One approach to distinguishing between these possibilities is to probe effective connectivity using functional neuroimaging data. We have begun to apply this approach by performing structural equation modelling based on the correlation matrix between the time courses of fMRI activity in various regions. In a study where we explored the full range of possible linear interactions between regions that were active in the left hemisphere during haptic shape perception with the right hand, we found evidence for both bottom-up and top-down paths in a network comprising foci in the postcentral sulcus (corresponding to Brodmann's area 2 of primary somatosensory cortex), intraparietal sulcus (IPS), and LOC (Peltier et al., 2007). This suggests a potential neural substrate for both the visual imagery and the multisensory representation explanations.

The visual imagery hypothesis receives some support from a recent behavioral study of cross-modal memory for objects (Lacey and Campbell, 2006a). In this study, participants encoded familiar and unfamiliar objects visually (without being able to feel them) or haptically (without being able to see them) whilst performing a concurrent visual, verbal, or haptic interference task. Visual interference consisted of dynamic visual noise; haptic interference was achieved by manipulating an object, and verbal interference involved listening to a recorded text. Visual and verbal interference during encoding significantly reduced subsequent cross-modal recognition of the unfamiliar objects, but not familiar ones. The haptic interference task had no effect. Since visual interference disrupted haptic encoding as much as visual encoding, this indicates that visual processes were active during encoding in either modality. The visual interference task used was one known to disrupt visual imagery mnemonics during recall of word-lists (Quinn and McConnell, 1996). These results suggest that encoding in the haptic as well as visual modality could depend on visual representations, and could potentially be mediated by covert verbal descriptions during encoding. Since visual and verbal interference affected unfamiliar but not familiar objects, the results of Lacey and Campbell (2006a) could also support the notion advanced earlier in this chapter, that visual cortical activation during tactile perception may be due to recruitment of visual imagery mechanisms triggered by unfamiliarity with the tactile stimuli. Thus, cross-modal memory for familiar objects may depend less on visual imagery because their representations are well-established in several formats – visual, verbal and tactile – either in a network of associated representations or a single multisensory representation.

Does tactile input result in a specifically visual representation, or are imagery mechanisms recruited in order to generate a spatial representation containing information available to both vision and touch, thereby enabling cross-modal memory even for unfamiliar objects? In the preceding study, the interference tasks did not explicitly demand spatial processing. To investigate the possibility of a common spatial representation, a comparison was made of the effects of spatial and non-spatial interference tasks, in the visual and haptic modalities, on cross-modal memory for unfamiliar objects (Lacey and Campbell, 2006b). The results were clear: whether these tasks occurred during encoding or during retrieval, spatial interference disrupted cross-modal memory whilst non-spatial interference had no effect. More importantly, the modality of the spatial interference, whether visual or haptic, made no difference. These two studies can be summarized as follows: In the first (Lacey and Campbell, 2006a), visual interference disrupted performance because, although it did not involve spatial processing, it tapped imagery mechanisms that were important to the cross-modal memory task. Haptic interference had no effect because it neither reflected these mechanisms nor attracted spatial processing. In the second study (Lacey and Campbell, 2006b), when the haptic interference task involved spatial processing, cross-modal performance was disrupted, to the same extent as that owing to visual spatial interference. This was because these tasks disrupted the processing of spatial information, common to vision and touch and crucial for visual imagery and other top-down processes, as well as bottom-up mechanisms. These studies, therefore, offer a means of reconciling the visual imagery and the multisensory representation accounts.

Neurophysiological and neuroanatomical studies in monkeys have also helped elucidate the basis for cross-modal recruitment of visual cortex. Some neurons in area V4 (a non-primary area in the ventral visual pathway) were found to be selective for the orientation of a tactile grating, but only when it served as a cue to be matched to a subsequently presented visual stimulus and not when the tactile grating was task-irrelevant; such responses were absent in primary visual cortex (V1) (Haenny et al., 1988). Since selectivity for tactile grating orientation depended on the tactile stimulus being relevant, it must have been derived from top-down rather than bottom-up inputs. This study fits both with the possibility of a multisensory representation for grating orientation, suggested by the imaging studies referred to earlier (Sathian et al., 1997; Sergent et al., 1992), and the idea that top-down mechanisms can engage such representations. Multisensory inputs have been demonstrated in early sensory cortical areas that are generally considered to be unisensory, including V1 (Falchier et al., 2002; Rockland and Ojima, 2003) and auditory association cortex (Schroeder et al., 2001, 2003; Schroeder and Foxe, 2002). Analysis of the laminar profile of these projections indicates the likely presence of both top-down (Falchier et al., 2002; Rockland and Ojima, 2003; Schroeder et al., 2003; Schroeder and Foxe, 2002) and bottom-up (Schroeder et al., 2003; Schroeder and Foxe, 2002) inputs. Thus, the weight of evidence favors the existence of multisensory representations that are flexibly accessible via both vision and touch, and involving interactions between bottom-up sensory inputs and top-down processes such as visual imagery.

4. CROSS-MODAL INVOLVEMENT OF VISUAL CORTICAL AREAS IN TACTILE PERCEPTION IN THE BLIND

One of the earliest reports of cross-modal plasticity in blind humans used PET scanning to show that occipital (visual) cortical areas were more metabolically active in the early blind compared to the late blind or sighted (Veraart et al., 1990). This was interpreted as an indication of greater synaptic activity in the early blind, perhaps due to incomplete synaptic pruning during development. Subsequently, numerous PET and fMRI activation studies have shown that visual cortical regions of blind subjects are recruited during Braille reading (Amedi et al., 2003; Büchel et al., 1998a; Burton et al., 2002a; Melzer et al., 2001; Sadato et al., 1996, 1998, 2002). In the first such study, a large territory of medial occipital cortex was found to be active in the blind, relative to a rest control, during reading Braille to discriminate words from non-words (Sadato et al., 1996). Activation of medial occipital cortex by Braille reading (relative to rest) occurs not only when the reading finger is moved over Braille text (Sadato et al., 1996, 1998) but even when Braille characters are presented to the passive finger for discrimination (Sadato et al., 2002); and is specific to early blind subjects (Cohen et al., 1999; Sadato et al., 2002), as compared to the late blind and sighted who deactivate these regions (Sadato et al., 2002).

A series of parallel studies established that visual cortex is functionally involved in Braille reading in the blind. TMS over medial occipital cortex impairs the ability of blind subjects to identify Braille or Roman letters, and also distorts subjective perception of the stimuli (Cohen et al., 1997). Sighted subjects were not affected on tactile identification of Roman letters by occipital TMS, but were more affected than the blind by TMS over the sensorimotor cortex contralateral to the stimulated hand. An early blind person has been reported, after an infarct of bilateral occipital cortex, to develop alexia for Braille despite otherwise normal somatosensory perception (Hamilton et al., 2000). Like the activation studies cited above, TMS over medial occipital cortex disrupted Braille-reading performance in the early blind but not late blind (Cohen et al., 1999), implying that visual cortical involvement in Braille reading depends on cross-modal plasticity that occurs during a critical period of visual development.

Does visual cortical involvement in Braille reading by the blind reflect processing of tactile sensory input or language? A number of studies have provided evidence for recruitment of visual cortex during linguistic tasks. A left occipito-temporal region was found (Büchel et al., 1998b) to be preferentially active for words compared to non-words during visual presentations to sighted subjects as well as Braille presentations to blind subjects (whether early or late blind). Wide swaths of occipital and occipito-temporal visual cortical areas are recruited in blind subjects during covert verb generation in response to nouns presented via Braille (Burton et al., 2002a) or hearing (Burton et al., 2002b), relative to feeling the Braille # sign and hearing non-words with matched auditory characteristics, respectively. Further, such activity is stronger during semantic than phonological processing (Burton et al., 2003), and more extensive in the early blind (Burton et al., 2002a,b, 2003; Burton, 2003). The strength of activity in occipital regions also increases with both semantic and syntactic complexity, in congenitally blind subjects listening to sentences to identify incorrect syntactic structures (Röder et al., 2002). Moreover, congenitally blind individuals recruit occipital cortex during a verbal memory task lacking sensory input during scanning, the magnitude of medial occipital activation correlating with verbal memory performance (Amedi et al., 2003). There may be some segregation of language function within reorganized visual cortex, with verbal memory and verb generation (in response to heard nouns) showing a preference for posterior (hierarchically lower-order) occipital regions, and Braille reading for (higher-order) LOC more anteriorly (Amedi et al., 2003).

The overarching conclusion from these studies is that visual cortex is active during language processing in the blind; whether it is also involved in true somatosensory functions is still unclear. Studies of tactile perception *per se*, as distinct from Braille reading, have been relatively limited, and the few studies in which this has been investigated have employed rest controls and thus could not distinguish between sensory and linguistic processes. There was less activation in occipital cortex when blind subjects discriminated angles or the width of grooves cut in homogeneous Braille fields, compared to when they read Braille (Sadato et al., 1996). These tasks also activated ventral occipital cortex in the blind but deactivated cortex around secondary somatosensory cortex (S2), whereas sighted subjects activated the S2 region and deactivated medial and ventral occipital regions (Sadato et al., 1996, 1998). A study of Braille reading that did attempt to control for linguistic processes, using an auditory word control, found that while early as well as late blind subjects activated superior occipital and fusiform cortex during Braille reading, medial occipital cortical activity occurred in the late blind but not the early blind (Büchel et al., 1998a). It was suggested that these findings could be due to visual imagery in the late blind. Category-selectivity during haptic perception of three-dimensional form, similar to that in sighted subjects, was found in inferotemporal cortex of blind subjects, although the category-selective voxels were located more ventrally in the blind compared to sighted subjects (Pietrini et al., 2004). More research is needed to address the nature of cross-modal recruitment of visual cortex during tactile perception in the blind.

5. EFFECT OF SHORT-TERM VISUAL DEPRIVATION ON TOUCH

A recent topic of investigation has been the consequences of short-term visual deprivation. Blindfolding sighted subjects not only enhances the excitability of visual cortex in as little as an hour, as tested using TMS and fMRI (Boroojerdi et al., 2000), but also improves performance on discrimination of grating orientation after just 90 minutes (Facchini and Aglioti, 2003), by a similar magnitude as blindness (van Boven et al., 2000). After two hours of blindfolding, subjects showed significant deactivation during tactile form discrimination and gap detection in regions that are intermediate in the hierarchy of visual shape processing (V3A and ventral IPS), as well as task-specific increases in activation in blindfolded relative to control subjects, favoring the form over the gap task, along the IPS and in regions of frontal and temporal cortex (Weisser et al., 2005). Over a longer period of blindfolding, five days, Braille character discrimination is improved (Kauffman et al., 2002) and occipital cortex becomes responsive during tactile discrimination of Braille characters and auditory tone discrimination, and TMS over occipital cortex becomes able to disrupt Braille reading (Pascual-Leone and Hamilton, 2001). These findings suggest that cross-modal plasticity does not necessarily require the formation of new connections, but could operate on pre-existing connectivity between areas representing individual sensory modalities. Thus, visual deprivation might amplify the range of cross-modal recruitment that has been demonstrated under conditions of normal vision. In future research, it will be important to further define the effects of short-term, long-term and congenital visual deprivation with respect to specific perceptual and cognitive domains, and relate these effects to findings in the normally sighted.

ACKNOWLEDGEMENTS

Research support to KS from the National Eye Institute, National Institute of Neurological Disorders and Stroke, and the Veterans Administration is gratefully acknowledged. Comments should be addressed to K. Sathian (E-Mail: krish.sathian@emory.edu; Tel: 404-727-1366; Fax 404-727-3157).

REFERENCES

Amassian, V. E, Cracco, R. Q., Maccabee, P. J., Cracco, J. B., Rudell, A., and Eberle, L., 1989, Suppression of visual perception by magnetic coil stimulation of human occipital cortex, *EEG Clin Neurophysiol* **74:**458-462.

Amedi, A., Jacobson, G., Hendler, T., Malach, R., and Zohary, E., 2002, Convergence of visual and tactile shape processing in the human lateral occipital complex, *Cereb Cortex* **12:** 1202-1212.

Amedi, A., Malach, R., Hendler, T., Peled, S., and Zohary, E., 2001, Visuo-haptic object-related activation in the ventral visual pathway, *Nature Neurosci* **4:**324-330.

Amedi, A., Raz, N., Pianka, P., Malach, R., and Zohary, E., 2003, Early 'visual' cortex activation correlates with superior verbal memory performance in the blind. *Nature Neurosci* **6:**758-766.

Blake, R., Sobel, K. V., and James T. W., 2004, Neural synergy between kinetic vision and touch, *Psychol Sci* **15:**397-402.

Boroojerdi, B., Bushara, K. O, Corwell, B., Immisch, I., Battaglia, F., Muellbacher, W., and Cohen, L. G., 2000, Enhanced excitability of the human visual cortex induced by short-term light deprivation, *Cereb Cortex* **10:**529-534.

Büchel, C., Price, C., Frackowiak, R. S. J., and Friston, K., 1998a, Different activation patterns in the visual cortex of late and congenitally blind subjects, *Brain* **121:**409-419.

Büchel, C., Price, C., and Friston, K., 1998b, A multimodal language region in the ventral visual pathway, *Nature* **394:**274-277.

Burton, H., 2003, Visual cortex activity in early and late blind people, *J Neurosci* **23:**4005-4011.

Burton, H., Diamond, J. B., and McDermott, K. B., 2003, Dissociating cortical regions activated by semantic and phonological tasks: A fMRI study in blind and sighted people, *J Neurophysiol* **90:**1965-1982.

Burton, H., Snyder, A. Z., Conturo, T. E., Akbudak, E., Ollinger, J. M., and Raichle, M. E., 2002a, Adaptive changes in early and late blind: An fMRI study of Braille reading, *J Neurophysiol* **87:**589-607.

Burton, H., Snyder, A. Z., Diamond, J. B., and Raichle, M. E., 2002b, Adaptive changes in early and late blind: An fMRI study of verb generation to heard nouns, *J Neurophysiol* **88:**3359-3371.

Cohen, L.G., Celnik, P., Pascual-Leone, A., Corwell, B., Faiz, L., Dambrosia, J., Honda, M., Sadato, N., Gerloff, C., Catala, M.D., and Hallett, M., 1997, Functional relevance of cross-modal plasticity in blind humans, *Nature* **389:**180-183.

Cohen, L. G., Weeks, R. A., Sadato, N., Celnik, P., Ishii, K., and Hallett, M., 1999, Period of susceptibility for cross-modal plasticity in the blind, *Ann Neurol* **45:**451-460.

Deibert, E., Kraut, M., Kremen, S., and Hart, J. Jr., 1999, Neural pathways in tactile object recognition, *Neurology* **52:**1413-1417.

Easton, R. D., Greene, A. J., and Srinivas, K., 1997a, Transfer between vision and haptics: memory for 2-D patterns and 3-D objects, *Psychon B Rev* **4:**403-410.

Easton, R. D., Srinivas, K., and Greene, A.J., 1997b, Do vision and haptics share common representations? Implicit and explicit memory within and between modalities, *J Exp Psychol: Learn Mem Cogn* **23:**153-163.

Epstein, C. M., Verson, R., and Zangaladze, A., 1996, Magnetic coil suppression of visual perception at an extracalcarine site, *J Clin Neurophysiol* **13:**247-252.

Epstein, C. M., and Zangaladze, A., 1996, Magnetic coil suppression of extrafoveal visual perception using disappearance targets, *J Clin Neurophysiol* **13:**242-246.

Facchini, S., and Aglioti, S. M., 2003, Short term light deprivation increases tactile spatial acuity in humans, *Neurology* **60:**1998-1999.

Falchier, A., Clavagnier, S., Barone, P., and Kennedy, H., 2002, Anatomical evidence of multimodal integration in primate striate cortex. *J Neurosci* **22:**5749-5759.

Feinberg, T. E., Rothi, L. J., and Heilman, K. M., 1986, Multimodal agnosia after unilateral left hemisphere lesion, *Neurology* **36:**864-867.

Freides, D., 1974, Human information processing and sensory modality: Cross-modal functions, information complexity, memory and deficit, *Psychol Bull* **81:**284-310.

Galletti, C., Battaglini, P. P., and Fattori, P., 1991, Functional properties of neurons in the anterior bank of the parieto-occipital sulcus of the macaque monkey, *Eur J Neurosci* **3:**452-461.

Grill-Spector, K., Kushnir, T., Hendler, T., Edelman, S., Itzchak, Y., and Malach, R., 1998, A sequence of object-processing stages revealed by fMRI in the human occipital lobe, *Hum Brain Mapp* **6:**316-328.

Haenny, P. E., Maunsell. J. H. R., and Schiller, P. H., 1988, State dependent activity in monkey visual cortex. II. Retinal and extraretinal factors in V4, *Exp Brain Res* **69:**245-259.

Hagen, M. C., Franzen, O., McGlone, F., Essick, G., Dancer, C., and Pardo, J. V., 2002, Tactile motion activates the human middle temporal/V5 (MT/V5) complex, *Eur J Neurosci* **16:**957-964.

Hamilton, R., Keenan, J. P., Catala, M., and Pascual-Leone, A., 2000, Alexia for Braille following bilateral occipital stroke in an early blind woman, *Neuroreport* **11:**237-240.

Heller, M. A., 1989, Texture perception in sighted and blind observers, *Percept Psychophys* **45:**49-54.

James, T. W., and Blake, R., 2004, Perceiving object motion using vision and touch, *Cogn Affective Behav Neurosci* **4:**201-207.

James, T. W, Humphrey, G. K., Gati, J. S., Servos, P., Menon, R. S., and Goodale, M. A., 2002, Haptic study of three-dimensional objects activates extrastriate visual areas, *Neuropsychologia* **40:**1706-1714.

Kauffman, T., Theoret, H., and Pascual-Leone, A., 2002, Braille character discrimination in blindfolded human subjects, *Neuroreport* **13:**571-574.

Kilgour, A. R., de Gelder, B., and Lederman, S., 2004, Haptic face recognition and prosopagnosia, *Neuropsychologia* **42:**707-712.

Kilgour, A. R., Kitada, R., Servos, P., James, T. W., and Lederman, S. J., 2005, Haptic face identification activates ventral occipital and temporal areas: An fMRI study, *Brain Cogn* **59:**246-257.

Kilgour, A. R., and Lederman, S., 2002, Face recognition by hand, *Percept Psychophys* **64:**339-352.

Klatzky, R. L., and Lederman, S. J., Reed, C., 1987, There's more to touch than meets the eye: the salience of object attributes for haptics with and without vision, *J Exp Psychol Gen* **116:**356-369.

Kosslyn, S. M., Pascual-Leone, A., Felician, O., Camposano, S., Keenan, J. P., Thompson, W. L., Ganis, G., Sukel, K. E., and Alpert, N. M., 1999, The role of area 17 in visual imagery: Convergent evidence from PET and rTMS, *Science* **284:**167-170.

Kosslyn, S. M., Thompson, W. L., Kim, I. J., and Alpert, N. M., 1995, Topographical representations of mental images in primary visual cortex, *Nature* **378:**496-498.

Lacey, S., and Campbell, C., 2006a, Mental representation in visual/haptic crossmodal memory: evidence from interference effects, *Q J Exp Psychol* **59:**361-376.

Lacey, S., and Campbell, C., 2006b, Modality-independent spatial representation in visual/haptic crossmodal memory, unpublished manuscript.

Malach, R., Reppas, J. B., Benson, R. R., Kwong, K. K., Jiang, H., Kennedy, W. A., Ledden, P. J., Brady, T. J., Rosen, B. R., and Tootell, R. B., 1995, Object-related activity revealed by functional magnetic resonance imaging in human occipital cortex, *P Natl Acad Sci USA* **92:**8135-8139.

Marks, D. F., 1973, Visual imagery differences in the recall of pictures, *Brit J Psychol* **64:** 17-24.

Mellet, E., Tzourio, N., Crivello, F., Joliot, M., Denis, M., and Mazoyer, B., 1996, Functional anatomy of spatial mental imagery generated from verbal instructions, *J Neurosci* **16:**6504-6512.

Melzer, P., Morgan, V. L., Pickens, D. R., Price, R. R., Wall, R. S., and Ebner, F. F., 2001, Cortical activation during Braille reading is influenced by early visual experience in subjects with severe visual disability: A correlational fMRI study, *Hum Brain Mapp* **14:**186-195.

Merabet, L., Thut, G., Murray, B., Andrews, J., Hsiao, S., and Pascual-Leone, A., 2004, Feeling by sight or seeing by touch? *Neuron* **42:**173-179.

Pascual-Leone, A., and Hamilton, R., 2001, The metamodal organization of the brain, *Prog Brain Res* **134:**427-445.

Pietrini, P., Furey, M. L., Ricciardi, E., Gobbini, M. I., Wu, W.-H. C, Cohen, L., Guazzelli, M., and Haxby, J. V., 2004, Beyond sensory images: Object-based representation in the human ventral pathway, *P Natl Acad Sci USA* **101:**5658-5663.

Prather, S. C., Votaw, J. R., and Sathian, K., 2004, Task-specific recruitment of dorsal and ventral visual areas during tactile perception, *Neuropsychologia* **42:**1079-1087.

Quinn, J. G., and McConnell, J., 1996, Irrelevant pictures in visual working memory, *Q J Exp Psychol* **49A:**200-215.

Reales, J. M., and Ballesteros, S., 1999, Implicit and explicit memory for visual and haptic objects: cross-modal priming depends on structural descriptions, *J Exp Psychol: Learn Mem Cogn* **25:**644-663.

Peltier, S., Stilla, R., Mariola, E., LaConte, S., Hu, X., and Sathian, K., 2007, Activity and effective connectivity of parietal and occipital cortical regions during haptic shape perception, *Neuropsychologia* **45:**476-483.

Reed, C. L., Shoham, S., and Halgren, E., 2004, Neural substrates of tactile object recognition: an fMRI study, *Hum Brain Mapp* **21:**236-246.

Rockland, K. S., and Ojima, H., 2003, Multisensory convergence in calcarine visual areas in macaque monkey, *Int J Psychophysiol* **50:**19-26.

Röder, B., Stock O, Bien, S. N. H., and Rösler, F., 2002, Speech processing activates visual cortex in congenitally blind humans, *Eur J Neurosci* **16:**930-936.

Sadato, N., Okada, T., Honda, M., and Yonekura, Y., 2002, Critical period for cross-modal plasticity in blind humans: a functional MRI study, *Neuroimage* **16:**389-400.

Sadato, N., Pascual-Leone, A., Grafman, J., Deiber, M-P., Ibanez, V., and Hallett, M., 1998, Neural networks for Braille reading by the blind, *Brain* **121:**1213-1229.

Sadato, N., Pascual-Leone, A., Grafman, J., Ibanez, V., Deiber, M-P., Dold, G., and Hallett, M., 1996, Activation of the primary visual cortex by Braille reading in blind subjects, *Nature* **380:**526-528.

Sathian, K., 2004, Modality, quo vadis? Commentary on "The emulation theory of representation: motor control, imagery, and perception" by R. Grush, *Behav Brain Sci* **27:**413-414.

Sathian, K., and Zangaladze, A., 1996, Tactile spatial acuity at the human fingertip and lip: bilateral symmetry and inter-digit variability, *Neurology* **46:**1464-1466.

Sathian, K., Zangaladze, A., Hoffman, J. M., and Grafton, S. T., 1997, Feeling with the mind's eye, *Neuroreport* **8:**3877-3881.

Schroeder, C. E., and Foxe, J. J., 2002, The timing and laminar profile of converging inputs to multisensory areas of the macaque neocortex, *Cognitive Brain Res* **14:**187-198.

Schroeder, C. E., Lindsley, R. W., Specht, C., Marcovici, A., Smiley, J. F., and Javitt, D. C., 2001, Somatosensory input to auditory association cortex in the macaque monkey, *J Neurophysiol* **85:**1322-1327.

Schroeder, C. E., Smiley, J., Fu, K. G., McGinnis, T., O'Connell, M. N., and Hackett, T. A., 2003, Anatomical mechanisms and functional implications of multisensory convergence in early cortical processing, *Int J Psychophysiol* **50:**5-17.

Sergent, J., Ohta, S., and MacDonald, B., 1992, Functional neuroanatomy of face and object processing. A positron emission tomography study, *Brain* **115:**15-36.

Stoeckel, M. C., Weder, B., Binkofski, F., Buccino, G., Shah, N. J., and Seitz, R. J., 2003, A fronto-parietal circuit for tactile object discrimination: an event-related fMRI study. *Neuroimage* **19:**1103-1114.

Stoesz, M., Zhang, M., Weisser, V. D, Prather, S. C., Mao, H., and Sathian, K., 2003, Neural networks active during tactile form perception: common and differential activity during macrospatial and microspatial tasks, *Int J Psychophysiol* **50:**41-49.

van Boven, R. W., Hamilton, R. H., Kauffman, T., Keenan, J. P., and Pascual-Leone, A., 2000, Tactile spatial resolution in blind Braille readers, *Neurology* **54:**2230-2236.

van Boven, R. W., and Johnson, K. O., 1994a, A psychophysical study of the mechanisms of sensory recovery following nerve injury in humans, *Brain* **117:**149-167.

van Boven, R. W., and Johnson, K. O., 1994b, The limit of tactile spatial resolution in humans: grating orientation discrimination at the lip, tongue and finger, *Neurology* **44:**2361-2366.

Veraart, C., De Volder, A. G., Wanet-Defalque, M.-C., Bol, A., Michel, C., and Goffinet, A. M., 1990, Glucose utilization in human visual cortex is abnormally elevated in blindness of early onset but decreased in blindness of late onset, *Brain Res* **510:**115-121.

Weisser, V., Stilla, R., Peltier, S., Hu, X., and Sathian, K., 2005, Short-term visual deprivation alters neural processing of tactile form, *Exp Brain Res* **166:**572-582.

Zangaladze, A., Epstein, C. M., Grafton, S. T., and Sathian, K., 1999, Involvement of visual cortex in tactile discrimination of orientation, *Nature* **401:**587-590.

Zhang, M., Weisser, V. D., Stilla, R., Prather, S. C., and Sathian, K., 2004, Multisensory cortical processing of object shape and its relation to mental imagery, *Cogn Affective Behav Neurosci* **4:**251-259.

Chapter 9

NEUROANATOMY OF THE PARIETAL CORTEX

Lutz Jäncke
Department of Psychology, Neuropsychology, University of Zurich, Switzerland

Abstract: In this chapter the basic principles of parietal cortex anatomy will be described. In this context it is emphasized that there is lack of knowledge about the human parietal cortex in terms of neuroanatomical subregions compared to what is known from the monkey brain. However, based on the current anatomical knowledge different parcellation schemes are described. In addition, the inter- and intrahemispheric connectiviy of the human parietal cortex to other brain regions is estimated on the basis of what is known from monkey data. Finally, basic functional principles of the parietal cortex will be discussed.

Key words: parietal lobe; neuroanatomy; connectivity; frontal lobe.

1. GROSS ANATOMICAL SUBDIVISION

The parietal lobe is one of the core regions for spatial processing. This area has undergone a major expansion during human evolution and occupies approximately one fourth of the human brain. It is the region of the cerebral cortex located between the frontal and occipital lobes. On the medial surface this region is roughly demarcated anteriorly by the central sulcus, ventrally by the subparietal sulcus, and posteriorly by the parieto-occipital sulcus. On the lateral surface the parietal lobe is separated from the frontal lobe by the central sulcus but there is no clear macroanatomical separation from the occipital and temporal lobes. On the lateral surface the parietal lobe consists of the postcentral gyrus, the inferior and superior parietal lobule. The inferior parietal lobe comprises the supramarginal and angular gyrus as well as the parietal operculum. Several attempts have been undertaken to delineate cytoarchtectonic and myeloarchitectonic maps of the human parietal cortex (Zilles et al., 2003). The most widely used map is the cytoarchitectonic map provided by Brodmann (1908a, 1908b). According to Brodmann (BA), the postcentral gyrus is covered by areas 1, 2, and 3. The superior parietal lobule includes areas 5 and 7, while the inferior parietal lobule is covered by areas 40 (supramarginal gyrus) and 39 (angular gyrus), and the parietal operculum (area 43).

On von Economo's and Koskinas' maps (1925), in which the parietal areas are labelled as PA (parietal area A), PB, and so forth, three posterior parietal areas (PE, PF, and PG) are described both for humans and monkeys. In this system PF is equivalent to the ventral part of BA 7 extending into BA 39 and the posterior part of the middle temporal gyrus. Area PG is roughly equivalent with the ventral part of BA 7, the ventral bank of the intraparietal sulcus, the posterior part of the middle temporal gyrus, and the anterior part of the angular gyrus (BA 40). PE is located on the dorsal surface of the parietal lobule and in principal similar with BA 7 (Figure 9-1).

The superior and inferior parietal lobules of the monkey and human brain are separated by the intraparietal sulcus. This complex sulcus consists of numerous inner branches and contains many areas. Although this sulcus has been intensely studied in the macaque brain, only few anatomists have provided detailed cytoarchitectonic descriptions of this sulcus. Interestingly, neither Brodmann nor von Economo provided distinct cytoarchitectonic parcellation of the human intraparietal sulcus. Using modern cytoarchitectonic mapping techniques in combination with objective statistical classification techniques Zilles and coworkers (2001) have characterized two areas located on the anterior ventral bank of the intraparietal sulcus. By using a neutral nomenclature, they called these areas ip1 and ip2 (ip for intraparietal). While ip1 is located in the depth of the intraparietal sulcus, ip2 is found on the lower bank of the intraparietal sulcus and extends onto the surface of the inferior parietal lobule. The stereotaxic location of areas ip1 and ip2 suggest that they correspond to the functionally defined areas AIP and VIP (see below for further description).

In terms of gross anatomical landmarks the inferior parietal lobe is clearly asymmetric. This mainly depends on the angulation of the Sylvian fissure. Unlike sulci, the Sylvian fissure is not formed by an infolding of the cortex, but results from an uneven growth of the outer cortex leading to the opercularization of inner (insular) structures. Unique to the human brain is the extensive degree of opercularization of frontal, parietal, and temporal brain regions, which form the banks of the Sylvian fissure. This anatomical

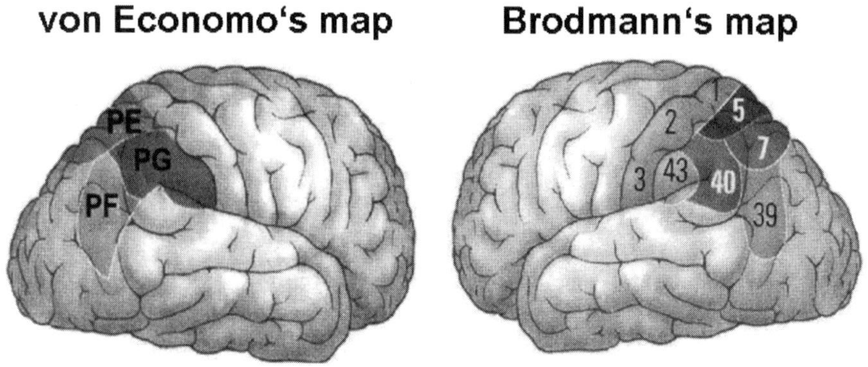

Figure 9-1. Von Economo's and Brodmann's cytoarchitectonic maps of the parietal cortex (figure modified according to Kolb and Whishaw (2006).

feature allows for great morphological variation among hemispheres, both between and within subjects. Sylvian fissure variability was noted early on and became the subject of several studies (for a summary see (Jäncke and Steinmetz, 2003). These studies have largely agreed in dividing the Sylvian fissure into a main horizontal segment (with an anterior and posterior portion, ASF and PSF; anterior and posterior Sylvian fissure) and four additional rami (anterior ascending ramus, anterior horizontal ramus, posterior ascending ramus, posterior descending ramus; abbreviated as AAR, AHR, PDR, PAR, respectively; Figure 9-2). According to the length of the posterior horizontal Sylvian fissure or the size of the posterior ascending and descending rami at least two subtypes of Sylvian fissure can be distinguished: (i) one which is more common on the left hemisphere and which is characterized by a long horizontal part of the Sylvian fissure, and (ii) a second type which is more common on the right hemisphere which typically is qualified by a relatively small posterior horizontal part of the Sylvian fissure. In some studies the ascending ramus was frequently absent on the left hemisphere (Steinmetz et al., 1990) or the right ascending ramus was directed more upward and forward (instead of upward and backward) (Ide et al., 1999).

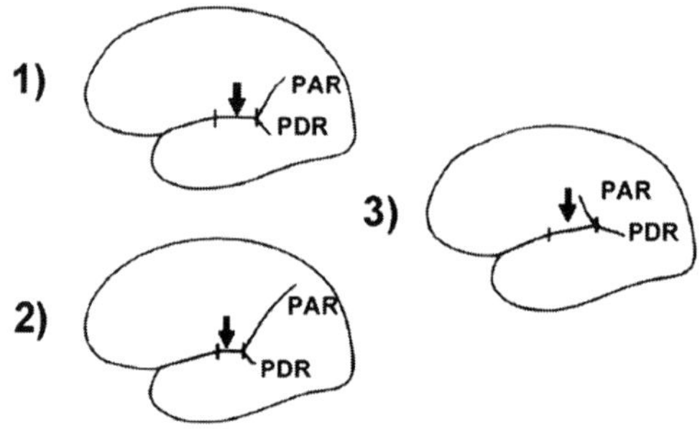

Figure 9-2 Three subtypes of Sylvian fissure shapes. Type 2 is predominantly found on the right hemisphere in right-handed subjects. The PAR determines the so called *planum parietale* which is the part of the supramarginal gyrus. The *planum parietale* is larger on the right hemisphere (PAR, PDR: posterior ascending ramus, posterior descending ramus of the Sylvian fissure).

In addition, the horizontal part of the Sylvian fissure is more strongly angulated upward on the right hemisphere while this part is more horizontally located on the left hemisphere. Most interestingly, when the two hemispheres of each subject are matched, no correspondence is observed between the fissurization pattern of one hemisphere and the other, indicating that fissurization develops independently in each hemisphere (Figure 9-2).

The brain region that has attracted by far the most attention with respect to cerebral asymmetry is the *planum temporale* which is located on the superior temporal gyrus (for a thorough discussion see Steinmetz, 1996) More interesting for the scope of the present paper is the so called *planum parietale* which is formed by the cortex covering the posterior bank of the posterior ascending ramus of the Sylvian fissure and which is larger on the right than on the left hemisphere. It remains to be shown whether the *planum parietale* may be a structural marker of human brain laterality measurable in vivo. Anatomically, the *planum parietale* is part of the inferior parietal lobule. Depending on the individually variable posterior extension of the Sylvian fissure, the *planum parietale* is covered mostly by the cytoarchitectonic area 40 or, in some hemispheres, area 39 of Brodmann. It is tempting to speculate that rightward *planum parietale* asymmetry may be related to right parietal specialization for aspects of non-verbal higher-order cognition. Thus, the inferior parietal lobule has been considered to be a nodal point of a network subserving directed attention or spatial working memory, functions that are lateralized to the right hemisphere. A further interesting finding of our own studies is that the direction of planum parietale asymmetry is basically unrelated to planum temporale asymmetry (Jäncke et al., 1994). This fits with recent morphological studies examining the shape and size of various parts of the Sylvian fissure (Ide et al., 1999). These findings altogether are also in concordance with neuropsychological data from lesion studies suggesting that the lateralizations of linguistic and spatial abilities are more or less statistically independent (Bryden et al., 1983). The cytoarchitecture of the perisylvian region is complex and its intersubject variation has not been intensively studied in humans. So far, no cytoarchitectonic asymmetries have been found for the areas covering the inferior parietal lobule (Schulze, 1960, 1962).

The most salient parietal landmark is the intraparietal sulcus that divides the parietal lobe into the superior and inferior parietal lobules. In humans the IPS is relatively long (~ 7 cm) and deep (~ 2cm). It extends posteriorly from the transverse occipital sulcus to the postcentral sulcus. In the monkey brain several subareas have been delineated along the intraparietal sulcus on the basis of single cell recordings comprising the V3a (in the occipital lobe), the caudal intraparietal sulcus (cIPS), the lateral intraparietal sulcus (LIP), the medial intraparietal sulcus (MIP), the ventral intraparietal sulcus (VIP), and the anterior intraparietal sulcus (AIP). Since the human neuroanatomy especially within the parietal lobule substantially differs from the monkey neuroanatomy, these regions can hardly be detected in the human brain. Several areas have been proposed in the human IPS region to be putative homologues of monkey intraparietal sulcus areas (e.g., the AIP, VIP, and the LIP).

2. CONNECTIONS OF THE PARIETAL CORTEX

The parietal cortex is part of the well-known dorsal information stream. However, the precise structural connectivity of the human parietal cortex is largely unknown. The reason for this apparent lack of knowledge is that axonal tracing techniques cannot be applied in the human brain. Although modern magnetic diffusion tensor imaging techniques are advancing these techniques they are still not powerful enough to demonstrate the anatomic connectivity at the required level of spatial resolution. Thus, our knowledge about the connectivity of the human parietal lobe is based mainly on axonal tracing studies in the macaque brain. However, the comparison of human and monkey parietal cortex is limited because there is only a rough similarity between the human and monkey parietal lobe. For example, referring to the classical maps of Brodmann the superior parietal lobe is covered by area 5 and a small part of area 7 (actually the most superior posterior region of area 7) in the monkey brain. Area 7 is even extending into the inferior parietal lobule of the monkey brain.

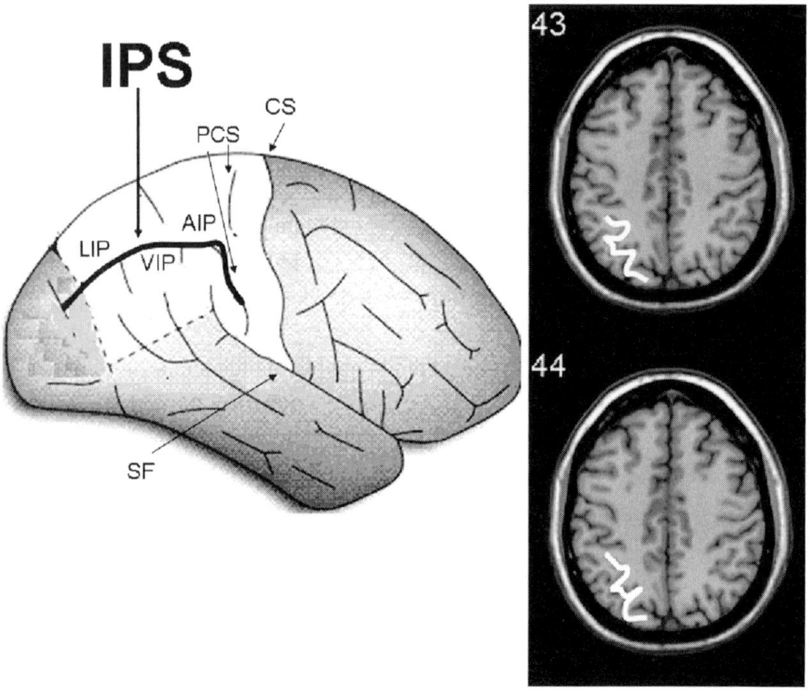

Figure 9-3. Schematic description of the intraparietal sulcus. On the left the intraparietal sulcus is indicated as bold line. This figure demonstrates the length of this complex sulcus extending from the dorsal part of the occipital lobe to the postcentral sulcus (PCS). Three subareas of the intraparietal sulcus are indicated according to their putative location. In addition, the central sulcus (CS) and the Sylvian fissure (SF) are shown. On the right panel two horizontal slides are shown at the level of z = 43 and z = 44 within the MNI stereotaxic space. In these slides the left-sided intraparietal sulcus is marked with a thick white line. The figure presented on the left has been adapted from a figure presented by Culham and Kanwisher (2001).

Whereas in the human brain, the superior parietal lobule is entirely covered by Brodmann areas 5 and 7 while the inferior parietal lobule is covered by the "human-specific" areas 39 and 40. However, the intraparietal sulcus can also be found in the monkey brain. Thus, the intraparietal sulcus is the crucial landmark for any valid comparison between the architectonic and connectivity structures of the monkey und human parietal cortex (Zilles et al., 2003). This view is supported by the fact that several human brain-imaging studies have recently described equivalents of monkey areas AIP and VIP in the human intraparietal sulcus (Binkofski et al., 1998; Jäncke et al., 2001). In the following I will present the connections of the parietal cortex on the basis of what is known from the monkey brain. Thus, one has to be cautious to directly transfer these findings to the human brain.

The monkey parietal cortex receives sensory input from the visual and somatosensory cortices and projects to the posterior motor areas (areas F1 to F5 in the monkey brain). The rostral motor areas F6 and F7 receive major input from the prefrontal cortex. Because the areas F1 to F5 are part of the motor system and give rise to the corticospinal tract, the function of this parieto-frontal system is sensory-motor transformation necessary for motor actions. In addition, there are also reciprocal connections from the primary sensorimotor area to the superior parietal lobule which have tactile recognition functions and are necessary for controlling limb movements. The intraparietal sulcus contains multimodal areas integrating visual, auditory, and somatosensory information. Some of these intraparietal sub-areas (VIP, AIP) are connected with the ventral premotor cortex (areas F4, F5) while others (MIP) are connected with the dorsal premotor cortex (e.g., F2 in the monkey brain) (see also Figure 9-2 for further information). In addition, there are also connections from monkey areas PF and PG to the orbitofrontal cortex and to the cingulum. There are back projections from the dorsolateral prefrontal cortex to area PE with collateral projections to the cingulum and the inferior temporal cortex including the hippocampus and paralimbic regions. These connections emphasize a close functional relation between the prefrontal cortex and the parietal cortex. This network is probably involved in controlling spatially guided behavior.

If one applies these finding with caution to the human brain the connections of the parietal cortex are as shown in Figure 9-3. Brodmann's area 5 is mainly a somatosensory area receiving most of its connections from the primary somatosensory areas (Brodmann's area 1, 2, and 3). The output of this area is going to the primary motor area (BA 4), and the mesial and lateral premotor areas (SMA, PMC). There are also connections to the inferior parietal lobe and to the posterior part of the temporal cortex (area PE in the monkey). This network is involved in controlling and guiding movements providing information about limb position.

3. FUNCTIONS CONTROLLED BY THE PARIETAL CORTEX

Many psychological functions are related to the parietal cortex. Some of them are presented and discussed in this book. In this paragraph I would like to summarize these functions in a broader context. The parietal cortex can be subdivided on a functional basis in at least two parts, the anterior and posterior parietal cortex. The anterior parietal cortex is covered by the postcentral gyrus (Brodmann areas 1, 2, and 3) and the parietal operculum (area subparietalis).

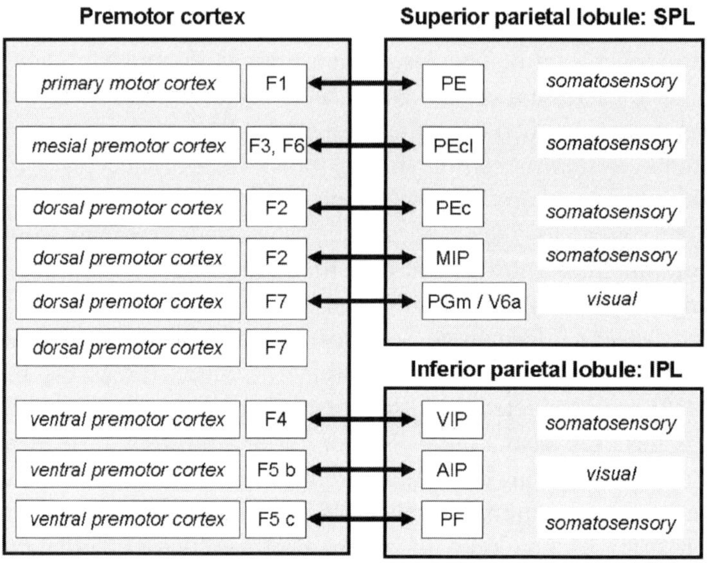

Figure 9-4. Schematic description of the parieto-frontal connections found so far in the monkey brain. F1 to F7 are acronyms describing subareas in the premotor cortex. The nomenclature used for describing the parietal subareas are taken from Economo's map. The entire figure was composed according to a figure published in Zilles et al. (2003).

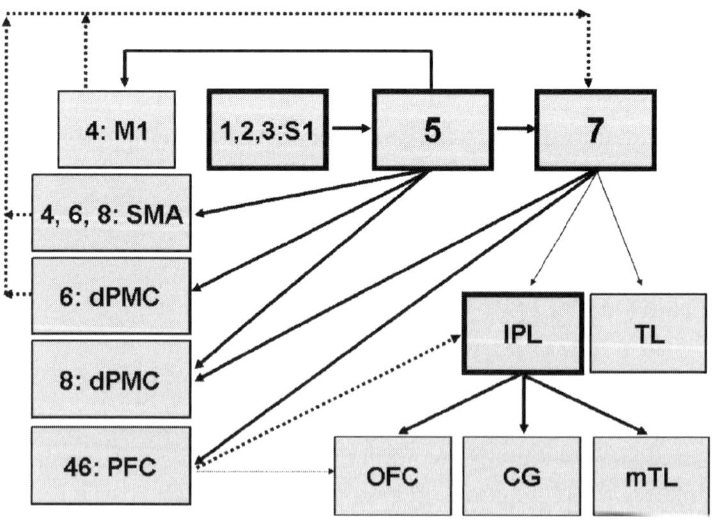

Figure 9-5. This figure represents the putative connections of the parietal areas in the human brain. Please note that these connections are estimated from findings obtained from the

monkey brain and do not refer to findings of the human brain. As mentioned in the text, there are currently no substantial data available for the human brain. The parietal areas are indicated as boxes with thick lines. Each box contains Brodmann area numbers and/or functional anatomical descriptions. (IPL: inferior parietal lobule, TL: temporal lobe, CG: cingulum, mTL: middle temporal lobe, OFC: orbitofrontal cortex, S1: primary somatosensory cortex, M1: primary motor cortex, SMA: supplementary motor area, dPMC: dorsal premotor cortex, PFC: prefrontal cortex). Dotted arrows indicate back projections to the parietal cortex from non-parietal areas.

The posterior parietal cortex is composed of the superior parietal lobule (BA 5 and 7) and parts of the inferior parietal lobule (BA 39 and 40). The anterior part processes somatic sensations and perceptions while the posterior part is specialized for the integration of sensory input from various sources (mostly visual, but also somatic and auditory) mostly for the control of movement. A further function, which is associated with the posterior parietal cortex, is the processing of spatial information.

The posterior parietal cortex is part of the dorsal visual information stream and thus receives information from the visual cortex. However, this information is coded in so-called retinotopic coordinates and is not applicable for movement control. Therefore it is necessary to transform the retinotopically organized visual information into body-centered information, which can be used for movement control. In addition, it is necessary to integrate visual information with the other senses to create a unified map of the external world. Therefore the parietal cortex can be conceived of as an integration centre necessary to map the external space within the brain mainly for orientation and movement. However, although this idea is compelling and mentioned in many textbooks about functional neuroanatomy, there is very little evidence for the existence of such a map in the brain. Rather, it seems likely that there is no single map, but a series of representations of space. The different representations vary in terms of behavioral needs and complexity levels of spatial information to deal with.

An interesting aspect is that parts of the parietal cortex are involved in spatial processing. However, spatial information about the location of objects in the world is needed to direct actions to those objects. For example, in the case of visuomotor control the location of an object and its local orientation and motion must be determined relative to the viewer. Furthermore, because the eyes, head, limbs, and body are constantly moving, computations about orientation, motion, and location must take place on-line. Details of object characteristics (e.g., color or texture) are irrelevant for visuomotor control in the context of viewer-centered movements. The results of many electrophysiological and brain imaging studies in humans and monkeys support the general idea that the posterior parietal cortex plays a significant role in directing movements in space and in detecting stimuli in space (Fogassi and Luppino, 2005; Naghavi and Nyberg, 2005; Rizzolatti and Matelli, 2003).

4. CONCLUSION

The human parietal cortex is a brain area, which is not well understood so far in terms of cytoarchitectonic composition and connectivity. Most available information about cytoarchitectonics and connectivity stems from studies on the monkey brain (Fig.9-4). Thus, a simple transformation from monkey findings to the human brain is not feasible. However, the parietal cortex plays a major role in interfacing sensory information with movement and action control (see Fig. 9-5, and the Chapter 17 of Gallese in this book).

REFERENCES

Binkofski, F., Dohle, C., Posse, S., Stephan, K. M., Hefter, H., Seitz, R. J., and Freund, H. J., 1998, Human anterior intraparietal area subserves prehension: a combined lesion and functional MRI activation study, *Neurology* **50**(5):1253-1259.

Brodmann, K., 1908a, Beiträge zur histologischen Lokalisation der Grosshirnrinde. VI. Die Cortexgliederung des Menschen, *Journal für Psychologie und Neurologie* **10**:231-246.

Brodmann, K., 1908b, Beiträge zur histologischen Lokalisation der Grosshirnrinde. VI. Die Cortexgliederung des Menschen, *Journal für Psychologie und Neurologie* **10** 287-334.

Bryden, M. P., Hecaen, H., and DeAgostini, M., 1983, Patterns of cerebral organization, *Brain Lang* **20**(2):249-262.

Fogassi, L., and Luppino, G., 2005, Motor functions of the parietal lobe, *Curr Opin Neurobiol* **15**(6):626-631.

Ide, A., Dolezal, C., Fernandez, M., Labbe, E., Mandujano, R., Montes, S., Segura, P., Verschae, G., Yarmuch, P., and Aboitiz, F., 1999, Hemispheric differences in variability of fissural patterns in parasylvian and cingulate regions of human brains, *J Comp Neurol* **410**(2):235-242.

Jäncke, L., Kleinschmidt, A., Mirzazade, S., Shah, N. J., and Freund, H. J., 2001, The role of the inferior parietal cortex in linking the tactile perception and manual construction of object shapes, *Cereb Cortex* **11**(2):114-121.

Jäncke, L., Schlaug, G., Huang, Y., and Steinmetz, H., 1994, Asymmetry of the planum parietale, *Neuroreport* **5**(9):1161-1163.

Jäncke, L., and Steinmetz, H., 2003, Anatomical brain asymmetries and their relevance for functional asymmetrie, in: *The asymmetrical brain*, K. Hugdahl and R. J. Daidson, eds., The MIT Press, Cambridge, Massachusetts, pp. 187-230.

Naghavi, H. R., and Nyberg, L., 2005, Common fronto-parietal activity in attention, memory, and consciousness: shared demands on integration?, *Conscious Cogn* **14**(2):390-425.

Rizzolatti, G., and Matelli, M., 2003, Two different streams form the dorsal visual system: anatomy and functions, *Exp Brain Res* **153**(2):146-157.

Schulze, H. A. F., 1960, Zur individuellen cytoarchitektonischen Gestaltung der linken und rechten Hemisphäre im Bereiche des Lobulus parietalis inferior, *Journal für Hirnforschung* **4**:486-517.

Schulze, H. A. F., 1962, Quantitative Untersuchungen zur Frage der individuellen Variation und der Hemisphärendifferenzen der corticalen Areale des unteren Parietalläppchens, *Journal für Hirnforschung* **5**:345-376.

Steinmetz, H., 1996, Structure, functional and cerebral asymmetry: in vivo morphometry of the planum temporale, *Neurosci Biobehav Rev* **20**(4):587-591.

Steinmetz, H., Furst, G., and Freund, H. J., 1990, Variation of perisylvian and calcarine anatomic landmarks within stereotaxic proportional coordinates, *AJNR Am J Neuroradiol* **11**(6):1123-1130.

von Economo, C., and Koskinas, G. N., 1925, *Die Cytoarchitektonik der Hirnrinde des erwachsenen Menschen*, Springer, Berlin.

Zilles, K., Eickhoff, S., and Palomero-Gallagher, N., 2003, The human parietal cortex: a novel approach to its architectonic mapping, *Adv Neurol* **93**:1-21.

Zilles, K., and Palomero-Gallagher, N., 2001, Cyto-, myelo-, and receptor architectonics of the human parietal cortex, *Neuroimage* **14**(1 Pt 2):S8-20.

Chapter 10

SPATIAL MAPS, FEATURE INTEGRATION AND PARIETAL FUNCTION: IMPLICATIONS FROM THE STUDY OF SPATIAL DEFICITS

Lynn C. Robertson
Veterans Administration Medical Research, Martinez, CA, and Department of Psychology and Helen Wills Neuroscience Institute, University of California, Berkeley, USA

Abstract: In this chapter I discuss some of what can and cannot be visually perceived without spatial awareness, how attentional selection of visual information is affected by damage to neural systems that support spatial processing, how spatial processing in turn is involved in binding surface features such as color and shape together and how multiple spatial maps may guide attention. Relevance of neurolopsychological patient studies is also examined.

Key words: attention; perception; Balint's syndrome; unilateral neglect; feature binding; cognitive neuroscience; spatial deficits.

1. INTRODUCTION

Clinical evaluations have long made it clear that spatial awareness of the external world can be lost without affecting the ability to perceive an individual object or part of an object. One reason this observation is so striking is that experience suggests that everything we see must inhabit some area of space separate from others. Surface features such as color and texture also appear within different spatial regions and in normal perception are bound within the contours or shapes that spatially define separate objects or groups of objects. Space perception seems so fundamental that it has led to philosophical conclusions such as the claim made by the influential 18th century philosopher Immanuel Kant: "Space is a necessary representation *a priori* which serves the foundation of all external intuition. We never can imagine or make a representation to ourselves of the non-existence of space." In fact, it seems intuitively obvious that without space, there is nothing to see, nothing to imagine, nothing to conceive. It turns out that Kant and our own intuitions are partly right but partly wrong. The conscious existence of space can be lost without losing perception of everything in it.

2. WHAT IS PERCEIVED WITHOUT SPATIAL AWARENESS?

There are two major syndromes in neuropsychology that disrupt spatial awareness, and both are associated with functional damage to posterior dorsal cortex in humans. These are unilateral spatial neglect (sometimes called hemi-inattention or simply left or right neglect) and Balint's syndrome (see Heilman et al., 1993). Neglect disrupts awareness contralateral to the side of the lesion and is most often associated with right parietal and less often with frontal or left hemisphere damage (Heilman et al., 1994; Mesulam, 1981). The second syndrome is produced by bilateral occipital-parietal damage and disrupts spatial awareness across the extent of external space (Balint, 1909; Holmes and Horax, 1919; Rafal, 1997). The common anatomical denominator is parietal cortex and underlying white matter, although the critical areas may be somewhat different depending on the nature of the spatial deficits involved (Rizzi and Vecera, 2002; Rafal, 2001; Robertson, 2004). Balint's syndrome produces the most severe visual-spatial deficits observed in neuropsychology, and spatial awareness of the external world for perception becomes all but absent. Fortunately, this syndrome is rare, as it leaves the patient functionally blind and in need of constant care.

A hallmark of Balint's syndrome is simultanagnosia (seeing only one object or part of one object at any given time) accompanied by the inability to locate that object (either by reaching, pointing, fixating or verbally reporting its position). In pure cases, primary visual functions are left intact, as are memory, language, problem solving and other cognitive abilities. Patients are often confused by what they see and may be hesitant to report it for fear of being "thought of as crazy" by family members, friends or caregivers. Instead, they may simply say they are blind. Although Balints syndrome has been historically associated with visual perception, recent evidence has demonstrated its cross modality properties (Phan et al., 2000).

Another classic symptom of the syndrome is what Balint called a "pseudoparalysis of gaze" in which the patient's eyes remain fixed in one position as if they are frozen, yet no primary movement problems exist (e.g., the eyes move in their sockets when the patient's head is turned, and patients can move their eyes in a given direction upon command). Primary motor movements remain intact, but the patient seems hesitant to break fixation.

RM, a patient with Balint's syndrome who we studied for about 8 years found it hard to describe everyday visual scenes and explained that it was quite difficult to provide the words that could express what the world looked like to him. As would be expected when spatial awareness is affected, he did not know which way to turn or where to look or where to reach for his telephone when it rang or how to find his bed. However, even more was involved. He explained that an object could appear in view for several minutes and then abruptly be replaced by another object, completely outside his control. In addition, an object such as a house might be seen to move or a safety pin could appear much larger than it should be, as if filling an entire visual space. Basic features such as size, motion, color could change or merge into forms that he knew did not exist.

Unlike blindness from primary visual damage, basic visual signals continued to be processed, but without the spatial functions of the parietal lobes, they had nowhere to go. The signals were not organized or integrated in normal ways. The result was visual chaos with features such as color, shape, size and motion appearing either bound within objects that he did see (sometimes correctly, sometimes incorrectly) or existing as a feature signal bound to nothing at all (see Bernstein and Robertson, 1998; Freidman-Hill et al., 1995; Robertson et al., 1997).

How does one begin to scientifically probe such a non-intuitive system? We cannot rely on our own perceptual phenomenology as we so often do when probing normal vision. In fact, one learns very fast from such cases how influential our own subjective visual experiences are when deriving hypotheses for experiments within visual sciences and how useless they are when trying to imagine what the world would look like without a spatial map. Nevertheless, the mechanisms that contribute to normal vision should limit ways in which perception breaks down, and to the extent that perceptual and attentional theories are correct, these can be used to begin to make inferences in cases where it is difficult or impossible to visualize the way the world must appear.

Feature Integration Theory (FIT) proposed by Treisman and Gelade (1980) predicts at least some of what RM reported seeing. According to this theory, basic features such as color, motion, size, orientation and shape are initially represented in separate "feature maps", while the integration of these features requires spatial attention. A red shoe among black and brown shoes is detected regardless of the number of shoes on the shelf because red is a distinctive feature that "pops out" in the display (i.e., automatically attracts attention). In terms of brain function, all that would be needed to perceive the presence of red among brown would be a signal from a feature map that represents different colors. The location of red need not be processed to manage the task; although there is no reason it should not be if spatial processing is intact. In normal observers attention is quickly attracted to the location of distinctive features or new objects (see Kim and Cave, 1995; Yantis, 1993). Our work with RM showed that if a location signal is absent a feature signal can still be processed and detected, although its location may not be. Conversely, a red shoe among red socks and brown shoes would require controlled search through the display for the integration of red and shoe shape. In such cases, spatial maps become necessary for guiding attention. These maps can be used to co-locate features in the display and bind the features together (e.g., red and the shape of a shoe are bound through spatial attention).

In one of our studies with RM, we investigated perception without spatial awareness by presenting feature and conjunction search displays and asked him to detect and then to locate targets in the different types of displays. In one case we showed him a target (red X among several green Xs (feature search) and found that like normal observers, he detected the distinctive red item quickly and easily independent of the number of distractors present in the display. However, when asked to report whether the target was in the upper, lower, leftward or rightward position in the display he performed at chance. This was true whether he gave a verbal report, pointed in a direction or was asked to touch the screen. In fact, he had to be encouraged to guess, struggled with his response and often reported that he did not know where the red X that he saw was.

Importantly, diplopia or double vision was ruled out, as there was no consistency between where the feature was located in the display and where RM reported it to be (whether tested monocularly or binocularly). A red X appeared to be there but with no location, consistent with his own verbal description. In other sessions, when he was again shown a display with a red X target but now with detractors of green Xs and red Os (conjunction search) he had tremendous difficulty detecting the presence or absence of the target. This was the case whether the displays were shown for 500 ms or 10 seconds and even with set sizes as small as 4 or 6 items (Robertson et al., 1997). As FIT predicted, features could be detected without spatial attention, but detecting the proper conjunction of two features in a multi item array required spatial attention.

These findings are also supported by studies of patients with unilateral visual neglect; i.e., spatial deficits impair detection of conjunction targets, while feature targets continue to pop out on the contralesional side of space (Brooks et al., 2005; Eglin et al., 1989, 1991, 1994; Esterman et al., 2000; Riddoch and Humphreys, 1987; Laeng et al., 2002; Pavlovskya et al., 2002). When two features must be bound together to detect the presence of a target, or whenever a serial search for a target in contralesional space is required, patients with left neglect are either dismally slow to find a target (often up to a minute or more) or they miss it entirely. Conversely, features continue to pop out from the neglected side (Esterman et al., 2000; Brooks et al., 2005). These observations fit well with clinical observations that visual search into the contralesional side of space is one of the main consistent deficits in unilateral neglect. Whether at bedside, in everyday life or in the laboratory, the major problem is orienting to the contralesional side of space. These patients do not attend to information on that side and tend to bump into items on their contralesional side. Detecting basic features on that side does not seem to help them avoid obstacles, consistent with basic features being detected but not in their proper locations and/or not properly bound.

This is not to say that the locations of features are never known to patients with unilateral neglect. Unlike Balint's syndrome, neglect is produced by unilateral damage (usually in the right hemisphere causing left sided deficits). There is a functioning dorsal system in the opposite hemisphere that is intact (or at least relatively intact). In fact, there is recent functional imaging evidence that the *undamaged* hemisphere produces *greater* activity in specific posterior areas that correlates with behavioral measures of neglect (Corbetta et al., 2005). This activity may help resolve spatial information, albeit less efficiently and more slowly.

3. SPATIAL REFERENCE FRAMES AND SPATIAL DEFICITS?

In patients with left neglect, attention is shifted to the right (Posner et al., 1984). However, the degree of shift is variable, and it is not the case that patients must detect only the rightmost items in a display and no leftward items in order to be given a diagnosis of neglect. Instead, one patient might detect an item somewhat more right than another, even though both patients could miss all items on the left side. Other patients may miss only one or two of the left most items.

Marcel Kinsbourne proposed a vector model of spatial attention to account for the ipsilesional shift, which was based on a balance between neurobiological activation and inhibition within and between hemispheres (see Kinsbourne, 1977). This balance is disrupted by right parietal damage such that the attentional vector swings rightward. For instance, if the right hemisphere inhibits the propensity for the left hemisphere to swing the attentional vector towards the right side, then right hemisphere damage will disrupt that inhibition and the left hemisphere will be unencumbered. This likely involves subcortical systems such as the superior colliculus and pulvinar as well (Robertson and Rafal, 2000).

However, attentional vectors alone do not explain why part of the left side of space can simply drop off as if it has disappeared in many patients, which is most often observed in the acute stages when swelling and diaschesis are most likely to affect large areas beyond the damaged regions. Normal observers do not lose consciousness of the left side of space when attention is shifted to the right. They continue to be aware of a global spatial map and will orient to a new location without difficulty. So there seems to be something more involved than attentional vectors.

The degree of spatial deficits varies widely between patients, and although many patients with neglect can be cued to the contralesional (neglected) side, there are cases in which cues are ineffective. The challenge is to account for the variability in the loss of spatial awareness as well as for the variability in attentional shifts. One approach that may account for at least some of this variability is modeled by a system of perceptual coordinates or spatial reference frames (Robertson, 2004). Experimental work in normal observers has demonstrated the importance of such frames in visual perceptual organization (see Palmer, 1989; Rock, 1990), and more recent work has extended their role to visual search (Rhodes and Robertson, 2002).

The space in which attention is allocated may be guided by any number of reference frames that are known to influence perception and attention (e.g., body centered, head-centered, environmental-centered, object-centered, part-centered, etc.). There is ample evidence that multiple spatial frames are available, both from cognitive (see Palmer, 1999) and biological studies (Gross and Graziano, 1995). These spaces likely contribute to the representation of the overall global spatial map we usually experience; a map that is reasonably isomorphic with external space and corresponds to the single, seamless space we perceive. Depending on the site and extent of brain damage, none, all, or only some frames might be affected (see Figure 1). For instance, if all frames were shifted or skewed rightward by a right hemisphere lesion, then there would be little or no neural signal to represent the left side of space (as if it had disappeared), but if only some frames were shifted, then spatial signals from the left would be weakened although present (with a corresponding stronger signal from the right), resulting in a global spatial map that is shifted and skewed but not as severely as if all frames were affected. Figure 10-1 presents examples that may help imagine how this could work.

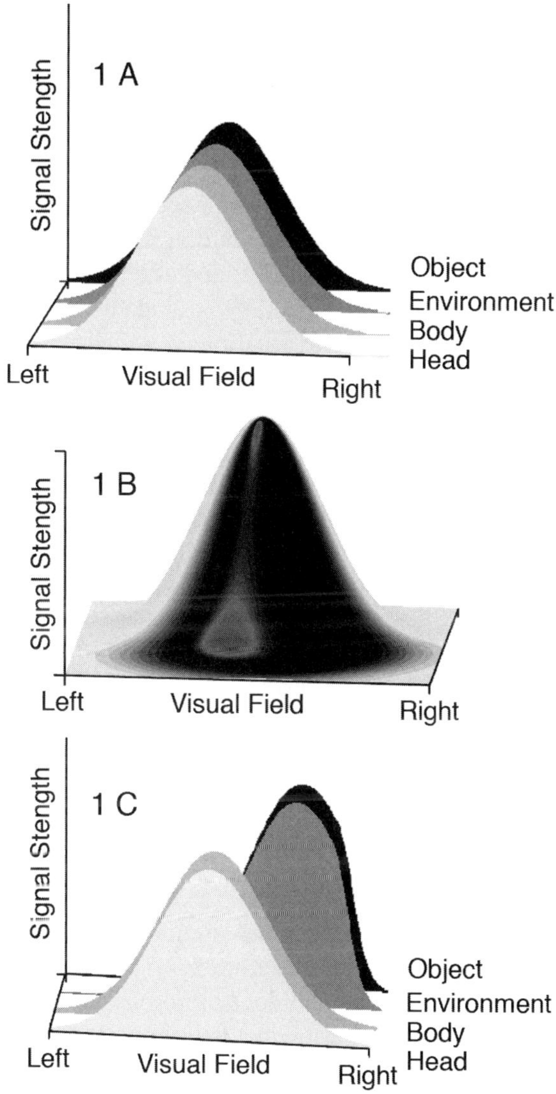

Figure 10-1. Example of underlying spatial maps (1A) that contribute to a global spatial reference frame that can be used to guide attention (1B). Example of skew that may occur in separate maps that result in one type of left neglect (1C).

Figure 10-1 B represents the distribution of attention (assuming central fixation) using an example with four reference frames (e.g., body, head, object, scene), any one of which could be selected to guide spatial attention. (I assume there are many more than four when considering the hierarchy of object/spaces in a real world display – see Robertson, 2004 – but four will do to make the point here). Figure 10-1 A represents a normal perceiver who is looking straight ahead with his/her body and head aligned with the overall display, and with both the scene and the object-based frames in line with fixation. Figure 10-1 B represents a global, unified spatial map that would emerge from the four frames in Figure 10-1 A. When searching for a certain part of a car, an object frame might be selected to guide visual search, while searching for clouds in the sky might utilize a scene-based spatial map. But since the frames are all aligned, the distribution of spatial attention would appear similar regardless of which frame guided attention.

What would happen if frames were skewed and shifted toward the right by a lesion? Figure 10-1 C shows a situation in which only the scene and object based frames of Figure 10-1A are shifted in this direction, but body and head-based frames remain centered. In this case, the attentional vector will be shifted to the right in each of the shifted frames but also in the global map of Figure 10-1 B. The degree of overall spatial omissions of left sided items will depend on how many spatial frames were shifted by the lesion, or when a shifted spatial frame has been selected to guide attention. If all four frames in Figure 10-1 A were shifted, the representation of perceptual space on the left would disappear completely. There would simply be no neural signal strong enough to carry spatial information from the left to contribute to attention. However, if only a subset of frames were shifted, then the remaining frames could be sufficient to represent and/or cue attention to the left, albeit more slowly. All combinations in between a shift of 1 frame and a shift of all frames would be possible, which could account for at least some of the variation in spatial deficits observed in the neglect syndrome.

This model of spatial shifts and reference frames in neglect is testable, which of course is an important element for any theory. For instance it predicts that detection of contralesional items in a display will vary depending on the number of reference frames affected by the lesion. One simple prediction is that neglect should be more severe in patients who exhibit both object and space based neglect than in patients who exhibit only one or the other. It also predicts that the pattern of neglect will differ depending on the reference frame selected for the task (as observed in a case reported by Humphreys and Riddoch, 1994) and whether the global map or other spatial frames are used in the task.

4. ATTENTION, PARIETAL FUNCTION AND BINDING

Whether the forgoing account is correct or not, functions that theoretically rely on a spatial map (e.g., serial search for the conjunction of two features) should be dramatically affected by spatial deficits, while those that do not rely on such maps should not (e.g., feature search). I have already discussed the evidence confirming this prediction for conjunction and feature search in neurological patients with spatial deficits. These results are consistent with feature integration theory's proposal that spatial attention is needed to bind features together to detect a target in a multi-item display. Treisman and Schmidt (1982) introduced another test of spatial attention's role in binding. When normal perceivers divert spatial attention and must report the color and shape of items in a 2 or 3 item display, perception of the shapes in the wrong colors occurs with some regularity. The wrong color is most often a color from another item in the display as opposed to a color that is not present but is part of the response set. In fact, confidence in the color of shapes that are perceived can be quite high even when these "illusory conjunctions" occur. Critically, Treisman and Schmidt (1982) found that illusory conjunctions were only observed when attention was diverted and displays were shown for a brief period of time (between 90 to 170 ms for different participants). They argued that under such conditions there was inadequate time for spatial attention to be allocated to the location of all the items, resulting in the miscombination of the shape and color in perceptual awareness. In order to bind the two features appropriately (whether under brief exposure or during conjunction search) spatial attention was involved (Treisman, 1988).

Given the brief exposure duration required to produce illusory conjunctions in normal observers, alternative explanations were suggested, such as memory confusions that could occur between offset of the display and participant report. Such alternatives have been ruled out by results in neurological patients by showing that illusory conjunctions can occur under free viewing conditions. For instance, the Balint's patient RM produced up to 38% illusory conjunctions when presented with only two letters in two different colors for several seconds. His illusory conjunction rate did not change significantly for display times that ranged from 500 ms to 10 seconds (Robertson et al., 1997). Without an adequate spatial map to guide spatial attention, illusory conjunctions were common for RM, and they were consistent with his poor conjunction search performance even in displays of only 4 or 6 items (see discussion in previous section). Such findings have since been replicated in other patients with Balint's syndrome (Humphreys et al., 2000).

The deficits in Balint's syndrome occur when dorsal occipital/parietal lobes of the angular gyrus and dorsal extrastriate areas of both hemispheres are functionally disrupted (see Hof, 1989; Rafal, 2001; Rizzo and Vecera, 2002; Robertson, 2004). RM's lesions were concentrated in these areas with some extension into more superior parietal regions in the left hemisphere.

Neither the calcarine cortex nor ventral occipital or temporal lobes were anatomically damaged (see Friedman-Hill et al., 1995 for reconstruction of his MRI). Consistently, RM's primary vision was intact (e.g., 20/20 in each eye, normal color vision, normal contrast sensitivity, etc.), suggesting that these areas were functionally intact as well.

Some investigators have argued that the deficits in binding and visual search are not a result of disrupting parietal spatial functions *per se*, but instead are due to the difficulty of conjunction vs feature search and the confusing visual signals that can occur in multi item arrays after brain injury. These arguments have been quieted by evidence acquired with functional imaging measures in groups of normal perceivers. For instance, in an fMRI study Donner et al. (2002) found that increased activity in parietal (and frontal) lobes in conjunction search could be attributed to increased difficulty, but some regions could not. Specifically, an area at the junction of the posterior inferior parietal sulcus and dorsal occipital lobe was active in tasks that required feature binding, independent of the search difficulty involved. Shafitz et al. (2002) also controlled for search difficulty to detect a target in conjunction and feature displays and linked spatial functions associated with parietal activation to binding errors and spatial processing.

In sum, there is converging evidence that the parietal lobes are involved in binding basic visual features that are encoded by specialized neural populations in the ventral stream of processing that have been associated with object formation (Ungerleider and Mishkin, 1982). Importantly, it is the spatial functions of the parietal lobe that seem to be involved. The data as a whole support predictions originally proposed by feature integration theory that spatial attention is important for accurate feature integration. They also demonstrate that feature binding in multi item displays does not take place normally without input from the parietal lobes.

5. WHERE IS ATTENTION WHEN SPACE PERCEPTION IS LOST?

Even when space disappears due to damage to both parietal lobes, there are areas outside these regions that are anatomically intact and respond to objects and basic features. Neurobiological evidence has documented object-based (ventral) and space-based (dorsal) cortical streams of processing (Ungerleider and Mishkin, 1982), with features such as color and shape encoded separately by the ventral (occipital-temporal), object-based systems (Zeki, 2003). Consistently, functional imaging data in humans has shown that feature search activates regions within the ventral stream in normal observers, while conjunction search activates regions within both (Corbetta, et al., 1995). As noted previously, differences in difficulty between feature and conjunction search cannot account for all these effects (see Donner et al., 2002: Shafritz et al., 2002).

The severe spatial deficits observed in Balint's syndrome make the case most convincingly. RM not only lost spatial awareness of the external world, but his visual system appeared to act as if it were a slave to single objects. He would report seeing one object (sometimes up to several minutes) that was abruptly replaced by another. Formal tests verified his object "simultanagnosia" but also showed that he was able to report multiple colors from a briefly presented display (unpublished data collected by Marcia Grabowecky and described in Robertson, 2004). When asked to indicate where a color he detected in a feature search was, he laughed and said that we knew he could not see where things were, although he would venture a guess to be cooperative. His guesses were generally at chance levels (see Robertson et al., 1997), and they were not systematic. For RM, signals that represented color continued to convey information, as did familiar objects (Robertson and Treisman, 2006). What was lacking was a neural signal that carried locations of the features and objects that he perceived.

It appears that one answer to the question of where attention is when spatial perception is lost is that it is where it has always been; on the neural signals that provide the basic ingredients for perceptual experience. If the neural signal for an object's location is absent, then attention to that signal cannot be accomplished, and if the signal is postponed (as it sometimes is in patients with neglect) then attention to the location signal will be delayed, possibly disrupting binding. However, this need not affect the signals that carry information about the features themselves (e.g., color, shape). This situation basically results in a feature that has nowhere to go in space and may be bound incorrectly in perception.

6. LESION STUDIES IN AN ERA OF BRAIN IMAGING

In a recent issue of the *Journal of Cognitive Neuroscience*, Anjan Chatterjee noted that, "As an example of the present-day peculiarities of conducting lesion studies, in a competitive renewal of a National Institutes of Health grant investigating unilateral spatial neglect, I was asked by a primary reviewer from a cognitive neuroscience study section to 'justify the use of patients to study the topic in question'" (p. 849). The implication is that studying cognitive problems in patients with lesions that are randomly inflicted by nature is no longer warranted in an era in which technological developments allow examinations in normal observers of activated brain regions during cognitive tasks, examination of coherence between brain areas, examination of changes in neural activity under various conditions, and stimulation of brain regions to transiently excite or inhibit brain function in selected regions of interest. I have also heard arguments that studies with patients are only justified when the goal is to better the quality of life of patients themselves rather than testing theories of brain-behavior relationships.

I find these arguments peculiar when considering how the study of parietal involvement in human attentional processing developed. In the 1980's a cognitive psychologist (Michael Posner) began collaborating with a behavioral neurologist (Robert Rafal) and, after studying a group of patients with unilateral visual neglect, they proposed that these patients had difficulty disengaging attention from their ipsilesional side. Both the scientific and clinical communities welcomed these findings, as they had relevance for theories of attention, theories of brain function *and* clinical management. Families and patients could be better informed about the disorder, and rehabilitation approaches were developed or revised based on scientific evidence. Many basic neuropsychological tests were changed to incorporate new evidence. Scientifically, these findings motivated a proliferation of electrophysiological and functional imaging studies of attention using "Posner cueing" methods that refined discussions of brain areas involved. They also served as a catalyst for increased interdisciplinary collaboration and the emergence of cognitive neuroscience as a discipline.

Similarly, despite the fact that Balint's syndrome has been observed in the clinical community for nearly 100 years, the idea of incorrectly binding features in free view after parietal insult was not reported until feature integration theory was proposed and specialized neural populations that encoded different features were discovered. It was assumed that if a patient reported seeing an object, its color was also seen, unless of course the patient was colorblind. There was no reason to suspect that a color in another part of the visual field could bind incorrectly with a perceived shape. Subsequently, functional imaging refined the extent and location of regions of the human parietal cortex that seem to be involved in detecting a conjunction target in a search task. The two methods together demonstrated the sufficiency and necessity of inferior parietal regions in normal feature integration and the sufficiency of ventral systems in feature detection. Without the lesion evidence the necessity of an activated area to a particular input in an imaging study is difficult if not impossible to evaluate.

Finally, the relief in patients is evident when they realize that medical staff can intelligently discuss perception in ways that touch the patient's own experience. Patients who have informed family members that they are blind to avoid telling them that objects are seen but they have no location suddenly realizes they are not alone. The family and the medical community can be educated to help them understand and to know better what to expect and how to cope. The reason to study lesions in humans is multifaceted. The same database can lead to better understanding about how normal cognition must work in order to break down in the particular ways that it does as well as to better lives for patients. These are not exclusive. The knowledge that results from the study of patients helps define limits on cognitive theories and it is often used to motivate designs for functional imaging, electrophysiological and brain stimulation studies by suggesting where to look and what to look for. At the same time, it helps clinical staff relate better to the patients and their families and to imagine new ways to examine, treat and manage cognitive disorders.

REFERENCES

Balint, R., 1909, Seelenlahmung des "Schauens", optische Ataxie, räumliche Störung der Aufmerksamkeit, *Monatsschrift für Psychiatrie und Neurologie* **25**:5-81 (Translated in *Cogn Neuropsychol* **12**:265-281, 1995).

Bernstein, L. J. and Robertson, L. C., 1998, Independence between illusory conjunctions of color and motion with shape following bilateral parietal lesions, *Psychol Sci* **9**:167-175.

Brooks, J. L., Wong, Y. and Robertson, L. C., 2005, Crossing the midline: Reducing visual extinction by re-establishing hemispheric balance, *Neuropsychologia* **43**:572-582.

Chatterjee, A., 2005, A madness to the methods in cognitive neuroscience?, *J Cognitive Neurosci* **6**:850-858.

Corbetta, M. C., Kincade, M. J., Lewis, C. and Sapir, A., 2005, Neural basis and recovery of spatial attention deficits, *Nat Neurosci* **8**:1603-1610.

Corbetta, M., Shulman, G., Miezin, F. and Petersen, S., 1995, Superior parietal cortex activation during spatial attention shifts and visual feature conjunctions, *Science* **270**: 802-805.

Donner, T. H., Kettermann, A, Diesch, E, Ostendorf, F, Villinger, A, Brandt, S., 2002, Visual feature and conjunction search of equal difficulty engage only partially overlapping frontoparietal networks, *Neuroimage* **15**:16-25.

Driver, J. and Veuilleumier, P., 2001, Perceptual awareness and its loss in unilateral neglect and extinction, *Cognition* **79**:39-88.

Eglin, M., Robertson, L. C. and Knight, R. T., 1989, Visual search performance in the neglect syndrome, *J Cognitive Neurosci* **4**:372-381.

Eglin, M., Robertson, L. C., and Knight, R. T., 1991, Cortical substrates supporting visual search in humans, *Cereb Cortex* **1**:262-272.

Eglin, M., Robertson, L. C., Knight, R. T., and Brugger, P., 1994, Search deficits in neglect patients are dependent on size of the visual scene, *Neuropsychology* **4**:451-463.

Esterman, M., McGlinchey-Berroth, R. and Milberg, W.P., 2000, Parallel and serial search in hemispatial neglect: Evidence for preserved preattentive but impaired attentive processing, *Neuropsychology* **14**:599-611.

Friedman-Hill, S., Robertson, L. C. and Treisman, A., 1995, Parietal contributions to visual feature binding: Evidence from a patient with bilateral lesions, *Science* **269**:853-855.

Graziano, M. S. A. and Gross, C. G., 1994, Mapping space with neurons, *Curr Dir Psychol Sci* **3**:164-167.

Gross, C. G. and Graziano, M. S. A., 1995, Multiple representations of space in the brain, *Neuroscientist* **1**:43-50.

Heilman, K. M., Watson, R. T. and Valenstein, E., 1993, Neglect and related disorders, in: *Clinical Neuropsychology, 3rd ed.,* K.M. Heilman and E. Valenstein, eds., Oxford University Press, London, pp. 279-336.

Heilman, K. M., Watson, R. T. and Valenstein, 1994, Localization of lesions in neglect and related disorders, in: *Localization and Neuroimaging in Neuropsychology*, A. Kertesz, ed., Academic Press: San Diego, pp. 471-492.

Hof, P. R., Bouras, C., Constantinidis, J., Morrison, J. H., 1989, Balint's syndrome in Alzheimer's disease: Specific disruption of the occipito-parietal visual pathway, *Brain Res* **493**:368-375.

Humphreys, G. W. and Riddoch, M. J., 1994, Attention to within-object and between-object spatial representations: Multiple sites for visual selection, *Cogn Neuropsychol* **11**:207-241.

Humphreys, G. W., Cinel, C., Wolfe, J., Olson, A. and Klempen, N., 2000, Fractionating the binding process: Neuropsychological evidence distinguishing binding of form from binding of surface features, *Vision Res* **40**:1569-1596.

Kim, M. S. and Cave, 1995, Spatial attention in visual search for features and feature conjunctions, *Psychol Sci* **6**:376-380.

Kim, M. S. and Robertson, L. C., 2001, Implicit representations of visual space after bilateral parietal damage, *J Cognitive Neurosci* **13**:1080-1087.

Laeng, B., Brennen, T., Espeseth, T., 2002, Fast responses to neglected targets in visual search reflect pre-attentive processes: An exploration of response times in visual neglect, *Neuropsychologia* **40**:1622-1636.

Mattingley, J. B., David, G. and Driver, J., 1997, Preattentive filling-in of visual surfaces in parietal extinction, *Science* **275**:671-674.

Mesulam, M. M., 1981, A cortical network for directed attention and unilateral neglect, *Ann Neurology* **4**:309-325.

Mort, D. J., Malhotra, P. Mannan, S. K., Rorden, C., Pambakian, A., Kennard, C. and Husain, M., 2003, The anatomy of visual neglect, *Brain* **126**:1986-1997.

Palmer, S. E., 1989, Reference frames in the perception of shape and orientation, in: *Object Perception: Structure and Process*, B. E. Shepp and S. Ballesterors, eds., Hilldale, Erlbaum, NJ, pp. 121-164.

Pavlovskaya, M., Ring, H., Groswasser, Z., and Hochstein, S., 2002, Searching with unilateral neglect, *J Cognitive Neurosci* **14**:74-756.

Phan, M. L., Schendel, K. L., Recanzone, G. H. and Robertson, L. R., 2000, Auditory and visual spatial localization deficits following bilateral parietal lobe lesions in a patient with Balint's syndrome, *J Cognitive Neurosci* **12**:583-600.

Posner, M. I., Walker, J. A., Friedrich, F. J. and Rafal, R. D., 1984, Effects of parietal injury on covert orienting of attention, *J Neurosci* **4**:1863-1877.

Rafal, R., 1997, Balint syndrome, in: *Behavioral Neurology and Neuropsychology*, T.E. Feinberg and M. J. Farah, eds., McGraw Hill, New York, pp. 337-356.

Rafal, R., 2001, Balint's syndrome in: *Disorders of Visual Behavior,* M. Behrmann, ed., Amsterdam, Elsevier Science, pp 121-141.

Rhodes, D. L. and Robertson, L. C., 2002, Visual field asymmetries and attention in scene-based coordinates, *Brain Cognition* **50**:95-115.

Riddoch, M. J. and Humphreys, G. W., 1987, Perception and action systems in unilateral visual neglect, in: *Neuropsychological and Neurophysiological Aspects of Spatial Neglect*, M. Jeannerod, ed., New Holland, Amsterdam.

Rizzo, M. and Vecera. S. P., 2002, Psychoanatomical substrates of Balint's syndrome, *J Neurol Neurosur Ps* **72**:162-178.

Robertson, L. C., 2003, Binding, spatial attention and perceptual awareness, *Nat Rev Neurosci* **4**:93-102.

Robertson, L. C., 2004, *Space, Objects, Minds and Brains*, Psychology Press (Francis and Taylor), Essays in Cognitive Science, New York.

Robertson, L. C. and Rafal, R., 2000, Disorders of visual attention, in: *The New Cognitive Neuroscience,* 2nd ed., M. Gazzaniga, ed., MIT Press, Cambridge, pp. 633-649.

Robertson, L. C. and Treisman, A., 2006, Attending to space within and between objects: Implications from a patient with Balints syndrome, *Cogn Neuropsychol* **23**:448-462.

Robertson, L. C., Eglin, M. and Knight, R. T., 2003, Grouping influences in unilateral visual neglect, *J Clin Exp Neuropsyc* **25**:297-307.

Robertson, L. C., Treisman, A., Friedman-Hill, S. and Grabowecky, M., 1997, The interaction of spatial and object pathways: Evidence from Balint's syndrome, *J Cognitive Neurosci* **9**:295-317.

Rock, I., 1990, The frame of reference, in: *The Legacy of Solomon Asch: Essays in Cognition and Social Psychology*, I. Rock, ed., Hillsdale, Lawrence Erlbaum, NJ, pp. 243-268.

Shafritz, K. M., Gore, J. C. and Marois, R., 2002, The role of the parietal cortex in visual feature binding, *P Natl Acad Sci* **99**:10917-10922.

Treisman, A. M., 1996, The binding problem, *Curr Opin Neurobiol* **6**:171-178.

Treisman, A., 2006, Consciousness and perceptual binding, in: *The Unity of Consciousness: Binding, Integration and Dissociation*, A. Cleeremans, ed., Oxford University Press, London, pp. 95-113.

Treisman, A. M. and Gelade, G., 1980, A feature-integration theory of attention, *Cognitive Psychol* **12**:97-136.

Treisman, A. M. and Schmidt, H., 1982, Illusory conjunctions in perception of objects, *Cognitive Psychol* **14**:107-141.

Treisman, A., 1988, Feature and objects: The fourteenth Bartlett memorial lecture, *QJ Exp Psychol* **40**(A):201.237.

Ungerleider, L. G. and Mishkin, M., 1982, Two cortical visual systems, in: *Analysis of Visual Behavior*, J. Ingle , M. A. Goodale and R. J. W Mansfield, eds., MA: MIT Press Cambridge, pp. 549-586.

Yantis, S., 1993, Stimulus-driven attentional capture, *Current Directions in Psychological Science* **2**:156-161.

Zeki, S., 2003, The disunity of consciousness, *Trends in Cognitive Sciences* **7**:214-218.

Chapter 11

PIGS IN SPACE[1]: HOW WE RECOGNIZE ROTATED OBJECTS

Michael C. Corballis[1], Branka Milivojevic[1] and Irina M. Harris[2]
[1]University of Auckland, New Zealand; [2]University of Sydney, Australia

Abstract: As organisms that move freely in space, we are adept at visually recognising objects regardless of their orientations. This facility probably does not depend on a "correction" mechanism, such as mental rotation, that might render an object in some canonical orientation. Rather, it is likely that an orientation-free description is extracted, probably by the ventral visual system. This suggests further that we can recognize the identity of an object before we can determine its orientation in space. This may depend in turn on the integration of shape information extracted by the ventral system with information about the space-occupying property of the object extracted by the dorsal system. The dissociation between identity and orientation may explain cases of "orientation agnosia," in which the patient can recognize common objects but cannot determine their orientations. Orientation free descriptions are nevertheless relatively crude. In order to distinguish between shapes that differ in more subtle ways, such as individual faces, or between shapes that are mirror-images of one another, a correction may be necessary, through either physical or mental rotation to the upright.

Key words: agnosias; face perception; mental rotation; mirror-image discrimination; object recognition; pattern recognition; what vs. where systems.

1. INTRODUCTION

People show remarkable flexibility in recognizing familiar objects or shapes. We can recognize our friends whether they are smiling or frowning, facing us or in profile, in sunshine or shadow, standing or sitting, in swimming gear or formal attire, in photographs or caricature. We can

[4] See Wu, Tillett, McFarlane, Xiangyang, Siebert, & Schofield (2004).

recognize a wheelbarrow, or a bicycle, or a letter of the alphabet, or even a pig, more or less regardless of its distance from us, its colour, its illumination, or its orientation in space. In the auditory domain, too, we can recognize tunes regardless of tempo or key, or spoken words regardless of who speaks them. Conversely, we can recognize the voices of individual speakers regardless of what they are saying. This extraordinary ability to separate out constant aspects of the environment from the individual circumstances in which they appear is the classic problem of pattern recognition.

This chapter is concerned with a particular aspect of this problem, our ability to recognize familiar visual objects when they are rotated in space, or when we ourselves are rotated relative to the objects. The problem of how we do this is especially challenging, since rotation of an object relative to the viewer radically alters the image on the retina, which poses the question of how the brain is able to extract sufficient information to find a match in memory, and so recognize it. Indeed, the number of possible retinal images generated by rotation, not to mention other parameters such as distance from the eye, is theoretically infinite, so one can rule out the possibility that the brain simply stores every possible view of the object, and is therefore always able to find a match. Such a mechanism would simply be too unwieldy to be practicable, even allowing for the vast storage potential in the human brain.

Although we can usually recognize objects when they are rotated away from their normal uprights, there are exceptions. One involves the recognition of faces. Usually we can recognize an upside-down face as a face, but it is often difficult to tell whose face it is. Figure 11-1 shows an example of a person we may recognize even when upside down, but rotating the picture will reveal a grotesque face uncharacteristic of the former British Prime Minister even during the worst governmental crises (after Thompson, 1980). The figure also shows a pattern that is not even recognizable as a face until rotated 180° (after Mooney, 1957). These simple illustrations show that correction for orientation may be less than optimal in some circumstances. Nevertheless we are generally able to identify rotated objects at least to the point of being able to categorize them, if not identify some of the subtleties that may distinguish one member of a category from another.

2. MENTAL ROTATION

2.1 Mental rotation and recognition

In the above examples, accurate recognition of inverted pictures was restored by rotating them back to the upright. In the natural world, of course, this is often impracticable in the physical sense, but one might suppose that rotation could be carried out *mentally*. That is, observers might imagine a rotated object as though it were upright, and then find a match among objects stored in their canonical orientations. Such a scheme would be parsimonious, in that only one version of an object need be stored. Mental rotation might therefore be one mechanism that would allow such a system to operate.

Mental rotation is a well-established phenomenon, in both humans (Shepard and Metzler, 1971) and monkeys (Georgopoulos et al., 1993). One classic study was that of Cooper and Shepard (1973), who timed people as they decided whether rotated alphanumeric characters were normal or backward (mirror-reversed). Their response times increased monotonically with the departure of the characters from their normal upright orientations. This suggested that the subjects mentally rotated the characters back to the upright in order to make their judgments, a conclusion supported by the subjects' own subjective reports. Subsequent studies, whether of rotation of alphanumeric characters, hands, human bodies, or nonsense shapes, have revealed a remarkably consistent pattern (see Corballis, 1982, for a review).

Figure 11-1. Two examples of upside-down faces. The face on the left may be recognizable, but changes dramatically when rotated 180 degrees. The picture on the right is not recognized even as a face until rotated 180 degrees.

The rate of mental rotation appears to be fairly constant over the range of different stimuli, at around 360 deg/s.[2] The estimated rate of mental rotation is considerably slower, though, when the task involves mentally rotating one object into conformity with another (e.g., Shepard and Metzler, 1971), probably because mental rotation under such conditions is undertaken in piecemeal fashion (Just and Carpenter, 1976).

People are often slower to name letters and digits (Corballis et al., 1978) or two-dimensional pictures of objects (e.g., Jolicoeur, 1985; Murray, 1995) when they are rotated away from the upright. This might be taken as evidence that mental rotation is indeed involved in "correcting" for rotation away from the upright. There is, however, a logical argument against this interpretation: It is not clear how the observer can mentally (or physically) rotate a shape to the upright unless one already knows what the shape is, for otherwise one would not know which way to rotate or where to stop (Corballis, 1988). It is logically possible, of course, that one might rotate all possible ways and stop when a match is found, but again this seems an unwieldy solution. Furthermore, people can recognize rotated alphanumeric characters (Corballis et al., 1978), at least, or categorize them as letters or digits (Corballis and Nagourney, 1978), much more quickly than they can decide whether they are normal or backward—a task that *does* require mental rotation.

The effect of orientation on recognition also tends to level out with repetitions of the same objects (Murray, 1995), leading some to propose that two processes are involved. Jolicoeur (1990), for example, argued that the initial registration of an object is accomplished with reference to a viewpoint-dependent internal representation, so that some correction is necessary when the object is rotated away from the upright, but that as viewers become more familiar with the objects on display recognition is accomplished by viewpoint-independent mechanisms, perhaps involving featural descriptions. This still raises the question, though, as to how observers can correct for rotation away from the upright on early trials unless they already know what the object is. An alternative is that the observer can indeed identify a rotated object without performing a correction, but may then mentally rotate the object to the upright as a check on the initial identification (Corballis, 1988). As the illustrations in Figure 11-1 show, the initial identification may be less than perfect, but may still

[2] One might suppose, perhaps frivolously, that this gave rise to the time unit we have come to know as the "second." Some time after the invention of the wheel, a group of savants may have decreed that the "second" be defined as the time it takes to imagine the wheel rotating through a full revolution.

suffice to indicate how the shape is to be rotated in order to check the initial impression, or to gain a more accurate identification. This check is no longer necessary when the observer becomes familiar with the objects used in the experiment.

Nevertheless, there are empirical reasons to suppose that mental rotation may not be involved at all, even in studies where recognition time increases with rotation of the object away from the upright—although there is an important exception, discussed in the next section. For one thing, it has long been observed that recognition time is not a monotonic function of angular departure from the upright. There is characteristically a dip at 180°, so that observers are faster to identify shapes that are completely inverted than to identify those rotated, say, 120° from the upright (e.g., Corballis et al., 1978; Murray, 1995). Mental-rotation functions, in contrast, typically show, if anything, an over-pronounced peak at 180°, a point discussed in some detail by Cooper and Shepard (1973). Another argument against the involvement of mental rotation has to do with the influence of a motion aftereffect. If observers watch a rotating disk for a while, and then look at a stationary object, that object seems to rotate in the opposite direction. If the observer is required to mentally rotate the object to the upright, the rate of rotation is slowed by an aftereffect in the direction opposite that of the mental rotation, and (to a lesser extent) speeded by an aftereffect in the same direction (Corballis and McLaren, 1982). There is no such effect, however, on the speed at which rotated objects are recognized, suggesting that recognition itself does not depend on mental rotation (Jolicoeur et al., 1998).

Further, brain-imaging studies show that the recognition and mental rotation of rotated objects activate different brain areas, which to some extent follow the distinction between the *ventral* and *dorsal* systems described by Ungerleider and Mishkin (1982). The dorsal system, sometimes called the "where" system, extends from the occipital lobes to the parietal lobe, and is responsible for locating objects in space. The ventral system, sometimes called the "what" system, extends from the occipital lobe to the inferotemporal lobe, and has to do with the identification of objects independently of where they are located. Brain-imaging studies suggest that mental rotation is dependent on areas in or close to the dorsal stream, while recognition depends on areas in the ventral stream (e.g., Gauthier et al., 2002). Areas within the dorsal stream that have been associated with mental rotation include the superior parietal lobule (Alivisatos and Petrides, 1997; Cohen et al., 1996; Ng et al., 2001; Tagaris et al., 1997), and the intraparietal sulcus (Harris and Miniussi, 2003; Harris et al., 2000), with possible involvement also of the inferior parietal lobule (Alivisatos and Petrides, 1997; Ng et al., 2001; Tagris et al., 1997) and precuneus (Ng et al., 2001). There may well be variation related to task demands (e.g., Harris et al.,

2000), and further study may be necessary to separate the regions involved in mental rotation from those involved in the registration of orientation itself.

There is also evidence that the frontal lobes are involved in mental rotation (Alivisatos and Petrides, 1997; Cohen et al., 1996; Ng et al., 2001; Tagriset al., 1997). This is especially evident in studies requiring the discrimination of left and right hands (e.g., Thayer et al., 2001; Tomasino et al., 2005) and tools (Vingerhoets et al., 2002), presented in different orientations. While this may reflect the mental rotation of body parts to match the stimuli, rather than mental rotation of the stimuli themselves, there is also evidence that both frontal and parietal regions may contribute in both cases. For example, Silberstein et al. (2003) found EEG synchronization between left frontal and right parietal regions when subjects mentally rotated the 3D shapes used by Shepard and Metzler (1971) in their original mental-rotation experiment. It seems likely that time-varying processes such as mental rotation recruit large populations of neurons, involving both parietal and frontal regions (Georgopoulos et al., 1993). The neural underpinnings of mental rotation are explored in more detail by Jäncke and Jordan (this volume).

As a final point against the role of mental rotation, the reader might observe that it is impossible to mentally rotate the two pictures shown in Figure 11-1 to the upright. In order to see Margaret Thatcher's grossly distorted visage, or to see the other pattern as a face, it is necessary to turn the images physically. *Mental* rotation fails precisely where it is needed most, probably because it cannot carry sufficient information for accurate decoding of a complex structure.

2.2 Mental rotation, parity, and bilateral symmetry

Despite the apparent dissociation of recognition and mental rotation, there is one limited domain in which mental rotation does play a role. The studies of mental rotation referred to above have virtually all been concerned with the discrimination of mirror-images (Corballis, 1982). In the study by Cooper and Shepard (1973), the task was to discriminate normal from mirror-reversed alphanumeric characters, while in the earlier study by Shepard and Metzler (1971) the task was to determine whether two 3D shapes were the same, and in trials where they were different they were in fact mirror-images. Mental-rotation studies have therefore exploited the fact that the left-right orientation, or *parity*, of a shape tends to be lost with rotation away from the upright. A picture of a rotated hand, for example, is readily seen *as a hand*, but it is difficult to tell whether it is a left or a right hand unless you can map one of your own hands onto it, by either a physical or a mental act of rotation and translation (e.g., Sekiyama, 1982; Thayer et al., 2001). Similarly, we can tell that the inverted letters shown in Figure 11-2 are Rs, but we can't easily tell which is normal and which is reversed without rotating them back to the upright, again either physically or mentally.

It follows that mental rotation *is* required in object identification when parity is critical to the identification itself. Sometimes we need to know whether a shoe is a left or a right shoe, or whether an automobile is a left- or a right-hand-drive. Although it is rare for objects to be named differently depending on parity, an important exception is the naming of the lowercase letters b, d, p, and q, which are notoriously confusing to children learning to read. When disoriented, these letters need to be rotated to the upright in order to be disambiguated and named correctly (Corballis and McLaren, 1984).[3]

The discrimination of parity, such as telling whether a shoe is a left or right shoe, depends on an asymmetry within one's own person, which is why mental rotation to some bodily reference is necessary. A perfectly bilaterally symmetrical individual would not be able to label mirror-image objects as distinct. To understand that this is so, imagine a perfectly symmetrical person, with a perfectly symmetrical brain. Imagine now that this unlikely creature is correctly able to say "left" when holding up a left shoe, and "right" when holding up a right shoe. Now imagine that this scene is mirror-reversed, as though through the looking glass. The person, being perfectly symmetrical, is quite unaltered by mirror reversal, but is now seen to be proclaiming the left shoe as "right", and the right shoe as "left." By *reductio ad absurdum*, then, the person would be unable to consistently label the shoes correctly. This is part of a more general rule, which is that a bilaterally symmetrical organism would be unable to tell left from right (Corballis and Beale, 1976).

Figure 11-2. Which of these letters is the right way round?

[3] Assuming the q to be represented without the give-away tail.

The point here is not entirely a frivolous one. Most organisms, including ourselves, are a close approximation to bilateral symmetry, and most do have difficulty telling left from right (see Corballis and Beale, 1976, for review). Confusion over left and right turns will often lead one up the wrong garden path. As mentioned earlier, children learning to read often have special difficulty with the pesky letters b, d, p, and q, and may confuse words like WAS and SAW. One of the first author's own sons, now a writer, came home from one of his early days at school with the entire alphabet neatly printed in reverse. Orton (1937) saw left-right confusion as the major factor underlying dyslexia, and related it to poorly established cerebral dominance. Nevertheless there is a positive aspect to left-right confusion, which we may also regard as left-right equivalence. In the natural world, the same object is often encountered in opposite mirror-image guises. The profile of any object reverses if we simply view it from the other side—the same face may be seen in left or right profile, the predator may attack from either side. We are therefore naturally equipped to treat mirror images as equivalent. Indeed the very symmetry of our brains and bodies is an adaptation to the lack of consistent left-right asymmetry in the natural world. To be sure, by evolving a degree of cerebral asymmetry we have captured an additional dimension, enabling us to read directional scripts, or observe consistent rules when driving, or kiss the appropriate cheek when greeting continental acquaintances of the opposite gender, but this is at the expense of the natural advantages bestowed by bilateral symmetry.

The problem of discriminating mirror images is something of a special case, but gives rise to two general points. First, since one must rotate an object to its canonical orientation in order to determine its parity, one must have first extracted a description that is independent of parity. Second, that description must be extracted regardless of its orientation. One knows that a shoe is a shoe regardless of its orientation, and the description does not tell you whether it is a left or a right shoe until it has been related to one's left or right foot. This suggests that the brain does indeed extract a description of a shape that is independent of both angular orientation and parity. Once this description finds a match in memory, one can then determine information as to its upright (or canonical) orientation. This information may in turn guide mental rotation to that canonical orientation, so that its parity may then be discerned (Corballis, 1988).

3. EXTRACTING SHAPE INFORMATION

It seems to be clear, then, that the brain does indeed extract descriptions of shapes that are orientation-free, and probably parity-free. As we have seen, this probably takes place in the so-called ventral system, or "what" system, in the occipital and temporal lobes. By stripping input of information about orientation and location in space, the brain not only eliminates information that is irrelevant to shape recognition itself, but also maximizes total storage capacity through the coding of information as economically as possible. This property has been named *sparseness* (Vinje and Gallant, 2000). The theoretical limit in sparseness occurs when every object is represented by a single cell, as illustrated by the concept of the "grandmother cell," famously suggested in 1967 by J. Y. Lettvin to refer to the possibility that even one's grandmother might find representation in the brain in the form of a single neuron (Gross, 2002). Although the suggestion may have been tongue-in-cheek, it has recently been shown that single cells in the human medial temporal lobe respond to well-known individuals, landmarks, or objects shown from different viewpoints (Quiroga et al., 2005). If not a grandmother cell, there does appear to be a cell representing Jennifer Aniston (who may indeed one day be a grandmother).

There is also evidence, though, that the ventral system does preserve both view-sensitive and view-independent representations of faces, at least, both in monkeys (Logothetis et al., 1995; Rolls, 2000) and in humans (Pourtois et al., 2005). This is perhaps not surprising, since processing begins with a representation on the retina that is totally view-dependent, and successive stage of processing must be required to produce representations that are independent of the particular view. In one study, view-independent activation to faces was found in the left fusiform gyrus of the temporal lobe in humans, but not in the homologous area on the right, suggesting that the brain deals with shapes in asymmetrical fashion (Pourtois et al., 2005). View-independent representations, such as that implied by the Jennifer Aniston cell, are probably closer to the conceptual than to the perceptual end of processing.

The independence of mechanisms that extract shape from those sensitive to orientation has also been demonstrated in humans through brain-imaging. Valyear et al. (2005) have shown that changing the orientation of objects elicits activity in the occipital part of the right intraparietal sulcus, regardless of whether the object identity was also changed, whereas changing identity regardless of orientation changes elicits activity in the right occipito-temporal area, posterior to the fusiform gyrus. These results suggest that, if shape information is extracted in the ventral stream independently of orientation, orientation itself seems to depend on the dorsal stream.

Neurological evidence also supports the idea that the recognition of a shape can be dissociated from the perception of its orientation. Some neurological patients are able to recognize objects correctly, but are unable to determine their orientations in space (Fujinaga et al., 2001). Another patient who showed this dissociation was nevertheless able to mentally rotate shapes to a canonical orientation, suggesting that mental rotation may also be dissociated from the perception of orientation (Turnbull et al., 2002). That is, even within the dorsal stream, there seems to be a dissociation between orientation itself and mental rotation. This is in line with the suggestion of a three way distinction between "what," "where," and "how" systems (Creem and Profitt, 2001), and mental rotation may be a matter of "how" rather than "where."

It is also possible to dissociate recognition from orientation experimentally, using a phenomenon known as *repetition blindness* (RB). If pictures are presented very rapidly, at rates of about 10 per s, observers typically fail to notice repetitions of the same picture, but report having seen it only once. One interpretation is that each picture may find a match in memory, but presentation is too fast for each picture to be recorded as a conscious event. That is, there is recognition of *type*, without *token* individuation (Kanwisher, 1987). Experiments have shown that RB occurs when the repeated pictures differ in orientation (Harris and Dux, 2005; Kanwisher et al., 1999). The same is true of uppercase letters shown in different orientations; recognition of individual letters is independent of orientation, and RB occurs regardless of the angular distance between two presentations of the same letter. RB also occurred among all pairs of the letters b, d, p, and q, suggesting that each was recorded as the same *type* (Corballis and Armstrong, in press). For letters, at least, shape may be registered independently of both orientation and parity, and it requires a further act of tokenization to distinguish among those letters that share the same basic shape.

In further unpublished experiments, we have also shown that, with rapid presentation of letters in different orientations, orientation probes (in the form of arrows representing letter orientation) presented either before or after the sequence result in very poor identification of the probed letter. Observers nearly always report one of the presented letters, but are almost as likely to report one of the unprobed letters as to report the probed one. Again, this suggests that the orientation is dissociated from the letter in the initial stages of registration.

Although object recognition seems to involve the extraction of information that is independent of orientation, it does not follow that recognition is itself independent of orientation. To take an obvious example, it is harder to recognize a person, even Jennifer Aniston, viewed from the back than from the front, simply because distinctive features of the face are no longer visible. Rotations in three-dimensional space may often occlude critical features, or introduce different features to the visible array. Studies have also shown that faces in three-quarter view are better recognized than those in frontal or side view, probably because of improved 3D cues (Logie et al., 1987; O'Toole et al., 1998). We have also seen that even 2D rotations in the picture plane may impair recognition. Rock (1973) noted that it is almost impossible to recognize cursive script when it is upside down, and that inverted faces are also notoriously difficulty to recognize—as we saw earlier from the illustrations in Figure 11-1. There is also evidence that pre-knowledge of orientation can improve the subsequent identification of objects (Graf et al., 2005). These points emphasize again that there are aspects of some shapes that are lost in orientation-free descriptions.

3.1 How is it done?

Although it seems clear that the ventral system extracts information about identity independent of orientation, and it is not clear how this is actually done. We might also conclude that the extracted information is independent of parity, since, as we have seen, people can usually identify a rotated object but are unable to distinguish it from its mirror image until it is rotated to the upright. Indeed parity is in a sense a special case of orientation, since one can change the parity of an n-dimensional object by flipping it in (n+1)-dimensional space. The 2D figures shown in Figure 11-3, for example, are mirror-images, but can be made identical by flipping one of them in depth. If you are unfortunate (or foolish) enough to buy two left-footed shoes, you can convert one of them to a right-footed shoe by flipping it in the fourth dimension—although the opportunity to do this seems to be denied to mere mortals.

An orientation-free description might be defined as one that makes no reference, either explicitly or implicitly, to such terms as *top*, *bottom*, *back*, *front*, *left*, or *right*. For example, the lowercase letter b might be (roughly) described as a line with a closed loop running from one end to a point roughly half-way along the line.

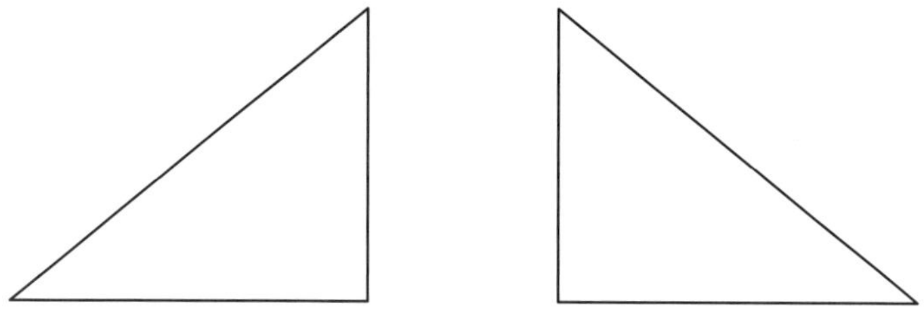

Figure 11-3. These triangles are mirror-images in 2D space, but identical in 3D space.

This description, though, does not distinguish between the letters b, d, p, and q, which might explain why all combinations of these letters give rise to RB. In orientation-free terms, they all have the same shape. In real-life settings, objects appear in arbitrary orientations, and the letters b, d, p, and q somewhat resemble a primitive hand-axe, with a sharpened slab of stone hafted to one end of a wooden handle. Our hominid forebears were no doubt able to recognize such objects regardless of their orientations.

The more general question of how orientation- and parity-free descriptions are extracted remains unknown, although there have been some speculations. Perhaps the most elegant solution was due to Deutsch (1962), who proposed in effect that shapes might be coded in terms of the internal distances between their contours. There are different ways in which this might work, but for illustrative purposes suppose we consider closed 2D shapes such as a circle or a triangle. Suppose then that we take each point on the contour and draw a line orthogonal to the contour to meet the contour on the opposite side, and measure its length. We then construct a frequency distribution of all of the line lengths. This distribution is then the description of the shape. The simplest such shape is the circle, since all lines will be of equal length. For a triangle, the lines will increase or decrease linearly as one proceeds along an edge. One can readily extend the idea to 3D closed surfaces, such as that of a pig (Wu et al., 2004). A description based on contours or surfaces has the appealing property that it is independent of any reference frame, and is therefore independent of orientation. It is also independent of parity. The same description would apply, for example, to the two mirrored triangles shown in Figure 11-3.

This approach works best for simple closed shapes, such as circles, rectangles, or triangles in 2D space, or cylinders, cones, or parallelepipeds in 3D space. It becomes unwieldy in the extreme for the more complex objects that inhabit our worlds. Biederman (1987), following earlier suggestions by Marr (1982), proposed a system in which shapes are reduced to combinations of what he called *geons*, short for *geometric ions*. The elementary shapes are precisely the sort of shapes that lend themselves to orientation-free description according to the scheme described by Deutsch. To the extent that objects can be understood as assemblages of geons, then, an orientation-free description may indeed be possible, at least insofar as the viewpoint allows perception of the individual geons. Biederman proposed a specific set of 36 geons. Although there has been little direct neurophysiological or psychophysical support for this or any other set of geons, it seems entirely reasonable to suppose that we can reduce most shapes to combinations of a finite set of idealized elements, in a sort of Legoland construction of the world.

This representation does not, however, capture many of the subtleties of objects. It does not, for example, clearly distinguish one face from another, or even one make of automobile from another. The illustrations in Figure 11-1 showed that subtle features of face recognition are not perceptually available in an upside-down image, and probably depend on a more holistic representation that does not in fact survive rotation away from the upright, at least if the rotation is extreme. At some level, then, face recognition depends on configurational cues, which are not apparent in upside-down faces (Valentine, 1988), and which do not survive mental rotation to the upright. It is these more subtle components that seem to be lost in orientation-free descriptions.

4. PERCEPTION OF ORIENTATION

If object recognition depends on the extraction of shape information independently of orientation, the next question is how the orientation of an object itself is perceived. The empirical evidence is unclear as to whether this depends on the dorsal or the ventral stream. As we saw earlier, there is evidence from brain-imaging suggesting the variations in orientation, regardless of identity, are associated with activity in the dorsal stream (Valyear et al., 2005). However, a number of studies have shown that identification of orientation or discrimination of orientations involves both intraparietal sulcus and middle occipital gyrus (Fias et al., 2002; Faillenot et al., 2001; Faillenot et al., 1999; Ng et al., 2001). Activity within the ventral stream seems to depend on the specifics of the task (Ng et al., 2001; Faillenot et al., 2001). It has been suggested that the involvement of the ventral processing stream is related to a comparison process required by orientation discrimination tasks (Faillenot et al., 1999; Faillenot et al., 2001). Alternatively, attention to the orientation features of the stimuli may modulate activity in the ventral stream as indicated by the finding that neurons in the monkey homologue of this part of the ventral stream, area V4, seem to be orientation-sensitive, and their activity is modulated by attention (McAdams and Maunsell, 1999).

Part of the difficulty is that, logically, perceiving the orientation of an object requires that one knows both what the object is, and where it is located in space. That is, one might expect both ventral and dorsal streams to play a role. The dorsal stream presumably does not "know" the actual identity of a given object, but it nevertheless registers the regions of three-dimensional space that it occupies. It has been shown, for example, that patients with object agnosia may nevertheless understand the shape of an object well enough to be able to insert it into an appropriately shaped slot

(Milner and Goodale, 1995). This system presumably does not "know" what the object actually is, but provides sufficient information to enable one to navigate in space, and to pick up objects and manipulate them. The ventral system presumably knows what an object is, but has no information as to where it is or how it is oriented. The question of how a particular object is oriented presumably requires that the object identity be somehow fitted to the space defined by the dorsal system.

Although the process of recognition strips the perceptual representation of orientation, it is sufficient to locate the stored information about the object and thus permit recognition itself. This stored information presumably includes information as to the internal axes, or upright orientation, of the recognized object, along with other information about the object, such as its name, function, and so forth. This orientation information would then be necessary in order to map the perceived object onto the information provided by the dorsal system. Once this is accomplished, the object has not only been recognized, but is also perceived in a particular orientation. This mapping might also be considered part of the process of tokenizing the object, recognizing it as part of an event in which it occurred in a particular orientation. The mapping process is not well understood, but is presumably prevented under conditions of rapid serial presentation (Harris and Dux, 2005; Kanwisher et al., 1999), and is also disrupted in patients with orientation agnosia, who are able to recognize objects without being able to discriminate their orientations (Fujinaga et al., 2005; Harris et al., 2001; Turnbull et al., 2002).

The mapping of information between dorsal and ventral systems is also part of the more general problem of perceptual binding, and has been little studied in this context. It has been suggested that binding involves synchronized firing of cells across different regions of the brain (Singer, 2001), but to our knowledge this has not been tested specifically with respect to the perception of rotated objects *as rotated.*

5. CONCLUSIONS

Vision serves two primary functions. First, a mobile organism needs to know what is out there in space. Animals need to distinguish food that is good to eat from that which is unpalatable or even poisonous. We humans need to recognize friends, enemies, and unscrupulous real-estate agents. This is the "what" system, and is contained largely in the occipital and temporal lobes of the brain. The second, perhaps more primitive system is required for navigation, enabling organisms to move around without colliding with other objects, and see and map paths through space toward goals. At some stage in evolution, it became important to be able to pick things up, especially foodstuffs, manipulate them, or bring them to the mouth for consumption. This system may require rather accurate perception of where things are, and precisely how they fill space. This is the "where" system, and is contained largely in the parietal lobes of the brain.

These two systems are complementary in that the "what" system functions by discarding the information extracted by the "where" system. For mobile organisms, important objects may appear anywhere in the surrounding space, so they need to be recognized wherever they are located relative to the organisms. In humans and other primates, there is loss of acuity away from the fovea, so we are equipped with a system of eye- and head-movements that can rapidly bring an object into foveal vision for more accurate inspection; peripheral vision is tuned more to relatively crude signals, such as movement or size, that might signal relevance without providing detail. Orientation, though, presents a different problem, since the eyes do not rotate in their sockets (except marginally as compensation for head-tilt), and the head and body are capable of only slow and limited rotation. This means that other mechanisms are needed to deal with the recognition of objects that are rotated relative to the head of the viewer.

The evidence reviewed in this chapter suggests that recognition of rotated objects depends on the extraction of information in the "what" system that is sufficient to match the object with stored information, and so reveal more information about the object. This is a "sparse" system that operates through the progressive elimination of information about orientation. In so doing, though, it may lose information of potential relevance to the viewer. For example, we may recognize animals or other humans rotated away from their canonical orientations, but fail to perceive information about facial expression or demeanour. The sparse description of a rotated animal may well indicate a quadruped but fail to accurately identify the species. The Thatcher illusion shown in Figure 11-1 indicates that facial expression may be grossly misperceived in an inverted face. Orientation-free descriptions therefore seem to be impoverished relative to the descriptions available in an upright image.

We also need to combine information between the "what" and "where" systems. For example, we need to know that it's the leopard on the right and the means of escape – a tree, say – on the left. Getting it round the wrong way plays into the hands, and mouths, of the enemy. We may also need to know how a particular object we recognize is oriented in space. A horizontal person may be asleep, or victim of a hit-and-run car accident. A predator lying down may be less of a threat than one that is upright. Since the "what" and "where" systems were designed to serve complementary functions, they have evolved separately in the brain, but at some point the information needs to be combined again. Relatively little is known as to how the visual scene, systematically dismantled by the visual systems of the brain, is put back together again. Brains must be able to do to a fallen Humpty Dumpty what all the King's horses and all the King's men were unable to do.

Mental rotation is perhaps one of the more recent visual functions to evolve, although present in other primates (Georgopoulos et al., 1993). It does serve a function in those rare cases, probably restricted to human experience, when we need to be able to distinguish between mirror images. The involvement of mental rotation in mirror-image discrimination may be simply a consequence of the fact that the "what" system extracts descriptions that are not only orientation-free, but are also parity-free. It probably did not evolve for this purpose, though. Rather, it may have been selected to enable an animal to imagine what the world would look like from another perspective. We may also use it to plan living spaces, as when we imagine what the living room would look like if the furniture were moved around.

The perception and understanding of visual scenes, and the array of objects contained in them, is not simply a matter of mental photography, in which images are simply transmitted from retina to brain for immediate interpretation. Rather, it is complex, involving distinct subprocesses that dissect the visual information, extract different aspects of the scene, and then somehow put it all back together again. This chapter is an attempt to identify just some of those subprocesses, and the manner in which they interact.

REFERENCES

Alivisatos, B. and Petrides, M., 1997, Functional activation of the human brain during mental rotation, *Neuropsychologia* **35**:111-118.

Biederman, I., 1987, Recognition by components: A theory of human image understanding, *Psychol Rev* **94**:115-147.

Cohen, M. S., Kosslyn, S. M., Breiter, H. C., DiGirolamo, G. J., Thompson, W. L. Anderson, A. K., Bookheimer, S. Y., Rosen, B. R., and Belliveau, J. W., 1996, Changes in cortical activity during mental rotation: A mapping study using functional magnetic resonance imaging, *Brain* **119**:89-100.

Cooper, L. A., and Shepard, R. N., 1973, Chronometric studies of the rotation of mental images, in: *Visual information processing*, W. G. Chase, ed., Academic Press: New York, pp. 76-176. New York.

Corballis, M. C., 1982, Mental rotation: Anatomy of a paradigm, in: *Spatial abilities: Developmental and physiological foundations*, M. Potegal, ed., Academic Press: New York, pp.173 198.

Corballis, M. C., 1988, Recognition of disoriented shapes, *Psychol Rev* **95**:115-123.

Corballis, M. C., and Armstrong, C., 2006, Repetition blindness is orientation blind, *Mem Cognition*, in press.

Corballis, M. C., and Beale, I. L., 1967, *The psychology of left and right*, NJ: Erlbaum, Hillsdale.

Corballis, M. C., and McLaren, R., 1982, Interaction between perceived and imagined rotation, *J Exp Psychol Human* **8**:215-224.

Corballis, M. C., and McLaren, R., 1984, Winding one's p's and q's: Mental rotation and mirror-image discrimination, *J Exp Psychol Human* **10**:318-327.

Corballis, M. C., and Nagourney, B. A., 1978, Latency to categorize disoriented alphanumeric characters as letters or digits, *Can J Psychology* **32**:186-188.

Corballis, M. C., Zbrodoff, N. J., Shetzer, L. I., and Butler, P. B., 1978, Decisions about identity and orientation of rotated letters and digits, *Mem Cognition* **6**:98-107.

Creem, S. H, and Proffitt, D. R., 2001, Defining the cortical visual systems: "What", "Where", and "How.", *Acta Psychol* **107**:43-68.

Deutsch, J. A., 1962, A system for shape recognition, *Psychol Rev* **69**:492-500.

Faillenot, I., Decety, J., and Jeannerod, M., 1999, Human brain activity related to perception of spatial features of objects, *Neuroimage* **10**:114-124.

Faillenot, I., Suneart, S., van Hecke, P., and Orban, G.A., 2001, Orientation discrimination of objects and gratings: an fMRI study, *Eur J Neurosci* **13**:585-596.

Fias, W., Dupont, P., Reynvoet, B., and Orban, G.A., 2002, The quantitative nature of a visual task differentiates between ventral and dorsal stream, *J Cognitive Neurosci* **14**:646-658.

Fujinaga, N., Muramatsu, T., Ogano, M., and Kato, M., 2005, A three-year follow-up of 'orientation agnosia', *Neuropsychologia* **43**:1222-1226.

Gauthier, I., Hayward, W. G., Tarr, M. J., Anderson, A. W., Skudlarski, P., and Gore, J. G., 2002, BOLD activity during mental rotation and viewpoint-invariant object recognition, *Neuron*, **34**:161-171.

Georgopoulos, A. P., Taira, M., and Lukashin, A., 1993, A cognitive neurophysiology of the motor cortex, *Science* **260**:47-52.

Graf, M., Kaping, D., and Bulthoff, H. H., 2005, Orientation congruency effects for familiar objects, *Psychol Sci* **16**:214-221.

Gross, C. G., 2002, Geneology of the "grandmother cell", *Neuroscientist* **8**:512-518.

Harris, I. M. and Dux, P. E., 2005, Orientation-invariant object recognition: Evidence from repetition blindness, *Cognition* **95**:73-93.

Harris, I. M., Egan, G. F., Sonkkila, C., Tochon-Danguy, H. J., Paxions G., and Watston, J. D, G., 2000, Selective right parietal lobe activation during mental rotation. A parametric PET study, *Brain* **123**:65-73.

Harris, I. M., Harris, J. A., and Caine, D., 2001, Object orientation agnosia: A failure to find the axis?, *J Cognitive Neurosci* **13**:800-812.

Harris, I. M., and Miniussi, C., 2003, Parietal lobe contribution to mental rotation demonstrated with rTMS, *J Cognitive Neurosci* **15**:315-323.

Jolicoeur, P., 1985, The time to name disoriented natural objects, *Mem Cognition* **13**:289-303.

Jolicoeur, P., 1990, Identification of disoriented objects: A dual-systems theory, *Mind Lang* **5**:387-410.

Jolicoeur, P., Corballis, M. C., and Lawson, R., 1998, The influence of perceived rotary motion on the recognition of rotated objects, *Psychon B Rev* **5**:140-146.

Just, M. A., and Carpenter, P. A., 1976, Eye fixations and cognitive processes, *Cognitive Psychol* **8**:441-480.

Kanwisher, N., 1987, Repetition blindness: Type recognition without token individuation, *Cognition* **37**:117-143.

Kanwisher, N., Yin, C., and Wojciulik, E., 1999, Repetition blindness for pictures: Evidence for the rapid computation of abstract visual descriptions, in: *Fleeting memories: Cognition of brief visual stimuli*, V. Coltheart, ed., MA: MIT Press, Cambridge, pp. 119-150.

Logie, R. H., Baddeley, A. D., and Woodhead, M. M., 1987, Face recognition, pose, and ecological validity, *Appl Cognitive Psych* **1**:53-69.

Logothetis, N. K., Pauls, J., and Poggio, T., 1995, Shape representation in the inferior temporal cortex of monkeys, *Curr Biol* **5**:552-563.

Marr, D., 1982, *Vision*, Freeman, New York.

McAdams, C. J., and Maunsell, J. H. R., 1999, Effects of attention on orientation-tuning functions of single neurons in macaque cortical Area V4, *J Neurosci* **19**:431-441.

Milner, A. D., and Goodale, M. A., 1995, *The visual brain in action*, Oxford University Press, New York.

Mooney, C. M., 1957, Age in the development of closure ability in children, *Can J Psychol* **11**:219–226.

Murray, J. E., 1995, The role of attention in the shift from orientation-dependent to orientation-invariant identification of disoriented objects, *Mem Cognition* **23**:49-58.

Ng, V. W. K., Bullmore, E. T., de Zubicaray, G. I., Cooper, A., Suckling, J., and Williams, S. C. R., 2001, Identifying rate-limiting nodes in large-scale cortical networks for visuospatial processing: an illustration using fMRI, *J Cognitive Neurosci* **13**:537-545.

Orton, S. T., 1937, *Reading, writing, and speech problems in children*, Norton, New York.

O'Toole, A. J., Edelman, S., and Bulthoff, H. H., 1998, Stimulus-specific effects in face recognition changes over changes in viewpoint, *Vision Res* **38**:2351-2363.

Pourtois, G., Schwartz, S., Seghier, M. L., Lazeyras, F., and Vuilleumier, P., 2005, Portraits or people? Distinct representations of face identity in the human visual cortex, *J Cognitive Neurosci* **17**:1043-1057.

Quiroga, R. Q., Reddy, L., Kreiman, G., and Fried, I., 2005, Invariant visual representation by single neurons in the human brain, *Nature* **435**:1102-1107.

Rock, I., 1973, *Orientation and form*, Academic Press, New York.

Rolls, E. T., 2000, Functions of the primate temporal lobe cortical visual areas in invariant visual object and face recognition, *Neuron* **27**:205-218.

Sekiyama, K., 1982, Kinesthetic aspects of mental representations in the identification of left and right hands, *Percept Psychophy* **32**:89-95.

Silberstein, R. B, Danieli, F., andNunez, P. L., 2003, Fronto-parietal evoked potential synchronization is increased during mental rotation, *Neuroreport* **14**:67-71.

Shepard, R. N., and Metzler, J., 1971, Mental rotation of three-dimensional objects, *Science* **171**:701-703.

Singer, W., 2001, Consciousness and the binding problem, *Ann NY Acad Sci* **929**:123-146.

Tagaris, G. A., Kim, S. G., Strupp, J. P., Andersen, P., Ugurbil, K., and Georgopolous, A. P., 1997, Mental rotation studies by functional magnetic resonance imaging at high field (4 Tesla): Performance and cortical activation, *J Cognitive Neurosci* **9**:419-432.

Thayer, Z. C., Johnson, B. W, Corballis, M. C., and Hamm, J. P., 2001, Perceptual and motor mechanisms for mental rotation of human hands, *Neuroreport* **12**:3433-3437.

Thompson, P., 1980, Margaret Thatcher: a new illusion, *Perception* **9**:483-484.

Tomasino, B., Borroni, P., Isaja, A., and Rumiati, R. I., 2005, The role of the primary motor cortex in mental rotation: A TMS study, *Cogn Neuropsychol* **22**:348-363.

Turnbull, O. H., Della Sala, S., and Beschin, N., 2002, Agnosia for object orientation: naming and mental rotation evidence, *Neurocase* **8**:296-305.

Ungerleider, L. G., and Mishkin, M., 1982, Two cortical visual systems, in: *The analysis of visual behavior*, D. J. Ingle, R. J. W. Mansfield, and M. S. Goodale, eds., MIT Press, Cambridge, pp. 549-586.

Valentine, T., 1988, Upside-down faces: A review of the effect of inversion upon face recognition, *Brit J Psychol* **79**:471-491.

Valyear, K. F., Culham, J. C., Sharif, N., Westwood, D., and Goodale, M. A., 2006, A double dissociation between sensitivity to changes in object identity and object orientation in the ventral and dorsal visual streams: A human fMRI study, *Neuropsychologia* **44**:218-228.

Vingerhoets, G., de Lange, F. P., Vandemaele, P., Deblaere, K., and Achten, E., 2002, Motor imagery in mental rotation: An fMRI study, *Neuroimage* **17**:1623-1633.

Vinje, W. E., and Gallant, J., 2000, Sparse coding and decorrelation in primary visual cortex during natural vision, *Science* **287**:1273-1276.

Wu, J., Tillett, R., McFarlane, N. Xiangyang, J., Siebert J. P. , and Schofield P., 2004, Extracting the three-dimensional shape of live pigs using stereophotogrammetry, *Comput Electron Agr* **44**:203-222.

Chapter 12

FUNCTIONAL NEUROANATOMY OF MENTAL ROTATION PERFORMANCE

Lutz Jäncke[1] and Kirsten Jordan[2]

[1]*Department of Psychology, Neuropsychology, University of Zürich, Switzerland;* [2]*Institute of Medical Psychology, University of Goettingen, Germany*

Abstract: Mental rotation is an important part of human spatial cognition. In the last decade a growing number of brain imaging studies have been undertaken to uncover the neural underpinnings of mental rotation. These studies demonstrated that several brain areas are involved in the control of mental rotation. In this chapter we will summarize these. Although the reviewed studies differ in terms of used stimuli, mental rotation procedure, or brain imaging method, there is consistency for the core regions, which are involved in mental rotation (superior parietal lobe and the intraparietal sulcus). However, frontal, temporal, and occipital areas are also included into mental rotation processes depending on various aspects including used cognitive strategy, task difficulty, measuring protocol, or concentration of sexual hormones.

Key words: mental rotation; functional neuroanatomy; parietal lobe; intraparietal sulcus; cognitive strategies.

1. DEFINITION AND CLASSIFICATION OF SPATIAL COGNITION AND MENTAL ROTATION

Spatial cognition may be defined as the ability to generate, retain, retrieve, and transform well-structured visual images. Spatial ability occupies a pivotal role in all models of human ability. For example, most models of human ability state that together with verbal ability spatial cognition captures more variance than any other dimensions in large, representative batteries of ability tests (Thurstone, 1938). Spatial ability has been measured with various tests such as form boards, block manipulation, paper-folding tasks, and mental rotation tests. Many of these tasks are used in contemporary intelligence tests as measures of performance or nonverbal intelligence. A further variant of spatial tests utilizes navigation tests, in which the subjects are required to find efficient routes between locations. However, these tests show at best moderate correlations with other measures of spatial abilities. Several factor analytic studies have shown that spatial ability is not a unitary construct as reflected in the fact that performance scores of many spatial tests consistently revealed several spatial factors. On the basis of extensive reviews, Lohman (1979) identified five major factors of spatial cognition: (1) spatial visualization, (2) speeded rotation, (3) closure speed, (4) closure flexibility, and (5) perceptual speed. Linn and Petersen (1985) on the other hand identified only three factors including (1) spatial perception, (2) mental rotation, and (3) spatial visualization. Several subsequent studies have substantiated this distinction of subtypes of spatial cognition, most supporting the three-factor solution. Closure speed, closure flexibility, and perceptual speed load also on other factors not directly related to spatial cognition (e.g., executive functions or perception), thus indicating that they may not be typical for spatial cognition. According to current view, the three-factor solution is supplemented by a fourth factor representing spatial navigation. A recent development in spatial cognition has been to differentiate the investigative approach into large-scale spatial tasks where the observer is part of the environment and cannot see the entire space from one viewpoint and small-scale spatial tasks where the spatial relations of objects can be seen (Golledge and Stimson, 1997).

This research has demonstrated that mental rotation is an important aspect of spatial cognition. We will focus in the present contribution on this special subcomponent of spatial cognition because of the considerable interest it has received over the last 20 years, especially in the context of brain imaging studies. Mental rotation includes the ability to imagine how objects will appear when they are rotated. Shepard and Metzler (1971) were the first to systematically explore this function. They did so by presenting pairs of solid geometrical objects that were in form either identical or mirror images of one another and timing participants as they judged whether the pairs were the same or different. The reaction times to make same judgments increased linearly as a function of the angle between the forms, suggesting that participants mentally rotated one form into congruence with the other before making the judgment.

Although this principal result has been replicated many times it has also been noted that there are individual differences in solving the mental rotation task. The most prominent difference is a gender difference, with men normally outperforming women. Environmental (experience-dependent, spatial activities, socialization) and biological (genetic, hormonal, evolutionary) factors are often discussed as possible causes for these sex differences. It is recognized that subjects prefer different strategies, known as "analog-holistic" or "verbal-analytic strategies, in mental rotation tasks. One of the more prevailing of possible explanations for the sex differences in mental rotation is that men and women solve mental rotation tasks by using different strategies. Men are thought to use the more successful visual-spatial holistic strategies, in which they picture the object in their mind and then mentally rotate the object. Women on the other hand prefer less efficient verbal or analytic strategies with a more "piecemeal" mental rotation process. Evidence for the role of environmental factors has been found in the frequently observed relation between a person's participation in spatial activities and his or her spatial ability, as well as in the fact that performance can be improved by training (Heil et al., 1998; Newcombe et al., 1989; Quaiser-Pohl and Lehmann, 2002; Voyer et al., 1995). Environmental factors may interact with the different strategies of the two sexes. Thus, a reasonable working hypothesis is that differences in strategies underlie differences in mental rotation performances of the sexes. However, it is unclear whether the different strategies in females and males depend on sex-specific brain development or on differential socialization (Halpern, 2000; Kimura, 1999). In terms of biological factors, explanations of sex differences in spatial performance are both indirect, as evidenced by evolutionary speculations about different selective pressures on males and females when navigating in the environment (Jones et al., 2003), and direct, as demonstrated by the important role played by sex hormones in organizing a variety of sexually dimorphic behaviors. For instance, sex hormones have been shown to be important in mental rotation performance, independently of differential experience (Hausmann et al., 2000). Hausmann and colleagues have demonstrated that testosterone and estradiol are able to modulate spatial cognition during the menstrual cycle, and Maki and Resnik (Maki and Resnick, 2001) recently pointed out that sex hormones like estrogen could influence cerebral blood flow and neuronal activation in women.. Sex differences in functional cerebral organization, which may reflect organizational hormone effects, have also been implicated in differential mental rotation performance of the sexes, albeit without any direct evidence and without any suggestions as to the nature of "functional cerebral organization". There is otherwise good evidence of the influence of sex hormones present during prenatal development on human sex-typed behavior (Cohen-Bendahan et al., 2005). For example, females suffering from congenital adrenal hyperplasia (CAH) display higher spatial ability than controls. Because of an enzymatic defect caused by a single gene, individuals with CAH produce high levels of adrenal androgens at a very early stage in gestation (Cohen-Bendahan et al., 2005). Besides the above-mentioned strategic differences in solving the mental rotation problem, Kosslyn et al (Kosslyn et al., 1998; Wexler et al., 1998) propose another difference in strategies which is not directly linked to gender: an internal strategy in which one anticipates the appearance of the object prior to physically manipulating it and an external strategy in which one visualizes the consequences of someone else or an external force moving the object. The internal strategy is also sometimes called the egocentric strategy. In the following we will summarize and discuss the current literature on mental rotation where its investigation has used modern brain imaging methods. We place the emphasis on describing and discussing the brain imaging results in the context of the following topics: (1) general functional neuroanatomy of mental rotation, (2) possible hemispheric asymmetries, (3) gender differences and (4) influence of different cognitive strategies on mental rotation and the concomitant cortical activation pattern.

2. FUNCTIONAL NEUROANATOMY OF MENTAL ROTATION

Several papers have been published in the last 10 years that dealt with exploring the functional neuroanatomy of mental rotation. These studies used positron emission tomography (PET) or functional magnetic resonance imaging (fMRI). This line of research began with the publications of Cohen et al. and Tagaris et al. in 1996 (Cohen et al., 1996; Tagaris et al., 1996). These and subsequent studies identified three brain complexes involved in mental rotation: (1) superior and inferior parietal regions, (2) extrastriate visual regions, (3) temporal lobe regions as part of the ventral stream, and (4) motor and/or premotor regions (Barnes et al., 2000; Carpenter et al., 1999a; Carpenter et al., 1999b; Cohen et al., 1996; de Lange et al., 2005; Dietrich et al., 2001; Harris et al., 2000; Jordan et al., 2001; Jordan et al., 2002; Just et al., 2001; Koshino et al., 2005; O'Boyle et al., 2005; Richter et al., 2000; Richter et al., 1997; Seurinck et al., 2004; Seurinck et al., 2005; Suchan et al., 2005; Tagaris et al., 1996; Tagaris et al., 1998; Unterrainer et al., 2000; Vanrie et al., 2002; Vingerhoets et al., 2002; Vingerhoets et al., 2001; Weiss et al., 2003). Thus, a distributed network of cortical activations has been identified in the context of mental rotation.

2.1 The parietal cortex

The parietal cortex has been identified most consistently in all brain imaging studies as being the core region involved in mental rotation. Some studies report activations centered more on the superior parietal lobe (SPL), while others emphasise the role of the intraparietal sulcus (IPS) as the core region. Most of the aforementioned brains imaging studies have used group analysis techniques to analyze the functional images statistically. In order to adjust for individual differences in brain anatomy, this analytical approach involves spatially smoothing the data using spatial kernels of 8-12mm radius. This can however diminish spatial precision and might result in spatially blurred activation clusters. Consideration of the spatial inaccuracy, of approximately 1 cm, that is inherent to this analytical approach is warranted when interpreting reported activation peaks. However, even if one does consider this inaccuracy, most studies do in fact report activation peaks centered around the IPS (Carpenter et al., 1999a; Gauthier et al., 2002; Jordan et al., 2002; Tagaris et al., 1996; Vingerhoets et al., 2002). In addition, several studies reported that the IPS activation linearly increased with greater angular disparity (Carpenter et al., 1999a; Tagaris et al., 1996). Because most research is consistent in this matter, it is plausible to assume a major role of the SPL and the IPS as computation centers for spatial transformations such as mental rotation.. The IPS might have a special role in this context. The anterior IPS (AIP) and adjacent areas has been shown to be involved in explicit, complex hand operations in humans, like grasping (Binkofski et al., 1999; Jancke et al., 2001). The more posterior part of the IPS (cIPS) is involved in a variety of mental operations beside mental rotation. Among them are (1) imagery of motor acts (Deiber et al., 1998; Jancke et al., 2001), (2) learning movement trajectories (Seitz et al., 1997), (3) controlling of self-determined finger movements (Schubert et al., 1998), (4) generating internal body representations (Bonda et al., 1995), (5) recognising objects in relation to actions (Faillenot Sakata et al., 1997; Faillenot Toni et al., 1997), (6) cross-modal matching tasks (Banati et al., 2000), and (7) during spatial attention as well as spatial working memory (Coull, 1998; Mattingley Corben et al., 1998; Mattingley Husain et al., 1998). From single cell recordings in monkeys and few recent human imaging studies (Merboldt et al., 2002; Tsao et al., 2003) we also know that the posterior part of monkey IPS (cIPS) is a higher centre of stereopsis, integrating various binocular disparity signals received from primary and secondary visual cortex. Thus, the cIPS possibly represents the neural code of three-dimensional features of objects to be sent to the anterior parts of the IPS (e.g., the AIP) for visual guidance of hand movements. The common denominator of the tasks described above that evokes IPS activations may be that they require the generation of a mental representation of a three- or two-dimensional object, which should be grasped, explored or manipulated (explicitly or implicitly). Obviously, this mental representation is established in the IPS and adjacent brain regions. These areas can be activated without any explicit visual input, thus serving as supramodal integration centres for mental images. Because these areas are part of the dorsal stream projecting to the frontal premotor areas, the mental images are prepared mostly for guiding human actions. This action-oriented object representation is therefore clearly different from the object recognition processes linked with the ventral stream (Goodale and Milner, 1992).

2.2 The frontal cortex

While all brain imaging studies are more or less consistent in the reported involvement of the parietal lobe in mental rotation processes, reports of frontal area involvement are highly inconsistent. Several studies report activations in premotor areas, however, within different subareas. For example, Thomsen et al. (2000) found bilateral activations in ventral regions, while others reported activations in more dorsal parts of the premotor cortex (Cohen et al., 1996; Gauthier et al., 2002; Jordan et al., 2002; Richter et al., 1997). Some of these studies also report unilateral (e.g., Gauthier et al., 2002) while others bilateral activations. Beside these activations there are also some studies reporting activations in mesial premotor areas (e.g., supplementary motor cortex) (Gauthier et al., 2002). However, the reason for these activations is so far unclear. Maybe these activations are associated with the also occasionally reported activations within the primary motor cortex (M1) and might represent the involvement of motor control processes during mental rotation.

That motor areas are active during mental rotation of visual objects is often related to subliminal or imagined movements, which might be activated either automatically or as a kind of strategy to solve the mental rotation task. This idea is supported by studies demonstrating that covert motor actions such as intending actions, learning by observation, but also imagining actions, lead to subliminal activations in the motor system (Jeannerod, 1995, 2001). In view of the preceding, the implication is that if people mentally rotate a visual mental image in the same way they rotate an actual object physically, the motor cortex is very likely recruited during mental rotation tasks. Kosslyn et al. (Kosslyn et al., 1998; Kosslyn et al., 2001) suggest that such a motor, or *egocentric strategy* when mentally rotating an object can explain the involvement of motor cortex and can be evoked implicitly by stimulus type or explicitly by instruction. Indeed, Kosslyn et al. (1998) provided supporting data for this hypothesis. In one experiment they compared mental rotation of drawings of hand with the classical Shepard-Metzler three-dimensional figures and found M1 activation limited to the hands condition. In a further experiment they showed that when subjects were instructed to use two different strategies for mental rotation of the Shepard-Metzler figures, M1 was only active when the subjects performed the mental rotation task by imagining the rotation of the objects as a consequence of physically turning them. When they mentally rotated the figures as a consequence of an external force (e.g., an electric motor), M1 was not activated. However, the premotor cortex was activated in both conditions. Thus, at least M1 activation would be a consequence of the used strategy to solve one aspect of the mental rotation problem.

One of the first to make the claim that motor control areas might be involved in mental rotation was Georgopoulous (Georgopoulos, 2000; Georgopoulos et al., 1989), on the basis of single cell recordings in monkeys. He found that neurons in M1 tuned for specific movement directions discharge before a monkey began to shift a lever in that specific direction. In addition, and more important for the scope of the present paper was the finding that movements composed of sub-movements that have to be executed in different directions "unfold" before movements that have to be performed in the opposite/reverse sequence to actual movements. Later on, Georgopoulous argued that the orientation-specific neurons in M1 might also be involved in mental rotation of visual figures. A further finding that emphasizes the importance of motor control processes for mental rotation is the study of Wexler et al. (1998). These authors showed that when participants performed mental rotation tasks while executing an unseen motor rotation in a given direction, motor rotation compatible with mental rotation resulted in faster responses than when the two rotations were incompatible. Thus, mental rotation of a visual image interacts with an explicit motor rotation. A third study about that topic has been published by Parsons (1994) in which participants had to identify a picture as being of a left or a right hand. Reaction times were longer when the pictures had to be mentally rotated in a way that would be awkward to perform physically.

Although these studies support the idea that motor control processes are involved in mental rotation processes, they do not necessarily support the notion that M1 is directly involved in mental rotation. Several brain imaging studies argue for anticipatory M1 activation to be present in mental rotation (Vingerhoets et al., 2002), because in typical mental rotation experiments it is necessary to indicate the *same* or *different* judgement by a motor response. This view has also been supported by a study of Windischberger et al. (2003) which adopts a specific statistical approach to analyze fMRI data (fuzzy cluster analysis). They showed that activation in M1 is exclusively related to the button presses, while parietal and premotor regions showed signal enhancement during the whole stimulus presentation period, reflecting cognitive processing underlying the mental rotation task. Similarly, Richter et al. (2000) reported that the left M1 was activated at the end of each mental rotation trial, reflecting neural activation caused by the button press with the right hand at the end of each trial. Thus, there is no strong evidence so far that M1 is inevitably involved in the mental rotation process. There is rather more evidence supporting the idea that M1 activations are related to subsequent motor control processes associated with the motor response.

However, although not all mental rotation studies report premotor activation (Jordan et al., 2001), there is stronger evidence that premotor areas are involved in mental rotation. This makes sense from a functional neuroanatomy perspective because there are strong functional and anatomical connections between posterior parietal and premotor cortex in humans. The parietal cortex (SPL and IPS) serves as an essential node generating information (transformation from retinotopic to body centered coordinates) which is projected to the dorsal premotor cortex for further processing. The dorsal premotor cortex is strongly involved in learning, storing, and controlling of sensorimotor associations (Hanakawa et al., 2002); thus, sensory cues are linked to motor programs in this area. This area is also activated during tool use (e.g., the tool as part of the body image), action observation (Buccino et al., 2004), and during planning of motor processes (Binkofski et al., 1999). In addition, parts of the dorsal premotor cortex are also involved in spatial working memory processes necessary to maintain and hold spatial information (Suchan et al., 2005). Beside dorsal premotor cortex activations some studies also report activations in the ventral premotor cortex in the vicinity of Broadmann's area 44 and 45 (Just et al., 2001; Thomsen et al., 2000). The ventral premotor cortex is an interesting region, which has been associated with several higher order cognitive functions. Among them are (1) action observation of movements made by someone else (Buccino et al., 2004), (2) imagination and preparation of forthcoming movements (Jancke et al., 2001), and a variety of language related functions (Longcamp et al., 2005). That the ventral premotor cortex is activated during mental rotation processes might depend on all of the above listed cognitive processes. Which of them are more important has to be examined in future experiments. However, the recruitment of this area and the associated cognitive processes might be a matter of used strategy or skill level in solving the mental rotation task.

2.3 The occipital cortex

A number of mental rotation studies report occipital cortex activation although there is considerable variation in reported exact location of activation and the stimulus and task condition under which these areas light up. Extrastriate cortex activation during mental rotation has been found in fMRI studies using the classical voxel-based analyses methods (Podzebenko et al., 2002; Vingerhoets et al., 2002). However, this activation has not been associated with cognitive processes directly related to mental rotation. It has been argued that this activation corresponds to the visuospatial attentional demands associated with the demanding mental rotation task. Interestingly, Seurink et al. (2004) recently demonstrated that extrastriate activations were only present during fixed-paced intervals. In this experimental condition each mental rotation trial appeared at fixed intervals regardless of the speed with which the subjects made a judgment. However, during self-paced trials (during which the next mental rotation trial was triggered by the subject's response) there was no activation in extrastriate cortex. Hence, this study clearly demonstrates that extrastriate activations do not reflect visual processing that is specific to mental rotation per se. The authors argue that during fixed-paced trials visual processing is ongoing after solving the mental rotation task. What kind of specific aspect of visual processing is at work cannot be determined, but possible candidates are perceptual learning, priming, or simply visual imagery.

A major problem in comparing the activation patterns across different studies is related to differences in used paradigms and statistical analysis methods. The latter point is quite problematic in the context of brain imaging studies. Most studies use whole-head voxel-based analyses requiring strict and conservative statistical tests. In addition, the voxel-based analyses are also biased by the inherent problem of inter-individual differences in terms of location of prominent anatomical landmarks and cytoarchitectonic borders. Thus, if there are small effects in particular areas (as it is the case in the occipital cortex during mental rotation) it will be difficult to detect signal changes in these areas due to reduced statistical power. An alternative method to analyze such data is to refer to regions of interest (ROIs) for pre-specified anatomical areas and to calculate for these areas average statistics like mean activation or mean number of significantly activated voxels (Bosch, 2000). Applying this technique, Koshino et al. (2005) found strong activations in occipital areas (especially in the inferior, superior extrastriate region and in the occipital pole) related to the mental rotation processes (see Fig. 12-1). Thus, there is considerable disagreement as to whether these areas are directly involved in mental rotation. Since the dorsal part of the occipital cortex projects to the dorsal stream it might be possible that these areas indeed play a role in mental rotation. This idea receives support from single-cell studies in monkeys showing that the causal part of the IPS (c-IPS) receives strong input from dorsally located cells in the occipital lobe specialized for the analyses of stereoscopic information (Sakata et al., 1997).

Thus, early spatial visual analysis might support the mental rotation process. However, this has to be proved in human brain imaging studies.

2.4 The temporal cortex

Several studies report also temporal lobe activations, mostly in the inferior temporal gyrus region, sometimes unilateral (Seurinck et al., 2004) or bilateral (Carpenter et al., 1999a; Jordan et al., 2002; Koshino et al., 2005). These activations are interpreted as reflecting participation of the so-called what system. Numerous functional imaging studies of visual object recognition have shown activation in inferior temporal cortex (Kanwisher et al., 1996; Kanwisher et al., 1997). In addition, it has also been shown that these regions (in association with the IPS) revealed increased activation in situations in which the recognition of line drawings is made more difficult by eliminating pictorial elements (Diwadkar et al., 2000). Whether these areas are also directly involved in mental rotation was unclear until a recent paper thoroughly investigated this issue in more detail (Koshino et al., 2005). By using a region of interest approach these authors found that the inferior temporal region is indeed part of a network involved in controlling mental rotation. However, these regions came into play with increasing stimulus complexity supporting the idea that more brain regions are bound together for the sake of mental rotation with increasing stimulus complexity.

2.5 Functional connectivity between brain areas

The preceding reviewed studies have highlighted the anatomical structures involved in mental rotation. Some areas are more involved in specific mental rotation processes than others, reflecting the relative specialization of cortical regions for specific mental rotation processes. A major problem of the papers referred to is that they have mostly used the so-called voxel-based analysis method to compare cortical activations obtained in different experimental conditions. Although this method is useful in detecting strong activation differences without a priori knowledge of the involved brain regions, this method is very conservative in that it only uncovers those brain regions, which are additionally, or more strongly activated than in the control conditions. This method has been shown to neglect the fact that both single isolated brain regions and, importantly, functionally coupled networks are involved in controlling a specific function. In order to overcome this obstacle, Koshino and colleagues (2005) conducted an fMRI study in which they used the hemodynamic responses obtained from predefined regions of interest to calculate the functional correlation between these regions of interest. In fact by using a factor analyses of the region of interest data, they uncovered a three-factor structure representing a large-scale cortical network, which was operative in all experimental conditions comprising mental rotation in the context of different stimulus complexities. The uncovered networks were (1) an executive control network, consisting mainly of the frontal areas, (2) a spatial information processing network, consisting primarily of the parietal regions, and (3) a lower level visual network for object recognition, consisting mainly of the occipital regions and, sometimes, the inferior temporal regions. Interestingly, the size (and number of involved regions) tended to be smaller with decreasing complexity, and smaller size was accompanied by relatively strong separation of the three networks. During more complex mental rotation conditions the networks became larger, sharing more brain areas and a lesser degree of separation. There was also a stronger functional correlation between the ventral and dorsal stream indicating stronger synchronization between the dorsal and the ventral systems as task difficulty increased. Thus, with increasing workload increasingly more brain areas are recruited to solve the mental rotation problem. Obviously, this is a typical strategy of the human brain dealing with complex cognitive problems, as has been shown in various brain imaging studies in the context of working memory, sentence comprehension, or attention. For future studies it will be interesting to investigate whether the size of each network as well as the interrelation between the different networks changes according to training levels or depends on or is related to individual differences in performance in solving the mental rotation.

3. LATERALISATION AND MENTAL ROTATION

The processing dominance of the right parietal lobe for spatial tasks has long been suggested as a kind of complement to the well known left-sided hemispheric dominance of language functions. A seminal review of Bryden et al. (1983) summarized the findings concerning spatial processing deficits as a consequence of right hemisphere (mostly parietal cortex) lesion in the early 80s. Although there was and is no doubt that the right hemisphere and especially the right parietal cortex plays a specific role in the control of spatial functions Bryden et al. already noted at that time, that the right-sided processing dominance for spatial functions is much smaller in effect than the left-sided dominance for language functions. This has also been substantiated in a recent review about laterality effects obtained in behavioural studies (Voyer, 1998; Voyer and Bryden, 1990).

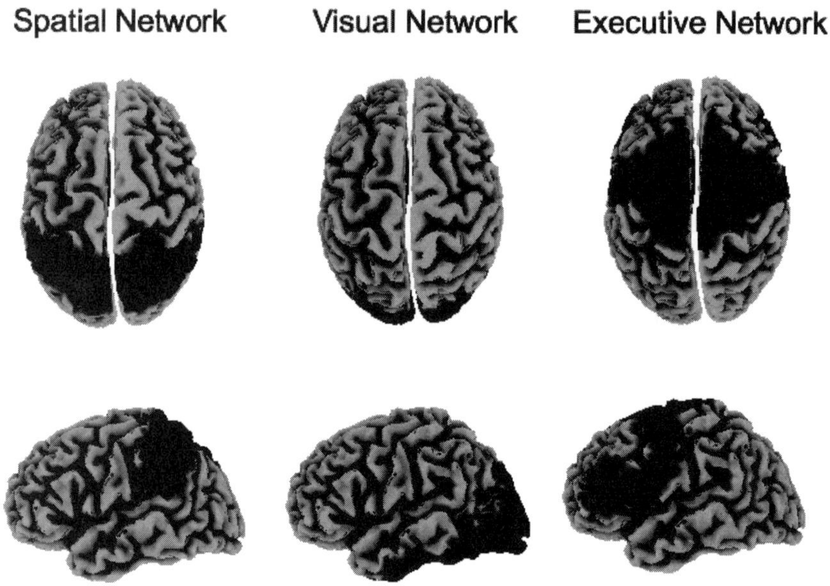

Figure 12-1. Schematic description of the different neural networks involved in mental rotation. The relative participation of each network depends on the kind of stimuli used for mental rotation, individual strategies to solve the mental rotation task, and different skill levels. This figure has been designed according to the interpretation given by Koshino et al. (2005).

Because mental rotation is regarded as being a substantial part of spatial processing, several investigators have argued that mental rotation performance relies on the integrity of the right-sided neural networks (including the parietal cortex) (Ditunno and Mann, 1990). However, earlier behavioral studies have also uncovered that the right-sided processing dominance is dynamic and depends on stimulus type and processing strategies. For example Corballis (1977, 1978, 1986) has proposed that bilaterality might increase with increasing task demands. The laterality issue has also received much attention in recent brain imaging studies employing mental rotation tasks. Before we discuss the current status about lateralization effects during mental rotation we will briefly describe the methodological problems associated with this kind of research.

Several investigators using PET, fMRI, or EEG methods have reported asymmetric activation changes associated with mental rotation (Cohen et al., 1996; Harris et al., 2000; Kosslyn et al., 1998; Muthukumaraswamy et al., 2003; Roberts and Bell, 2003; Rosler et al., 1995; Rosler et al., 1993;

Tagaris et al., 1996; Tagaris et al., 1998; Yoshino et al., 2000). In the case of PET or fMRI studies, claims of asymmetric activation were made on the basis of voxels in one hemisphere exceeding a statistical threshold while homologous voxels in the opposite hemisphere did not. However, such an analytic strategy, while typical, tests only for the main effects of the condition. To demonstrate an actual difference between the two hemispheres, it is necessary to test the Condition X Hemisphere interaction (see for a similar discussion Davidson and Irwin (1999) about asymmetric activations in the context of emotion studies). The fact that such tests are rarely performed is largely a function of the fact that appropriate software is not commercially available to perform such analyses for the entire brain volume (e.g., within the SPM, BrainVoyager, or MedEx packages). An alternative method, which is often used in the context of anatomical but also functional studies, is to flip the brains horizontally and compare the flipped with the unflipped brain, resulting in a between-hemisphere comparison. Although this approach seems compelling, it is weakened by the very fact of anatomical differences between both hemispheres, particularly very strong a priori anatomical differences in the inferior parietal cortex (Ide et al., 1996; Jancke et al., 1994). However, the most convenient way to conduct interaction analyses is to define regions of interest (ROI) in homologue brain areas of both hemispheres and to calculate mean activations for these ROIs. It is important for statistical reasons that one should not use statistical thresholds for this kind of analyses in order to obtain a reliable estimate of the activation of the particular areas (see for a more detailed discussion of this critical topic Bosch, 2000). If the interaction calculated with this data is not significant, any claim to have an asymmetric finding is unfounded since the lack of a significant interaction means only that the changes found in one hemisphere are not significantly different from those observed in the other, even if the effects were independently significant in one hemisphere but not in the other. Moreover, it is possible for significant interactions to arise in the absence of any significant main effects. To the best of our knowledge there are no reports in the literature of asymmetric effects where the Condition X Hemisphere interaction has also been tested properly. Thus, we have to be cautious in interpreting reported asymmetries in the context of mental rotation processes.

Most brain imaging studies examining mental rotation processes have however reported bilateral activations in parietal and frontal areas, thus supporting the view that both hemispheres are involved in controlling mental rotation. On the other hand, there is growing evidence that both hemispheres are differently involved in mental rotation depending on the stimuli, applied strategies, and the reference frames used for mental rotation. For example, mental rotation of one's body (Bonda et al., 1995; Creem et al., 2001) is typically associated with increased left-sided parietal activations. These studies argue in favour of the left SPL as being at the core of processing of egocentric or viewer-based spatial information. In line with this argument, a recent rTMS study of Harris and Miniussi (2003) found that inhibition of the right SPL at 400-600 ms after stimulus presentation disrupted mental rotation of letters and digits while imagined rotation of one's body was unaffected by this stimulation. As Parsons (2003) mentioned, this result might indicate that the right SPL is involved in various aspects of "visual-spatial processing that relate the object to viewer-based or scene-based frames of reference". This interpretation is partly supported by a study of Zacks et al. (2003) using a chronically implanted electrode grid placed on 14 sub-regions in the vicinity of the right parietal lobe of an epileptic patient. The authors found that electric stimulation of one particular sub-region of the right parietal lobe diminished mental rotation performance, thus supporting, an anatomical specialization within the right SPL for mental rotation processes from an egocentric viewer-based perspective.

This view has been challenged by a recent fMRI study of Wolbers et al. (2003). Using an elegant design, these authors revealed that the side of parietal activations strongly depends on the strategy used for mental rotation. For example, when the subjects were instructed to mentally imagine grasping and rotating the objects to be rotated with their own right hand, left-sided activations were found in the SPL and IPS. However, when the left hand was used for motor imagery, the activation peaks were stronger on the right parietal lobe. Thus, the activation in the parietal lobe was contralateral to the imagined hand involved in the mental rotation process. This finding has been corroborated by a recent lesion study, which showed that patients with lesion in the right parietal lobe (RH patients) were more accurate than patients with left-sided parietal lesions (LH patients) when they imagined physically rotating the stimuli with their dominant right hand (Tomasino and Rumiati, 2004). On the contrary, LH patients outperformed RH patients when they imagined the stimulus as rotating in the visual space. Hence, when the RH patients can use their unaffected left-sided parietal and motor areas for solving the mental rotation task by using the "mental motor strategy" they show no performance deterioration. On the other hand, LH patients cannot use the left-sided parietal-frontal network for mental rotation and can therefore no longer use the "mental motor strategy" (or with less proficiency). But they can use a more visual strategy, which makes use of a more right-sided parietal-frontal network.

Hemispheric differences were also found in the motor and premotor areas. For example, Vingerhoets et al. (2002) used pictures of hands and tools in their fMRI study and uncovered bilateral activations in parietal and premotor areas during rotation of hands while rotation of pictures of tools revealed bilateral parietal and left-sided activations in premotor areas. This finding corresponds with a more recent study that demonstrated predominantly left-sided activation in premotor and motor areas when "motor strategies" are used for mental rotation (Wraga et al., 2005; Wraga et al., 2003).

In summary, the lateralisation patterns reported so far are much more complicated than simple dichotomies might suggest (right = spatial and left = verbal). One problem common to all brain imaging methods is that the time resolution is poor. Thus, it is possible that there is a dynamic lateralisation pattern but that it cannot be resolved by the poor time resolution of fMRI and the even poorer resolution of PET. First findings supporting a dynamic pattern of lateralization during mental rotation have been demonstrated in several EEG studies calculating event related potentials (ERPs). These studies have shown that early processing stages at 200-300 ms post stimulus presentation were mostly symmetric in the parietal lobe while later stages at 400-500ms are often stronger on the right parietal lobe (Muthukumaraswamy et al., 2003; Yoshino et al., 2000). It would therefore be most interesting to combine the superior spatial resolution of electrophysiological methods like EEG with fMRI or PET in order to bring to light the sequence of cognitive processes and modules involved in mental rotation.

4. GENDER DIFFERENCES

As mentioned above mental rotation is one of the few cognitive abilities for which men have been shown to consistently outscore women (Linn and Petersen, 1985; Peters, 1995; Peters et al., 1995; Voyer and Bryden, 1990; Voyer et al., 1995). Some studies addressed this issue using functional imaging methods (Dietrich et al., 2001; Halari et al., 2005; Jordan et al., 2002; Seurinck et al., 2004; Tagaris et al., 1996; Thomsen et al., 2000). The main question of these studies is, whether the sex differences in behavior are visible at the neuronal level. There is also growing interest to disentangle whether gender specific cortical activation are related to gender-specific cognitive strategies or whether there are substantial hormonal influences affecting not only the mental rotation task but also the associated cortical activation.

Although both sexes show in principle similar activation patterns during mental rotation with activations in the above-mentioned networks, there are some gender-specific differences. For example, Thomsen et al. (2000) stronger activations in parietal areas for men while women exhibited stroneger activations in inferior frontal areas. The authors suggest that males and females may differ in the used strategies and men would use a "gestalt" strategy and women a "serial" reasoning strategy, reminiscent of the speculations about "holistic" and "analytic" strategies (see also Weiss et al. 2003 for similar findings and interpretations). Dietrich and colleagues (2001) were interested to study whether oestrogen exerted a modulatory effect on the hemodynamic responses during mental rotation. In fact they found that the intensity and spatial extend of cortical activation in women during mental rotation depends on the oestrogen phase with lower activation during the low oestrogen phase and much higher overall level of cerebral hemodynamic response during the high oestrogen phase of women. Interestingly, although there were substantial differences in cortical activations in women depending on the oestrogen phase, there was no corresponding difference in mental rotation performance. This study elegantly demonstrates that other factors than cognitive involvement or cogntitive strategies influence cortical activations and thus, challenge all fMRI studies on gender differences in mental rotation.

A very critical point in gender studies on mental rotation is that men consistently outperform women in this task. Thus, gender differences in cortical activation might also be influenced by performance differences. Therefore, it is necessary to compare men and women with similar mental rotation performance. This has been done in one of our studies (Jordan et al., 2002) in which we found different cortical activation patterns for women and men although there was no performance difference. Women showed stronger bilateral activation within the intraparietal sulcus and an additional activation in the inferior temporal lobe. On the other hand, the left motor cortex and the parieto-occipital sulcus were stronger activated in men. This pattern might be due to different cognitive strategies with a more analytical strategy in women and a more visual or motor strategy in men. Recently Seurinck et al. (2004) reported that men and women use a very similar "motor" strategy during egocentric mental rotation of hands and tools with a potential gender-specific accent. As Weiss et al. (2003) and Jordan et al. they tested a men and women with approximately similar performance. Gender differences were modest and limited to the mental rotation of pictures of hands with a stronger activation of the left ventral premotor cortex in women and a stronger involvement of the lingual gyrus in men. The authors supposed that women rely more strongly on imitation or use more perceptual comparisons whereas men would prefer a more extensive semantic or early visual processing. Recently Halari et al (Halari et al. 2005) conducted a very well controlled study in which the female subjects were chosen because they didn't take any contraceptives and who reported having regular menstrual cycles. In addition, all women were tested during the low-oestrogen phase of the menstrual cycle. Using this strict protocol they found stronger hemodynamic responses in the left middle temporal gyrus and the right angular gyrus in men while there were no additional activation in women compared to men.

Summarizing these studies it can be suggested that both men and women exhibit very similar activation patterns during mental rotation. However, there are subtle gender differences in the superior and inferior parietal lobe, often comprising the intraparietal sulcus. Stronger gender differences are found in the inferior frontal gyrus, the inferior temporal gyrus and the primary motor cortex. The reason for these differences are currently not well understood. One possibility is that macroanatomical differences determine functional differences. There is a growing literature supporting the idea that there are substantial anatomical gender differences. For example, Luders et al. (2004) found greater cortical complexity (stronger gyrification) in women within frontal and parietal regions. Increased cortical complexity implies more cortical surface area, which may offset gender differences in brain volume and account for behavioral sex differences. Other authors described a significantly different cortical cytoarchitecture with larger neuropil volumes in women than in men, but higher neuronal densities in men without any differences in neuronal soma size and astrocytic volumes (Rabinowicz et al., 1999; Rabinowicz et al., 2002). Another often proposed explanation is that women and men might use different during mental rotation. Thomsen et al. (2000) and Weiss et al. (2003) assumed a "serial" reasoning strategy in women (more executive control) and a "gestalt" strategy in men (spatial and visual processing). Jordan et al. (2002) hypothesized that women might use a more "analytical" strategy, e.g. that women place more effort on the mental rotation tasks by picturing the figures in their mind. In this context it may be that women rely more on spatial working memory in such tasks than men. But in none of these studies the used individual strategies were explicitly measured (e.g., by asking the participants which strategy they have used). Besides the known gender-typical strategies strong interindividual differences in problem solving strategies could serve as one reason for the inconsistent results. Measuring strategies not only with questionnaires but for example using eye movements could be one step in uncovering the varieties between the studies (de'Sperati, 2003). Another possible source of variance is the used spatial reference system. At present it is not known whether men or women differ in terms of the applied references system during mental rotation.

A further point, which has not been looked upon in detail, is how and whether sexual hormones might modulate spatial cognition. Behavioural studies have clearly shown that different concentration levels of sexual hormones substantially influence spatial cognition (Sherwin, 2003). As mentioned above oestrogen could act via vascular and/or neuronal pathways

(Dietrich et al. 2001). Oestrogen as an vasodilator may act on the endothelial nitric oxide synthase (eNOS) which produces the vasodilatory molecule Nitric oxide (NO). Sexual hormones are also known to act on various neurotransmitter systems, like the serotonin-pathway, the catecholaminergic, the cholinergic and the GABA-pathway via genomic and nongenomic actions (Sherwin, 2003). According to the progesterone-decoupling hypothesis increasing progesterone concentration in the luteal phase of the menstrual cycle decreases non-NMDA and increases GABA receptor activation. These effects lead, via a decrease of transcallosal neuronal activation, to hemispheric decoupling, which then results in lesser functional asymmetries (Hausmann and Gunturkun, 2000). Hausmann et al. (2000) showed that a lesser degree of functional asymmetry could be associated with a decrease of the performance in a figural comparison task. Even though the dimension of the influence of sexual hormones on cognition is under discussion, it is know clear, that testing women with variable levels of sexual hormones could enlarge the variance in the behavioural as well was in the hemodynamic data. Thus, the search for gender specific activation pattern during mental rotation is still ongoing.

5. INDIVIDUAL STRATEGIES

As mentioned above individual differences to solve the mental rotation problem is a matter of intensive discussion. Kosslyn et al. (1998) supposed two different strategies, which can be called up in mental rotation: an *internal* and an *external* strategy. Using the *internal* strategy the subject would anticipate what one would see if he/she were to physically manipulate the objects. Thus, the internal strategy would rely on the participation of motor control processes. Using an *external* strategy the subject would visualize the consequences of someone else or an external force moving the object. Thus, there would be no participation of motor control centres during mental rotation. In fact several studies have shown M1 and/or premotor cortex activation during mental rotation. Most of the brain imaging studies reporting activations in motor areas during mental rotation (indicative of internal strategies) have asked the subjects to mentally rotate drawing of body parts, their own hands, hands of others, or tools (Kosslyn et al., 1998; Vingerhoets et al., 2002; Wraga et al., 2003 and Seurinck et al., 2004). Thus, it is thought that mentally rotating these objects automatically evoke egocentric strategies. In this context egocentric refers to the application of an egocentric frame of reference for both action and space. During mental rotation the subject imagines mental rotation from the first-person-perspective and experiences oneself as the agent of mental rotation. That this is the case has been shown in an elegant experiment published by Wraga et al. (2003). In this study two groups of subjects were tested. In one group the subjects performed mental rotations of pictures of hands followed by mental rotations of the classical Shepard-Metzler objects (*Hand-Object* group). The other group performed mental rotations of the Shepard-Metzler objects in both conditions (*Object-Object)*. It was found that the motor areas (M1 and premotor areas) were active during mental rotation of Shepard-Metzler figures only in the *Hand-Object* group supporting the idea that the implicitly evoked egocentric strategy during mental rotation of pictures of hands was transferred to the mental rotation of the Shepard-Metzler figures.

However, although mental rotation of body parts and objects might implicitly evoke egocentric strategies, it is also possible that egocentric strategies are used as a kind of individual strategy for all kinds of objects. This might explain that some studies have found motor and/or premotor activations even during mental rotation of non-body objects like the Shepard-Metzler figures. A typical example is the study of Cohen et al. (1996) who found that half of the studied subjects revealed activation in motor areas during mental rotation of the classical Shepard-Metzler objects. Interestingly, when the subjects are explicitly instructed to mentally rotate these objects by imagining that their dominant hand rotates the objects, strong activations were found in the motor areas (Kosslyn et al., 1998). As described in previous parts of this it has been argued that men might apply a more holistic strategy while women would apply a piecemeal strategy. However, these strategies have not been substantiated in the context of brain imaging strategies. Interestingly Ark and colleagues (2003) varied the task instructions and marked the classical 3D-cubes with features to examine differences between a possible "mental rotation / holistic strategy and a "feature / analytic strategy". For the precuneus they found a stronger right than left activation when subjects used the "mental rotation / holistic" strategy while a stronger left-sided activation was found for the "feature / analytic" strategy.

Mathematically gifted adolescents also seem to use different strategies than matched controls to solve the mental rotation task. Compared to the control subjects mathematically gifted adolescents additionally activated the right anterior cingulate, the left superior temporal gyrus and the left premotor cortex (O'Boyle et al., 2005). According to O'Boyle et al. these regions represent components of a larger anterior neural system thought to mediate a variety of cognitive control processes relating to general intelligence including spatial attention, working memory, the parsing of executive processes into strategic and evaluative functions, error detection, conflict resolution and the online monitoring of performance. Taken together it is meanwhile accepted that different cognitive strategies are used to solve the mental rotation task depending on the kind of stimuli, the complexity of the task, the used reference system, the individual experience and gender.

REFERENCES

Ark, W. S., Stiles, J., and Hollan, J. (2003). *Ventral Temporal Processing Further Distinguishes Mental Rotation and Feature Matching.* Paper presented at the 9th International Conference on Functional Mapping of the Human Brain, New York.

Banati, R. B., Goerres, G. W., Tjoa, C., Aggleton, J. P., and Grasby, P., 2000, The functional anatomy of visual-tactile integration in man: a study using positron emission tomography, *Neuropsychologia* **38**(2):115-124.

Barnes, J., Howard, R. J., Senior, C., Brammer, M., Bullmore, E. T., Simmons, A., Woodruff, P., and David, A. S., 2000, Cortical activity during rotational and linear transformations, *Neuropsychologia* **38**(8):1148-1156.

Binkofski, F., Buccino, G., Posse, S., Seitz, R. J., Rizzolatti, G., and Freund, H., 1999, A fronto-parietal circuit for object manipulation in man: evidence from an fMRI-study, *Eur J Neurosci* **11**(9):3276-3286.

Bonda, E., Petrides, M., Frey, S., and Evans, A., 1995, Neural correlates of mental transformations of the body-in-space, *Proc Natl Acad Sci U S A* **92**(24):11180-11184.

Bosch, V., 2000, Statistical analysis of multi-subject fMRI data: assessment of focal activations, *J Magn Reson Imaging* **11**(1):61-64.

Bryden, M. P., Hecaen, H., and DeAgostini, M., 1983, Patterns of cerebral organization, *Brain Lang* **20**(2):249-262.

Buccino, G., Lui, F., Canessa, N., Patteri, I., Lagravinese, G., Benuzzi, F., Porro, C. A., and Rizzolatti, G., 2004, Neural circuits involved in the recognition of actions performed by nonconspecifics: an FMRI study, *J Cogn Neurosci* **16**(1):114-126.

Carpenter, P. A., Just, M. A., Keller, T. A., Eddy, W., and Thulborn, K., 1999a, Graded functional activation in the visuospatial system with the amount of task demand, *J Cogn Neurosci* **11**(1):9-24.

Carpenter, P. A., Just, M. A., Keller, T. A., Eddy, W. F., and Thulborn, K. R., 1999b, Time course of fMRI-activation in language and spatial networks during sentence comprehension, *Neuroimage* **10**(2):216-224.

Cohen, M. S., Kosslyn, S. M., Breiter, H. C., DiGirolamo, G. J., Thompson, W. L., Anderson, A. K., Brookheimer, S. Y., Rosen, B. R., and Belliveau, J. W., 1996, Changes in cortical activity during mental rotation. A mapping study using functional MRI, *Brain* **119 (Pt 1)**:89-100.

Cohen-Bendahan, C. C., van de Beek, C., and Berenbaum, S. A., 2005, Prenatal sex hormone effects on child and adult sex-typed behavior: methods and findings, *Neurosci Biobehav Rev* **29**(2):353-384.

Corballis, M. C., 1986, Is mental rotation controlled or automatic?, *Mem Cognit* **14**(2): 124-128.

Corballis, M. C., Nagourney, B. A., Shetzer, L. I., and Stefanatos, G., 1978, Mental rotation under head tilt: factors influencing the location of the subjective reference frame, *Percept Psychophys* **24**(3):263-273.

Corballis, M. C., and Zalik, M. C., 1977, Why do children confuse mirror-image obliques?, *J Exp Child Psychol* **24**(3):516-523.

Coull, J. T., 1998, Neural correlates of attention and arousal: insights from electrophysiology, functional neuroimaging and psychopharmacology, *Prog Neurobiol* **55**(4):343-361.

Creem, S. H., Downs, T. H., Wraga, M., Harrington, G. S., Proffitt, D. R., and Downs, J. H., 3rd, 2001, An fMRI study of imagined self-rotation, *Cogn Affect Behav Neurosci* **1**(3):239-249.

Davidson, R. J., and Irwin, W., 1999, The functional neuroanatomy of emotion and affective style, *Trends Cogn Sci* **3**(1):11-21.

de Lange, F. P., Hagoort, P., and Toni, I., 2005, Neural topography and content of movement representations, *J Cogn Neurosci* **17**(1):97-112.

de'Sperati, C., 2003, Precise oculomotor correlates of visuospatial mental rotation and circular motion imagery, *J Cogn Neurosci* **15**(8):1244-1259.

Deiber, M. P., Ibanez, V., Honda, M., Sadato, N., Raman, R., and Hallett, M., 1998, Cerebral processes related to visuomotor imagery and generation of simple finger movements studied with positron emission tomography, *Neuroimage* **7**(2):73-85.

Dietrich, T., Krings, T., Neulen, J., Willmes, K., Erberich, S., Thron, A., and Sturm, W., 2001, Effects of blood estrogen level on cortical activation patterns during cognitive activation as measured by functional MRI, *Neuroimage* **13**(3):425-432.

Ditunno, P. L., and Mann, V. A., 1990, Right hemisphere specialization for mental rotation in normals and brain damaged subjects, *Cortex* **26**(2):177-188.

Diwadkar, V. A., Carpenter, P. A., and Just, M. A., 2000, Collaborative activity between parietal and dorso-lateral prefrontal cortex in dynamic spatial working memory revealed by fMRI, *Neuroimage* **12**(1):85-99.

Faillenot, I., Sakata, H., Costes, N., Decety, J., and Jeannerod, M., 1997, Visual working memory for shape and 3D-orientation: a PET study, *Neuroreport* **8**(4):859-862.

Faillenot, I., Toni, I., Decety, J., Gregoire, M. C., and Jeannerod, M., 1997, Visual pathways for object-oriented action and object recognition: functional anatomy with PET, *Cereb Cortex* **7**(1):77-85.

Gauthier, I., Hayward, W. G., Tarr, M. J., Anderson, A. W., Skudlarski, P., and Gore, J. C., 2002, BOLD activity during mental rotation and viewpoint-dependent object recognition, *Neuron* **34**(1):161-171.

Georgopoulos, A. P., 2000, Cognition: mental rotation, *Am J Psychiatry* **157**(5):695.

Georgopoulos, A. P., Lurito, J. T., Petrides, M., Schwartz, A. B., and Massey, J. T., 1989, Mental rotation of the neuronal population vector, *Science* **243**(4888):234-236.

Golledge, R. G., and Stimson, R. J., 1997, *Spatial behavior: a geographic perspective.*Guilford, New York.

Goodale, M. A., and Milner, A. D., 1992, Separate visual pathways for perception and action, *Trends Neurosci* **15**(1):20-25.

Halari, R., Sharma, T., Hines, M., Andrew, C., Simmons, A., and Kumari, V., 2005, Comparable fMRI activity with differential behavioural performance on mental rotation and overt verbal fluency tasks in healthy men and women, *Exp Brain Res*:1-14.

Halpern, D. F., 2000, *Sex differences in cognitive abilities.*Lawrence Erlbaum Associates, Publishers., Mahawah, New Jersey, London:.

Hanakawa, T., Honda, M., Sawamoto, N., Okada, T., Yonekura, Y., Fukuyama, H., and Shibasaki, H., 2002, The role of rostral Brodmann area 6 in mental-operation tasks: an integrative neuroimaging approach, *Cereb Cortex* **12**(11):1157-1170.

Harris, I. M., Egan, G. F., Sonkkila, C., Tochon-Danguy, H. J., Paxinos, G., and Watson, J. D., 2000, Selective right parietal lobe activation during mental rotation: a parametric PET study, *Brain* **123**:65-73.

Harris, I. M., and Miniussi, C., 2003, Parietal lobe contribution to mental rotation demonstrated with rTMS, *J Cogn Neurosci* **15**(3):315-323.

Hausmann, M., and Gunturkun, O., 2000, Steroid fluctuations modify functional cerebral asymmetries: the hypothesis of progesterone-mediated interhemispheric decoupling, *Neuropsychologia* **38**(10):1362-1374.

Hausmann, M., Slabbekoorn, D., Van Goozen, S. H., Cohen-Kettenis, P. T., and Gunturkun, O., 2000, Sex hormones affect spatial abilities during the menstrual cycle, *Behav Neurosci* **114**(6):1245-1250.

Heil, M., Rauch, M., and Hennighausen, E., 1998, Response preparation begins before mental rotation is finished: evidence from event-related brain potentials, *Acta Psychol (Amst)* **99**(2):217-232.

Ide, A., Rodriguez, E., Zaidel, E., and Aboitiz, F., 1996, Bifurcation patterns in the human sylvian fissure: hemispheric and sex differences, *Cereb Cortex* **6**(5):717-725.

Jancke, L., Kleinschmidt, A., Mirzazade, S., Shah, N. J., and Freund, H. J., 2001, The role of the inferior parietal cortex in linking the tactile perception and manual construction of object shapes, *Cereb Cortex* **11**(2):114-121.

Jancke, L., Schlaug, G., Huang, Y., and Steinmetz, H., 1994, Asymmetry of the planum parietale, *Neuroreport* **5**(9):1161-1163.

Jeannerod, M., 1995, Mental imagery in the motor context, *Neuropsychologia* **33**(11): 1419-1432.

Jeannerod, M., 2001, Neural simulation of action: a unifying mechanism for motor cognition, *Neuroimage* **14**(1 Pt 2):S103-109.

Jones, C. M., Braithwaite, V. A., and Healy, S. D., 2003, The evolution of sex differences in spatial ability, *Behav Neurosci* **117**(3):403-411.

Jordan, K., Heinze, H. J., Lutz, K., Kanowski, M., and Jancke, L., 2001, Cortical activations during the mental rotation of different visual objects, *Neuroimage* **13**(1):143-152.

Jordan, K., Wustenberg, T., Heinze, H. J., Peters, M., and Jancke, L., 2002, Women and men exhibit different cortical activation patterns during mental rotation tasks, *Neuropsychologia* **40**(13):2397-2408.

Just, M. A., Carpenter, P. A., Maguire, M., Diwadkar, V., and McMains, S., 2001, Mental rotation of objects retrieved from memory: a functional MRI study of spatial processing, *J Exp Psychol Gen* **130**(3):493-504.

Kanwisher, N., Chun, M. M., McDermott, J., and Ledden, P. J., 1996, Functional imagining of human visual recognition, *Brain Res Cogn Brain Res* **5**(1-2):55-67.

Kanwisher, N., McDermott, J., and Chun, M. M., 1997, The fusiform face area: a module in human extrastriate cortex specialized for face perception, *J Neurosci* **17**(11):4302-4311.

Kimura, D., 1999, *Sex and Cognition*.MIT Press., Cambridge, Massachusetts, London, England.

Koshino, H., Carpenter, P. A., Keller, T. A., and Just, M. A., 2005, Interactions between the dorsal and the ventral pathways in mental rotation: an fMRI study, *Cogn Affect Behav Neurosci* **5**(1):54-66.

Kosslyn, S. M., DiGirolamo, G. J., Thompson, W. L., and Alpert, N. M., 1998, Mental rotation of objects versus hands: neural mechanisms revealed by positron emission tomography, *Psychophysiology* **35**(2):151-161.

Kosslyn, S. M., Thompson, W. L., Wraga, M., and Alpert, N. M., 2001, Imagining rotation by endogenous versus exogenous forces: distinct neural mechanisms, *Neuroreport* **12**(11):2519-2525.

Linn, M. C., and Petersen, A. C., 1985, Emergence and characterization of sex differences in spatial ability: a meta-analysis, *Child Dev* **56**(6):1479-1498.

Lohman, D. F. (1979). *Spatial ability: A review and reanalysis of the correlational literature.* Stanford, CA: Stanford University.

Longcamp, M., Anton, J. L., Roth, M., and Velay, J. L., 2005, Premotor activations in response to visually presented single letters depend on the hand used to write: a study on left-handers, *Neuropsychologia* **43**(12):1801-1809.

Luders, E., Narr, K. L., Thompson, P. M., Rex, D. E., Jancke, L., Steinmetz, H., and Toga, A. W., 2004, Gender differences in cortical complexity, *Nat Neurosci* **7**(8):799-800.

Maki, P. M., and Resnick, S. M., 2001, Effects of estrogen on patterns of brain activity at rest and during cognitive activity: a review of neuroimaging studies, *Neuroimage* **14**(4): 789-801.

Mattingley, J. B., Corben, L. A., Bradshaw, J. L., Bradshaw, J. A., Phillips, J. G., and Horne, M. K., 1998, The effects of competition and motor reprogramming on visuomotor selection in unilateral neglect, *Exp Brain Res* **120**(2):243-256.

Mattingley, J. B., Husain, M., Rorden, C., Kennard, C., and Driver, J., 1998, Motor role of human inferior parietal lobe revealed in unilateral neglect patients, *Nature* **392**(6672): 179-182.

Merboldt, K. D., Baudewig, J., Treue, S., and Frahm, J., 2002, Functional MRI of self-controlled stereoscopic depth perception, *Neuroreport* **13**(14):1721-1725.

Muthukumaraswamy, S. D., Johnson, B. W., and Hamm, J. P., 2003, A high density ERP comparison of mental rotation and mental size transformation, *Brain Cogn* **52**(2):271-280.

Newcombe, N., Dubas, J. S., and Baenninger, M. A., 1989, Associations of timing of puberty, spatial ability, and lateralization in adult women, *Child Dev* **60**(1):246-254.

O'Boyle, M. W., Cunnington, R., Silk, T. J., Vaughan, D., Jackson, G., Syngeniotis, A., and Egan, G. F., 2005, Mathematically gifted male adolescents activate a unique brain network during mental rotation, *Brain Res Cogn Brain Res* **25**(2):583-587.

Parsons, L. M., 1994, Temporal and kinematic properties of motor behavior reflected in mentally simulated action, *J Exp Psychol Hum Percept Perform* **20**(4):709-730.

Parsons, L. M., 2003, Superior parietal cortices and varieties of mental rotation, *Trends Cogn Sci* **7**(12):515-517.

Peters, M., 1995, Some Factors of Importance in Mental Relation Performance, *Brain Cogn* **28**(1):104-105.

Peters, M., Laeng, B., Latham, K., Jackson, M., Zaiyouna, R., and Richardson, C., 1995, A Redrawn Vandenberg and Kuse Mental Rotations Test - Different Versions and Factors That Affect Performance, *Brain Cogn* **28**(1):39-58.

Podzebenko, K., Egan, G. F., and Watson, J. D., 2002, Widespread dorsal stream activation during a parametric mental rotation task, revealed with functional magnetic resonance imaging, *Neuroimage* **15**(3):547-558.

Quaiser-Pohl, C., and Lehmann, W., 2002, Girls' spatial abilities: charting the contributions of experiences and attitudes in different academic groups, *Br J Educ Psychol* **72**(Pt 2): 245-260.

Rabinowicz, T., Dean, D. E., Petetot, J. M., and de Courten-Myers, G. M., 1999, Gender differences in the human cerebral cortex: more neurons in males; more processes in females, *J Child Neurol* **14**(2):98-107.

Rabinowicz, T., Petetot, J. M., Gartside, P. S., Sheyn, D., Sheyn, T., and de, C. M., 2002, Structure of the cerebral cortex in men and women, *J Neuropathol Exp Neurol* **61**(1): 46-57.

Richter, W., Somorjai, R., Summers, R., Jarmasz, M., Menon, R. S., Gati, J. S., Georgopoulos, A. P., Tegeler, C., Ugurbil, K., and Kim, S. G., 2000, Motor area activity during mental rotation studied by time-resolved single-trial fMRI, *J Cogn Neurosci* **12**(2):310-320.

Richter, W., Ugurbil, K., Georgopoulos, A., and Kim, S. G., 1997, Time-resolved fMRI of mental rotation, *Neuroreport* **8**(17):3697-3702.

Roberts, J. E., and Bell, M. A., 2003, Two- and three-dimensional mental rotation tasks lead to different parietal laterality for men and women, *Int J Psychophysiol* **50**(3):235-246.
Rosler, F., Heil, M., Bajric, J., Pauls, A. C., and Hennighausen, E., 1995, Patterns of cerebral activation while mental images are rotated and changed in size, *Psychophysiology* **32**(2):135-149.
Rosler, F., Roder, B., Heil, M., and Hennighausen, E., 1993, Topographic differences of slow event-related brain potentials in blind and sighted adult human subjects during haptic mental rotation, *Brain Res Cogn Brain Res* **1**(3):145-159.
Sakata, H., Taira, M., Kusunoki, M., Murata, A., and Tanaka, Y., 1997, The TINS Lecture. The parietal association cortex in depth perception and visual control of hand action, *Trends Neurosci* **20**(8):350-357.
Schubert, T., von Cramon, D. Y., Niendorf, T., Pollmann, S., and Bublak, P., 1998, Cortical areas and the control of self-determined finger movements: an fMRI study, *Neuroreport* **9**(14):3171-3176.
Seitz, R. J., Canavan, A. G., Yaguez, L., Herzog, H., Tellmann, L., Knorr, U., Huang, Y., and Homberg, V., 1997, Representations of graphomotor trajectories in the human parietal cortex: evidence for controlled processing and automatic performance, *Eur J Neurosci* **9**(2):378-389.
Seurinck, R., Vingerhoets, G., de Lange, F. P., and Achten, E., 2004, Does egocentric mental rotation elicit sex differences?, *Neuroimage* **23**(4):1440-1449.
Seurinck, R., Vingerhoets, G., Vandemaele, P., Deblaere, K., and Achten, E., 2005, Trial pacing in mental rotation tasks, *Neuroimage* **25**(4):1187-1196.
Shepard, R. N., and Metzler, J., 1971, Mental rotation of three-dimensional objects, *Science* **171**(972):701-703.
Sherwin, B. B., 2003, Estrogen and cognitive functioning in women, *Endocr Rev* **24**(2): 133-151.
Suchan, B., Botko, R., Gizewski, E., Forsting, M., and Daum, I., 2005, Neural substrates of manipulation in visuospatial working memory, *Neuroscience*.
Tagaris, G. A., Kim, S. G., Strupp, J. P., Andersen, P., Ugurbil, K., and Georgopoulos, A. P., 1996, Quantitative relations between parietal activation and performance in mental rotation, *Neuroreport* **7**(3):773-776.
Tagaris, G. A., Richter, W., Kim, S. G., Pellizzer, G., Andersen, P., Ugurbil, K., and Georgopoulos, A. P., 1998, Functional magnetic resonance imaging of mental rotation and memory scanning: a multidimensional scaling analysis of brain activation patterns, *Brain Res Brain Res Rev* **26**(2-3):106-112.
Thomsen, T., Hugdahl, K., Ersland, L., Barndon, R., Lundervold, A., Smievoll, A. I., Roscher, B. E., and Sundberg, H., 2000, Functional magnetic resonance imaging (fMRI) study of sex differences in a mental rotation task, *Med Sci Monit* **6**(6):1186-1196.
Thurstone, L. L., 1938, *Primary mental abilities*.
Tomasino, B., and Rumiati, R. I., 2004, Effects of strategies on mental rotation and hemispheric lateralization: neuropsychological evidence, *J Cogn Neurosci* **16**(5):878-888.
Tsao, D. Y., Vanduffel, W., Sasaki, Y., Fize, D., Knutsen, T. A., Mandeville, J. B., Wald, L. L., Dale, A. M., Rosen, B. R., Van Essen, D. C., Livingstone, M. S., Orban, G. A., and Tootell, R. B., 2003, Stereopsis activates V3A and caudal intraparietal areas in macaques and humans, *Neuron* **39**(3):555-568.
Unterrainer, J., Wranek, U., Staffen, W., Gruber, T., and Ladurner, G., 2000, Lateralized cognitive visuospatial processing: is it primarily gender-related or due to quality of performance? A HMPAO-SPECT study, *Neuropsychobiology* **41**(2):95-101.

Vanrie, J., Beatse, E., Wagemans, J., Sunaert, S., and Van Hecke, P., 2002, Mental rotation versus invariant features in object perception from different viewpoints: an fMRI study, *Neuropsychologia* **40**(7):917-930.

Vingerhoets, G., de Lange, F. P., Vandemaele, P., Deblaere, K., and Achten, E., 2002, Motor imagery in mental rotation: an fMRI study, *Neuroimage* **17**(3):1623-1633.

Vingerhoets, G., Santens, P., Van Laere, K., Lahorte, P., Dierckx, R. A., and De Reuck, J., 2001, Regional brain activity during different paradigms of mental rotation in healthy volunteers: a positron emission tomography study, *Neuroimage* **13**(2):381-391.

Voyer, D., 1998, On the reliability and validity of noninvasive laterality measures, *Brain Cogn* **36**(2):209-236.

Voyer, D., and Bryden, M. P., 1990, Gender, level of spatial ability, and lateralization of mental rotation, *Brain Cogn* **13**(1):18-29.

Voyer, D., Voyer, S., and Bryden, M. P., 1995, Magnitude of sex differences in spatial abilities: a meta-analysis and consideration of critical variables, *Psychol Bull* **117**(2): 250-270.

Weiss, E., Siedentopf, C. M., Hofer, A., Deisenhammer, E. A., Hoptman, M. J., Kremser, C., Golaszewski, S., Felber, S., Fleischhacker, W. W., and Delazer, M., 2003, Sex differences in brain activation pattern during a visuospatial cognitive task: a functional magnetic resonance imaging study in healthy volunteers, *Neurosci Lett* **344**(3):169-172.

Wexler, M., Kosslyn, S. M., and Berthoz, A., 1998, Motor processes in mental rotation, *Cognition* **68**(1):77-94.

Windischberger, C., Barth, M., Lamm, C., Schroeder, L., Bauer, H., Gur, R. C., and Moser, E., 2003, Fuzzy cluster analysis of high-field functional MRI data, *Artif Intell Med* **29**(3):203-223.

Wolbers, T., Weiller, C., and Buchel, C., 2003, Contralateral coding of imagined body parts in the superior parietal lobe, *Cereb Cortex* **13**(4):392-399.

Wraga, M., Shephard, J. M., Church, J. A., Inati, S., and Kosslyn, S. M., 2005, Imagined rotations of self versus objects: an fMRI study, *Neuropsychologia* **43**(9):1351-1361.

Wraga, M., Thompson, W. L., Alpert, N. M., and Kosslyn, S. M., 2003, Implicit transfer of motor strategies in mental rotation, *Brain Cogn* **52**(2):135-143.

Yoshino, A., Inoue, M., and Suzuki, A., 2000, A topographic electrophysiologic study of mental rotation, *Brain Res Cogn Brain Res* **9**(2):121-124.

Zacks, J. M., Gilliam, F., and Ojemann, J. G., 2003, Selective disturbance of mental rotation by cortical stimulation, *Neuropsychologia* **41**(12):1659-1667.

Chapter 13

SPATIAL ORIENTATION AND NAVIGATION IN MICROGRAVITY

Charles M. Oman
Man Vehicle Laboratory, Massachusetts Institute of Technology, Cambridge, USA

Abstract: This chapter summarizes the spatial disorientation problems and navigation difficulties described by astronauts and cosmonauts, and relates them to research findings on orientation and navigation in humans and animals. Spacecraft crew are uniquely free to float in any relative orientation with respect to the cabin, and experience no vestibular and haptic cues that directly indicate the direction of "down". They frequently traverse areas with inconsistently aligned visual vertical cues. As a result, most experience "Visual Reorientation Illusions" (VRIs) where the spacecraft floors, walls and ceiling surfaces exchange subjective identities. The illusion apparently results from a sudden reorientation of the observer's allocentric reference frame. Normally this frame realigns to local interior surfaces, but in some cases it can jump to the Earth beyond, as with "Inversion Illusions" and EVA height vertigo. These perceptual illusions make it difficult for crew to maintain a veridical perception of orientation and place within the spacecraft, make them more reliant upon landmark and route strategies for 3D navigation, and can trigger space motion sickness. This chapter distinguishes VRIs and Inversion Illusions, based on firsthand descriptions from Vostok, Apollo, Skylab, Mir, Shuttle and International Space Station crew. Theories on human "gravireceptor" and "idiotropic" biases, visual "frame" and "polarity" cues, top-down processing effects on object orientation perception, mental rotation and "direction vertigo" are discussed and related to animal experiments on limbic head direction and place cell responses. It is argued that the exchange in perceived surface identity characteristic of human VRIs is caused by a reorientation of the unseen allocentric navigation plane used by CNS mechanisms coding place and direction, as evidenced in the animal models. Human VRI susceptibility continues even on long flights, perhaps because our orientation and navigation mechanisms evolved to principally support 2D navigation.

Key words: vision; vestibular; spatial disorientation; navigation; inversion illusion; visual reorientation illusion; spacecraft architecture; head direction cells; place cells; height vertigo.

1. INTRODUCTION

In our normal lives on Earth, gravity furnishes a ubiquitous sensory cue that helps us keep the various self- and world-fixed coordinate frames we use for spatial perception, imagery, and actions in proper registration. We naturally locomote on two dimensional surfaces in a gravitationally upright orientation. Arguably the human nervous system has become somewhat specialized for terrestrial conditions, since – as reviewed in this chapter - astronauts and cosmonauts regularly experience occasional three dimensional orientation and navigation problems while in weightlessness, even long after the initial two to thee day period of susceptibility to space motion sickness has passed. The routine orientation and navigation problems experienced by astronauts probably have more to do with CNS spatial processing, imagery and perception – the central themes of this book – than they do with adaptation in the vestibular end organs or changes in vestibulo-ocular or vestibulo-spinal reflexes.

Dozens of crewmembers have served as subjects in various neurovestibular experiments in orbit. After their missions, all crewmembers are routinely debriefed on their operational experiences. However, only few are questioned in detail about orientation and navigation problems, and some are reluctant to raise the issue. Inevitably much of what is currently known is based on anecdotal but detailed descriptions provided by several dozen crewmembers, many of them scientist astronauts. Weightlessness is a unique environment. Though the reports are anecdotal, there is a great deal that can be learned from them that is of interest to neuroscientists. However some are unpublished, and others are scattered across the scientific and popular literature. My purpose in writing this chapter is to assemble and interpret them, including where possible many direct quotes, though preserving anonymity when required.

The organization of this chapter is straightforward: First, the two principal illusions of weightlessness - the "Visual Reorientation Illusion" (VRI) and the "Inversion Illusion"- are described. Next, related extravehicular activity (EVA, or spacewalking) disorientation, height vertigo and 3D navigation problems are discussed. The final sections review several theories and experiments that provide insight into visual and body axis spatial orientation cue interaction, and the mechanisms of reorientation and 3D navigation. Based on evidence in animal models, it is argued that the exchange in subjective identity of floors, ceilings, and walls one of the unique hallmarks of a Visual Reorientation Illusion - occurs when the CNS navigation reference plane coding azimuth and place erroneously aligns with the wrong spacecraft surface, due to the absence of gravity. Continued susceptibility to VRIs reflects the terrestrial heritage of human orientation, re-orientation and navigation mechanisms. Nonetheless, the crew reports and research reviewed here suggests ways to further reduce spatial orientation and navigation problems through improved spacecraft design and virtual reality based crew training.

1.1 Visual Reorientation Illusions

When an astronaut floats within the cabin of an orbiting spacecraft, the notion of a "gravitational down" is meaningless. Crew typically speak of the "visual down" reference defined by the orientation of surrounding wall, ceiling and floor surfaces, typically comprised of labeled racks and panels, readily recognizable from prior experience in ground simulators. In order to know which way to look or reach for remembered objects, or to move about in the cabin, astronauts must visually recognize landmark objects and surfaces, and correctly infer their self-orientation with respect to the cabin. Normally this process is automatic and effortless when they work with their feet oriented towards the familiar cabin floor. However, Skylab (Cooper, 1976; Johnston and Dietlein, 1977) and Spacelab astronauts (Oman et al., 1984, 1986, 87) reported that when moving about the cabin they frequently experienced disorientation. Two of the most common situations were when working upside down (relative to their normal 1-G orientation in training), or when floating right side up but viewing another crewmember floating upside down in the cabin. In either case, crew often experienced the striking illusion that the surrounding walls, ceiling, and floors had somehow exchanged identities. In the first situation, whichever surface was closest to their feet seemed like a generic floor. Surfaces approximately parallel to their body now seemed like walls, and overhead surfaces were perceived as ceilings. In the second situation, the orientation of the inverted crewmember determined the direction to the "floor". In both cases, it was as if an internal mental coordinate frame responsible for perception of surface identity had rotated into a new orientation determined by available visual cues. Since the crew often felt "right side up" after such visual reorientations, we (Oman et al, 1987) termed these phenomena "Visual Reorientation Illusions" (VRI), in order to distinguish them from the less commonly experienced "Inversion Illusion", detailed in Sect 2.2, wherein crew always feel continuously gravitationally upside down.

Sometimes the reorientation illusions were subtle, and crew were not aware of them till they reached or looked for a remembered object, or turned in the wrong direction. More often, the change in orientation perception was dramatic. One Skylab crewmember described it this way: "It was a strange sensation. You see brand-new things…It's really like a whole new room that you walk into…with the lights underneath your feet, and it's just an amazing situation to find yourself in". Another noted "All one has to do is to rotate one's body to [a new] orientation and whammo! What one thinks is up *is* up". "It's a feeling as though one could take this whole room and, by pushing a button, just rotate it around so that the ceiling up here would be the floor. It's a marvelous feeling of power over space – over the space around one" (Cooper, 1976). A third said: "Being upside down in the wardroom made it look like a different room than the one we were used to. After rotating back to approximately 45 degrees or so of the attitude which we normally called "up", the attitude in which we had trained, there was a sharp transition in the mind from a room which was sort of familiar to one which was intimately familiar... "We observed this phenomenon throughout the whole flight." Another commented: "I can move into a given room sideways or upside down and not recognize it. You would tend to get locked into one frame of reference. When you rotated your body to another one, it took a little time for the transition to occur" (Johnston and Deitlein, 1977).

Areas of Skylab that had locally incongruent visual vertical cues also triggered VRIs, depending on where the astronaut was working or directed their visual attention. For example, the Skylab Multiple Docking Adapter (MDA) tunnel had a cylindrical interior, and control panels to operate telescopes and other systems mounted on the walls in a variety of different orientations. The MDA was deliberately designed this way because it provided an efficient use of wall space, and to determine whether crews could get along without a single visual vertical (Cooper, 1976). The almost unanimous verdict was that crews disliked working there. One said: "It is one of the biggest mysteries in the world when you go in there to find something." Another commented "There's been some thought about mounting some furniture on the floor, some on the walls, some on the ceiling, but this doesn't work out. You tend to orient yourself when you're in a room, even though you're in zero-g, and when you orient yourself, you should find everything is the same." (Johnston and Deitlein, 1977).

In the early 1980s, our MIT laboratory began to develop experiments on vestibular function, spatial disorientation and motion sickness which ultimately flew on four US Shuttle Spacelab missions between 1983 and 1993. The science crew of the first mission included a Skylab astronaut (O. Garriott) who introduced us to these during illusions during repeated intervals of weightlessness on parabolic training flights. At our request, the crews made detailed notes on pocket voice recorders once they reached orbit. They documented for us in considerable detail the numerous circumstances that triggered orientation illusions aboard the Shuttle, emphasizing the previously unrecognized contribution of the illusions in causing space motion sickness. We summarized their reports in a series of papers (Oman et al., 1984, 1986, 1987, 1988). Our crews noted that the change subjective identity of surrounding surfaces was a perceptually distinct event. In this respect VRIs resemble other types of figure-ground illusions, except that what is being reinterpreted is the identity of surrounding surfaces, and implicitly, the viewer's own allocentric orientation with respect to an unseen environment beyond. VRIs typically occur spontaneously, but as with figure ground illusions, onset depends on visual attention and is therefore under cognitive control. One commented: "If you really want a surface to be "down", you can just look at it and decide that it is". Architectural symmetries and prior visual experience were important predisposing factors. For example, they noted that they frequently experienced VRIs when in the Shuttle flight deck, while the mid-deck beneath, or in the tunnel connecting the mid-deck to the Spacelab laboratory module. However, in the laboratory module, it usually required deliberate effort to make the ceiling and floor reverse, and making a wall seem like a compelling ceiling or floor was even more difficult. The crew noted that the mid-deck and tunnel areas had strong architectural symmetries, and that the science members of the crew had they received far less preflight training in the flight deck, mid-deck, and tunnel than the Spacelab laboratory module. They were intimately familiar the arrangement of the laboratory interior from two years of ground training in a high fidelity mockup. The implication was that visual vertical and surface identity cues are not entirely physically intrinsic, but depend on prior visual experience and familiarity with the spatial layout. Usually the only way to spontaneously experience a VRI while in the Spacelab laboratory was to float with feet towards the ceiling, or view a crewmember who was working that way.

Views of the Earth through the windows also provided powerful orienting cues. One of our Spacelab crewmembers commented: "Generally the visual verticals [in the laboratory module] kept me upright and oriented, and if I were to go and look out the window generally I would move myself around so that the Earth was down below me just so it was easier to see and understand where we were. If I was upside down I would come away from there and for several seconds look around. The first time you think: things are kind of strange and misplaced, like the air lock is sitting on the floor or on the side or something. But as soon as [I] saw one familiar thing like the airlock then I was able to figure out where I was in relation to the Spacelab." This crewmember added that "working on the [inward] slanting panels on the upper half [of the laboratory module walls], let's say you are...pulling out a [stowage] box to get things out, and it wasn't more than a couple of seconds than the [upper panel] would become vertical to me, and I would look down and I'd see the [lower panel] wall.coming out at an angle, slanting in towards me, and it was a very strange sensation."

Figure 13-1. Visual Reorientation Illusions are labile when visual cues to the vertical are ambiguous or conflict. Whether you feel upright or upside down depends on visual attention. Does your interpretation of the photograph change if you look at the face of the blue shirted crewmember ? If you turn both the page and your head upside down ? (NASA photo)

To some degree, non-astronauts can appreciate the VRI phenomenon simply by viewing rotated photographs (e.g. Figures 13-1 – 13-3), but astronauts and parabolic flight participants who have experienced 0-G VRIs firsthand say that when the gravity cue is truly physically absent, and the scene is real, the perceptual change in surface identity is far more distinct and the perceived self-orientation change far more compelling than when simply viewing a photograph.

As detailed in Section 4.6, visual reorientation, path integration and place recognition are the fundamental modes underlying navigation in humans and many animals. In our everyday lives on Earth, our gravireceptive organs provide an absolute vertical orientation reference. Our semicircular canals contribute to our sense of direction, but cannot provide a corresponding absolute azimuth reference. Hence our sense of direction and place ultimately must be updated - reoriented - by visual cues. We all occasionally experience "direction vertigo" (Viguer, 1882; Jonsson, 2002), for example when we emerge from a subway, realize we are not facing in the expected direction, and reorient. However, living on Earth all our visual reorientations can occur only in azimuth since gravity anchors our perceptions of pitch and roll. Our sense of azimuthal direction reorients, but ceilings and floors do not change subjective identity the way they do for astronauts. At most we say the wall we thought faced west actually is actually the one facing east.

The focus of this chapter is on spatial orientation, not motion sickness. However, the early Shuttle crews made the important observation that VRI onset could trigger an immediate increase in nausea and sometimes even cause vomiting during the first several days in weightlessness. We (Oman et al., 1986) noted that VRIs are caused by a sudden change in perceived allocentric orientation, and that this happens without concurrent movement commands or vestibular and proprioceptive cues. Hence one would expect VRIs to be provocative, based on the sensory conflict theory for motion sickness. One Shuttle pilot recalled awakening in his seat, removing the cockpit window shades, seeing the Earth in an unexpected location above rather than below, experiencing a sudden change in spacecraft orientation - and therefore in self orientation - and vomiting moments later. Several Spacelab crew described sudden vomiting episodes after seeing a nearby crewmember floating upside down. One commented "[Early in the mission] I really needed … a good optical "down". It was really distressing when [a second crewmember] came floating into the [Spacelab] module upside down and tumbling and things – that didn't sit too well with my own perception of Spacelab. I felt like I needed a real visual "down", and it was the floor…and I didn't really have one of my own." Subsequent Shuttle crews have noted that after reaching orbit, when the entire crew remove their orange launch/entry space suits and leave them floating about the cabin prior to stowage, the resulting visual environment is extremely disorienting to everyone.

One astronaut deliberately created VRIs to obtain nausea relief: "When I went into the mid-deck..and I didn't feel really well, I knew a method how to get better by vomiting…I [went into the connecting tunnel] and turned around, just to make sure that I didn't know the orientation of Spacelab or the mid-deck, and then I'd close my eyes, [float back into Spacelab], open my eyes, and see something I didn't expect." Several crewmembers suggested that VRIs and space sickness could be reduced by the practique of deliberately ignoring visual landmarks, relaxing, and not trying to control their own orientation. Others noted that belting into a seat or standing up against a bungee harness seemed to reduce spontaneous VRIs and nausea.

It is interesting that Skylab crews had described orientation illusions but had not noted a relationship between them and space sickness, whereas to our subsequent Shuttle crews the causal relationship seemed unequivocal. Possibly this is because two of the three Skylab crews were largely confined to their seats in small Apollo ferry vehicles during the first days in weightlessness, when their susceptibility to space sickness was highest. The Shuttle crews could roam about freely immediately. It seemed clear that although VRIs caused disorientation, their nauseogenic potential was evident primarily during the first days in flight. Because of this, Oman et al (1984) first recommended that during the first several days while Shuttle crew are at risk of space sickness, for the good of all onboard, *all* crewmembers - symptomatic or not – should remain in a locally visually upright orientation.

Subsequent Shuttle crews have followed this dictum, and operationally confirmed its efficacy.

With the exception of a questionnaire study (Kornilova, 1995, 1997) no comparably detailed descriptions of visual orientation illusions have yet appeared in the Soviet and Russian scientific literature. However, cosmonauts and designers of the Salyut, Mir and Russian ISS modules were clearly aware of the importance of providing at least a local visual frame of reference. Most Russian modules have a longitudinal floors, a rectangular interior cross section, ceiling lighting, and brown floors, tan walls and lighter tan or blue ceilings to help establish a locally consistent visual vertical in work areas within a module (Gurovskiy et al., 1980). In the ISS Zvezda module, the floor and ceiling rack labeling is symmetrically oriented about the surface midline. This way labeling on adjacent floors, walls, and ceiling are easily readable by crewmembers working upright. However, the local visual verticals of certain multi-compartmented modules (e.g. the Mir Priroda module) have adjacent work areas with oppositely oriented visual vertical cues.

NASA's 1995 Man Systems Integration Standard (MSIS 3000 Sect 8.4.3) for all future NASA spacecraft mandated that all NASA spacecraft be designed so color, lighting and equipment orientation provided unambiguous visual vertical cues that were consistent. The standard cited both Skylab and Spacelab experience and supporting ground research (e.g. Coss et al., 1989). However the rules were eventually changed for the non-Russian portions of ISS. Some early ISS designs Kitmacher (2002) featured large modules with parallel, "bologna slice" decks with globally congruent visual verticals, reminiscent of Skylab. Ultimately, however NASA designers opted for Shuttle payload bay sized modules with hatches at each end, connected by smaller "node" modules with hatches facing in all six directions. The interior of each NASA module have a square cross section, formed by four rows of superficially similar equipment racks running longitudinally. The front surfaces of these racks form the floors, walls and ceiling. The orientation and labeling of equipment mounted in the racks effectively creates dual visual verticals, oriented 90 degrees apart. One is defined by the equipment mounted on the walls, and the other by equipment on the floor and ceiling. To know which is the "true" visual vertical, the crew has to be familiar with the relative arrangement of specific equipment racks and permanently mounted equipment, or look for larger scale architectural cues. For example a row of lights running longitudinally between the ceiling and the walls outlines the ceiling. When viewed upright, the hatch openings at the ends of the module form a "**U**" rather than an "**∩**", and the text on prominent emergency egress signs around the hatches (e.g. "TO NODE 1") appears upright. Dual local visual verticals were not consistent with the general

MSIS standard, but NASA adopted a special standard for ISS (NASA SSP5005, 1999), which deleted the troublesome requirements. Engineers apparently determined that disorientation due to architectural factors introduced only short-term medical and habitability problems, and that rack commonality and stowage volume efficiency should have priority. Early ISS crews noted the potential problem, and mounted movable equipment (e.g. laptop computers and foot restraints) so they worked upright with respect to the "floor" defined by their experience in ground training modules.

Early ISS astronauts also noted that the hatches in the node modules could sometimes be difficult to distinguish. Also, the Russian modules had smaller openings, necessitating a change in body orientation when transiting. If crews experienced VRIs when entering a node, they reoriented using remembered equipment items or signs as landmarks. One early ISS crewmember described how he had been detailed to mount an emergency egress placard on a US node hatch leading to the docked shuttle. He had a VRI after entering the node, and inadvertently attached the placard to an unused hatch leading to space. Fortunately another crewmember discovered the error.

Although space station crews eventually become intimately familiar with the interior of their spacecraft, it is clear that some degree of VRI susceptibility remains even after many months in orbit. Crews learn to live with the VRI phenomenon, and (as described in Section 3) rely on landmark and route strategies for navigation. One US astronaut who lived on Mir for six months recalled that in some areas where he routinely preferred to work visually upside down, VRIs actually seemed easier to get: "When I rolled upside down, I didn't have to wait till my feet got near the floor before the ceiling became down. It happened even before I reached 90 degrees.". Another US Astronaut who lived for months on ISS astronaut wrote: "There really isn't an up or down anywhere else here, but there is a direction we think of as the floor and a direction we think of as the ceiling in each module. Most of the labeling on panels and equipment is written so that it is right side up assuming this orientation, and also most of the lights are on the "ceiling" so they cast light "downwards." To add to the effect, there is a simulator back on Earth [where] we spent a lot of time in where we got used to one direction as the floor and the opposite direction as the ceiling….This isn't true for the Progress [cargo vehicle]. Since it is just a cargo container, we don't have a simulator that we have trained in on the ground, and since it is spherical there really isn't a flat surface to call the floor. So that means that work inside Progress can be kind of disorienting. This is especially true if doing close-up work on something (say unbolting a piece of equipment). In weightlessness your body may shift position without you realizing it while you are intently working, so that when you pop your head back up after finishing you may find yourself in a totally different orientation than when you started. I recall looking out the hatch and being momentarily surprised to see [another crewmember] in the Service Module running on the treadmill on the ceiling! Actually, it was me that had flipped upside down.... In space you need to remember that you aren't limited like you are on the ground to having your feet on the floor - they can just as easily be on the wall or on the ceiling. I find that when I am working in a tight space, I don't really think about any particular direction as up or down, but when out in an open space like in the middle of a module I do. If for instance I am up on the ceiling, by concentrating I can make myself think of the ceiling as the floor. I really think it is a matter of just familiarity what you call up or down. An example is the area around our weightlifting exercise equipment, which is located on the ceiling of the Node module. I've gotten so used to spending time there in that orientation that I am more comfortable there upside down. I've also gotten used to looking towards the Service Module while I am working out and seeing [my colleague] upside down - or at least the opposite way since from his viewpoint - it is me that is upside down" (Lu, 2005).

1.2 0-G Inversion Illusions

Many astronauts have some flying background and are familiar with the somatogravic illusions of aerobatic flight, including a sensation of flying upside down during an aerobatic pushover due to the associated "eyeballs up" acceleration. (Cheung, 2004) Since the US Shuttle thrusts into orbit in an inverted attitude, crewmembers experience "eyeballs-in and up" acceleration, it is not surprising that crewmembers report feeling upside down during the launch phase.

Immediately after main engine cutoff and the onset of weightlessness, almost all US and Russian crews experience momentary somersaulting sensations, and thereafter frequently feel upside down for a period of time ranging from seconds to several minutes (Gazenko, 1964; Yuganov et al., 1966; Oman et al., 1986). Cosmonaut Titov reported: "the weight vanished as quickly as Vostok separated from the booster...and I felt suddenly as though I were turning a somersault and then flying with my legs up!.... Fortunately the sensation lasted only seconds"(Titov and Caidin, 1962). A similar illusion has been reported at the onset of the weightless phase of parabolic flight (Lackner, 1992). Almost all blindfolded subjects making their first flight experience somersaulting, or sometimes a paradoxical sensation of inversion without pitch. If vision is available, the incidence of the illusion is lower.

Figure 13-2. Persistent 0-G Inversion Illusion – both self and vehicle seem upside down relative to an unseen external gravitational reference frame. The illusion is often reported in the Shuttle mid-deck, where walls of rectangular stowage lockers make up and down ambiguous (NASA photo)

For a small minority of astronauts and cosmonauts, 0-G inversion illusions are more persistent, sometimes lasting for hours, and return sporadically during the first several days in orbit. Once crewmember said: "The only way I can describe it is that though I'm floating upright in the cabin in weightlessness, both the spacecraft and I seem to somehow be flying upside down". Rolling upside down in the cabin does not eliminate the inversion sensation - only the spacecraft seems right side up. A Spacelab crewmember said: " I knew I was standing upright … in the normal way with respect to the orbiter, and nevertheless I felt upside down … despite the fact that everything was normally oriented around me. This gave me the intellectual interpretation that the orbiter was flying upside down….I just interpreted intellectually that the orbiter has to be upside down because you feel upside down and yet you see are the right way up.." Some have reported that the inversion sensation was more noticeable in the visually symmetrical Shuttle mid-deck than when on the flight deck. Certain of the afflicted have found that inversion illusion can be momentarily eliminated by standing in bungee cords, or looking at their own face in a mirror. However such methods have little practical appeal to busy crewmembers. Since 1978, Russian crews have worn "Penguin" suits that use elastic cords to load their bodies along the head-foot axis as a countermeasure against muscle and bone deterioration. In the early 1980s they also evaluated two other disorientation countermeasures, a cap that applied a load between the top of the head and the shoulders, and also sandals with insoles that could be inflated, applying pressure to the feet. The latter became known as "Cuban Boots" after the Cuban cosmonaut who first tried them. However, the extent to which these devices can reproduce haptic gravitational cues is unclear, and users reportedly still experienced illusions and space sickness (Reschke et al., 1994b). Artificial cues that reinforce the perception that "down" is beneath the feet may render freely moving cosmonauts more susceptible to VRIs whenever they float inverted.

1.3 Distinguishing VRIs from 0-G Inversion Illusions

Prior to the first detailed descriptions provided by Skylab and Spacelab astronauts, VRIs were not distinguished from inversion illusions in the scientific literature. For example, Graybiel and Kellogg (1967) reviewed Titov's early account of 0-G inversion illusion after orbital insertion (Sect. 2.2) but assumed Titov's inversion sensation corresponded to the VRI produced by slowly rolling inverted in the cabin of an aircraft in parabolic flight. Prior to Skylab, crew accounts in the US and Soviet programs typically came from hurried debriefings or written questionnaires where terminology was rarely defined and or discussed. In some cases nuances were lost in language translation. Also, early investigators typically focused exclusively on perceived orientation with respect to the gravitational vertical, ignoring the changes in surface identify highlighted by Skylab and Shuttle crews. Consequently the distinction between inversion illusions and VRIs was overlooked in several otherwise comprehensive 1990s reviews (e.g. Lackner, 1992; Reschke et al., 1994a,b). However as confirming descriptions of VRIs emerged from Shuttle (e.g. Mukai and Igarashi, 1995), Mir and ISS (e.g. Liu, 2003), the scientific and operational medical significance of the distinction between VRIs and inversion illusions has become more widely appreciated. Although both illusions are clearly influenced by visual and interoceptive cues, we (Oman et al., 1984, 1986; Mittelstaedt, 1986; Oman, 2003) have noted that inversion illusions and VRIs differ in many important respects.

As summarized in Table 13-1, the hallmark of a VRI is a visual attention dependent change in the perceived identity of surrounding surfaces, resulting from an angular reorientation of the internal mental allocentric reference frame used as the basis for perception of orientation and place. When an astronaut floats upside down in the cabin and then looks at their own feet, the ceiling surface beyond suddenly seems like a "floor", and there is a corresponding illusory change in the identity adjacent surfaces. If the same astronaut is floating upright, but looks at a second astronaut floating inverted nearby, the surface beneath the second astronaut's feet is often suddenly the "floor". By contrast, a crewmember experiencing a persistent 0-G inversion illusion correctly perceives the identity of surrounding surfaces but feels continuously inverted with respect to an external gravitational reference frame, regardless of body orientation. Hence when floating inverted, such a person feels upside down in a gravitationally upright cabin. When floating upright, the entire cabin seems gravitationally upside down. Note that the orientation of the external gravitational reference frame is determined by body orientation, and not by the actual relationship to the unseen Earth. The inversion sensation continues even when body orientation is changed, and persists after the eyes are closed, whereas VRI sensations generally do not. Inversion Illusion sensations are difficult to reverse, whereas VRIs only occur with eyes open, are labile, and like figure-ground illusions are easily cognitively manipulated by redirecting visual attention. Crew typically look around for a known architectural visual landmark, reorient to it, and the surface identity illusion disappears after a few seconds.

Table 13-1. Characteristics of VRIs vs.Persistent 0-G Inversion Illusions

Character	VRI	0-G Inversion Illusion
Allocentric reference frame	Spacecraft	External gravitational frame
Perceived orientation	Usually feet towards "floor", but can vary	Always gravitationally inverted
Role of visual cues	Required	Not essential
Duration	Seconds	Many minutes
Lability	Easily cognitively reversed	Reversible with haptic cues
Incidence	Almost universal	< 25% of crew
Prevalence	Can occur throughout flight	Rare after second day
Paradoxical sensation	Momentary	Continuous
Allocentric reference frame	Spacecraft	External gravitational frame

Almost all crew experience a brief tumbling illusion upon reaching orbit, at the moment of booster engine cutoff. Persistent inversion illusions are far less common and prevalent only during the first several days of flight. The incidence of persistent inversion illusions among crewmembers is difficult to estimate from conventional crew reports. However, among twelve carefully debriefed science astronauts who understood the distinction, only two described persistent inversion illusions, and only during their first two days in orbit. By contrast, almost all astronauts admit to experiencing changes in subjective surface identity, and though susceptibility probably eventually diminishes, VRIs have been described throughout the duration of six month long missions aboard orbiting space stations.

VRIs create a momentary change or uncertainty in perceived orientation and place. Crews say VRIs are nauseogenic only during their onset. They are a significant space motion sickness stimulus only because VRIs occur often, as when crew leave their seats and move about in all degrees of freedom.

The afflicted describe persistent 0-G inversion illusions as continuously nauseogenic, since they feel continuously gravitationally inverted, regardless of their relative orientation with respect to the vehicle. Both of our science crewmembers who reported inversion illusions experienced space motion sickness, including vomiting.

2. EVA DISORIENTATION AND HEIGHT VERTIGO

Shuttle, Mir and ISS crewmembers have also sometimes experienced spatial disorientation episodes while performing spacewalks ("Extra Vehicular Activity", EVA). EVA astronauts typically move about using handrails, trailing a backup safety tether. They stabilize their body with one hand while working with the other, or install foot restraints and use both hands. Since the body tends to drift while working, crews must remain conscious of their orientation, and be careful not to inadvertently bump antennae, optics, or other sensitive equipment. They must avoid thruster keep-away-zones. Working upright within the Shuttle payload bay is disorienting, since the area can be illuminated with flood lights, and the floor and side walls define a convenient visual reference frame. However when crews work on the rounded exteriors surfaces of the Mir or ISS modules, fewer global visual landmarks are available, particularly during the dark portion of each orbit, when the only lighting comes from helmet mounted lamps. Crews prepare for EVA by memorizing landmarks and routes during their preflight underwater neutral buoyancy training. NASA EVA teams also train using an interactive immersive virtual reality display system. Once in orbit they use a laptop computer graphics program to review anticipated translation paths (Homan, 2001; Walz, 2002). Nonetheless most EVA crews admit they occasionally become disoriented and sometimes must even radio for advice, or await daylight.

Some EVA astronauts have described a 0-G form of height vertigo, apparently triggered by a VRI. Early reports came from Shuttle astronauts working in the open payload bay while it faced earthward. If the astronaut happened to float into an inverted orientation, looked toward their feet and saw the Earth moving by rapidly by several hundred kilometers away, their mental allocentric reference frame apparently jumped from the payload bay to the surface of the Earth below. Perceived orientation suddenly changed from floating inverted in the payload bay with the globe of the Earth "above" to hanging from a handrail with the surface of the Earth far "below" (Fig. 13-3).

Figure 13-3. Floating inverted in Shuttle payload bay can cause EVA Height Vertigo (NASA photo)

Height vertigo reports have also come from astronauts egressing from an ISS airlock through an Earthward facing hatch, or while standing in foot restraints on the end of the Shuttle remote manipulator arm, or while hanging on the end of a crane used on the Mir station to transfer crew from one module to another (e.g. Linenger, 2000).

Some of the afflicted have extensive parachuting or rock climbing background, so it is hard to think that acrophobia is a contributing factor. In many respects the phenomenon resembles physiological height vertigo (Brandt et al., 1980) that people describe on Earth when standing at the edge of a cliff or the roof of a tall building. However, some astronauts say they also experience enhanced awareness of the spacecraft's orbital motion, and the sensation that both they and the entire vehicle are falling toward Earth. In some cases, the compulsion to "hang on for dear life" for fear they will fall to Earth is disabling. The most common etiologic factor is that the Earth's surface is perceived as beneath the body, rather than as a blue planet floating above. Veterans say the best defense against EVA height vertigo is to look at their hands, and concentrate on the vehicle as the frame of reference. Changing relative body position so the Earth is "above" should also be effective (Oman, 2002).

3. 3D NAVIGATION PROBLEMS

Navigation problems deriving from the peculiar visual architectural relationships between the interiors of docked modules on Apollo, Skylab, Mir and ISS have been consistently reported. A common theme is that crews transiting between modules are momentarily disoriented when the visual verticals in the modules transited are not coaligned. Spatial relationships between non-aligned modules are apparently difficult for the crew to visualize.

The first reports came during Apollo, where astronauts in the Command Module (CM) on their way to the Moon normally sat facing the docked Lunar Module (LM). The primary visual axes of the LM cockpit were pitched back 90 degrees and yawed 90 degrees right with respect to the CM cockpit. One Apollo crewmember recalled: "..whenever I went from one spacecraft to the other through the connecting tunnel between the CM and the LM I was visually disoriented until I looked at a familiar spacecraft panel. Instantly my mind reoriented itself and I went about my business. In this case my mind apparently had a "learned" orientation from lying on my back during training in the command module simulators that was 90 degrees different from that learned while standing on my feet during training in the lunar module simulators" (Schmitt and Reid, 1985).

Figure 13-4. Russian Mir Space Station. In this picture, the Core module is behind, pointing upwards. The Priroda module is to the left, also attached to the central node mudule, and opposite the Kristall and the orange Shuttle docking module. Opposite the Core is a docked Soyuz. The cockpit interior of the docked vehicle is oriented at 45 degrees to the Core-Soyuz axis. (NASA Photo)

Skylab astronauts encountered similar problems. One noted "I get you know, [one local vertical] embedded in my mind, and I whistle [out of the workshop] through the docking adapter and into the command module [docked facing the other Skylab modules] and zingy ! All of a sudden it's upside down..." He felt the disorientation might be dangerous, since an astronaut might throw a switch the wrong way." (Cooper, 1976).

Disorientation and navigation problems were also common on the Russian Mir space station, due to its complex three dimensional architecture (Fig. 13-4). Mir research modules were connected at 90 degree angles to a central, 6 ported spherical "node". The visual verticals of many Mir modules were not co-aligned. For example the visual vertical in the Priroda science module was opposite to that in the Core module station control center. The visual vertical in the Kristall science module was oriented at 90 degrees to the Core. Crewmembers said that even though they intellectually knew the physical arrangement of the modules, and though had a small physical model of the Mir exterior onboard, the interior arrangement was so complicated that they could not readily mentally visualize it. Several observed they could not point in the direction of familiar interior landmarks in other modules the way they knew they could in their homes on Earth. When moving between modules they learned to use landmarks and rules to navigate. One crewmember recalled: "I learned that to go into Priroda, I needed to leave the [Core module] upright, go through the hatch to my left, and then immediately roll upside down so that Priroda would be right side up". Another said: "Even though you knew the modules went in six different directions, it felt like the node was a vestibule in a single story house.... You eventually just learned what to look for and do to get to your destination." A third said: "After I first boarded Mir, I decided to go back to the Shuttle, but discovered I didn't know which way to go, since I hadn't left behind any bread crumbs!". To assist Shuttle visitors, Mir crew fashioned red velcro arrows, and positioned them on the walls pointing toward the Shuttle.

In 1997, Mir crews successively had a fire, a near collision and a collision with a Progress robot resupply vehicle. The collision caused a depressurization and power loss. In both cases when collisions were imminent, crew tried to locate the inbound spacecraft visually, but could not readily keep track of its allocentric direction when moving from module to module and window to window. The power loss required the crew to reorient the entire station using thrusters on a docked Soyuz spacecraft. Crew in the Mir Core control center discovered they had great difficulty mentally visualizing the orientation of another crewmember in the differently oriented Soyuz cockpit, and performing the 3D mental rotations required to formulate appropriate verbal control instructions (Burrough, 1998) These events convinced the space agencies that in could be critical in certain emergency situations for crew to be able to maintain their allocentric orientation and be able to make complex three dimensional spatial judgments.

Fortunately no comparably serious emergencies have yet occurred on the ISS. The primary Russian and US modules orbited so far (2006) are arranged in a straight line. Although some modules have multiple visual verticals (Sect 2.1), their principal visual verticals - as defined by the crew's gravitational orientation during training in ground simulators - are coaligned. However crews have reported difficulties visualizing spatial relationships between these principal modules and other vehicles which often dock at 90 degrees to the main plane of ISS, such as the Shuttle, Soyuz crew vehicle, and also the Progress and Multipurpose Logistics Modules which deliver supplies. Other modules and vehicles will eventually be added at 90 degree orientations. As on Mir, visiting Shuttle crews are vulnerable to becoming lost. In emergencies, crewmembers must plan to leave ISS in different directions, since each of the Soyuz vehicles can only accommodate three people in custom fit couches, and visiting Shuttle crews must leave through a different hatch. Small relocatable luminescent signs (Smart et al., 2001) point the way to the various docked vehicles, but in conditions of reduced visibility, crew must remain oriented and be able to find their way. Since the ground mockups are not all connected in the actual flight configuration, egress routes and landmarks cannot be fully rehearsed. ISS crews and visitors do partial walkthroughs on the ground, and sometimes rehearse in orbit. Laptop based VR emergency egress trainers for 3D egress route rehearsal under simulated impaired visibility are also under development (Aoki et al., 2006).

4. RELATED THEORIES AND EXPERIMENTS

The concluding sections of this chapter review the physiologic and cognitive factors known to influence human perception of the gravitational vertical and surface identity, and the mechanisms of visual reorientation and 3D navigation. Notional models for sensory cue interaction based on ground laboratory experiments of Mittelstaedt, Howard, as well as more recent results from human and animal experiments conducted in parabolic and orbital flight. Taken together, these results account for many of the phenomena described in previous sections. It is argued that astronauts remember landmarks within spacecraft modules relative to a 3D allocentric coordinate frame that in terrestrial life defines a 2D navigation plane and thus the identities of floors, walls, and ceilings. The changes in perceived surface identity that occur during a human VRI are the direct result of a rotation of the astronaut's internal local allocentric frame. If the navigation planes of adjacent spacecraft modules are incongruently aligned, inter-module navigation and spatial judgment abilities are impaired.

4.1 Gravireceptor bias

On Earth, human perceptions of static tilt result from a synthesis of gravireceptor, body axis, and visual cues. Mittelstaedt (1987, 1997) showed that gravireceptor cues originate not only from the vestibular otolith organs in the inner ear, but also from receptors located in the trunk (e.g. kidneys and cardiovascular system). He also noted that when a person lies horizontal, the gravitational component acting along the head and body axis is eliminated. However a residual gravireceptor bias evidently remains in either a head ward or foot ward direction: Subjects with a head ward bias do not feel horizontal in darkness unless their body axis is tilted a few degrees head upward. The bias may originate in the saccular otoliths or in truncal receptors. Mittelstaedt argued that the perceptual effects of the residual gravireceptor bias should also be manifest in orbit. Those astronauts with a head ward bias in 1-G should experience persistent 0-G inversion illusions and be more susceptible to space sickness. Of five astronauts Mittelstaedt tested in 1-G, the two who had head ward biases reported inversion illusions in space. However, 1-G bias did not predict acute inversion illusions in brief parabolic flight (Glasauer and Mittelstaedt, 1992). Also, half of a large control population had a head ward bias, so evidently the 1-G bias over-predicts the incidence of persistent inversion illusions actually reported in orbit with eyes open. Nonetheless, it makes sense that a net gravireceptor bias acting along the body long axis could determine susceptibility to persistent inversion illusions in orbital flight. Perhaps fluid shift effects fully manifest only in orbital flight alters the effective gravireceptor bias from that measured in 1-G (Oman, 2002). The associated sensations of head fullness from fluid shift resemble those from whole body inversion in 1-G. Fluid shift typically begins even before launch, since crew typically sit on the launch pad with feet elevated, sometimes for hours.

4.2 Body Axis and Visual cues

On Earth, if subjects lying horizontal in a dark room are asked to rotate a luminous line to the gravitational vertical, they will set the line tilted about 30 degrees in the direction of their foot to head axis – the well-known Aubert illusion (Aubert, 1861). If the room lights are turned on so the gravitational vertical cue is supplemented by visual vertical and horizontal cues from the room, the Aubert effect is much reduced, but still present.

Conversely, if the entire surrounding visual environment is tilted with respect to gravity, subjects feel compelling illusions of self tilt. Witkin et al. (1948) showed that when a person sits upright in a darkness and views even a tilted, dimly lit square frame, the perceived vertical is biased away from true gravitational vertical in the direction of the frame axis of symmetry. The magnitude of the effect was shown to be an personal characteristic, and formed the basis of Witkin's well known "Rod-and-Frame" test of visual field dependency.

Mittlestaedt (1983) referred to the tendency for the perceived vertical to align with the body as an "idiotropic" effect, and introduced the idea of using weighted vectors to represent the visual, body axis and gravireceptor cues involved. Young et al. (1986) and Parker and Harm (1993) advocated similar models. Subsequent experiments (e.g. Mittelstaedt, 1989; Dyde et al., 2006) have shown that the magnitude and interaction of the visual and body axis vectors derived from experimental data depends on how the perceived vertical direction is measured. Though vector models provide a conceptually useful way of describing the relevant stimuli, the cue interaction is arguably (Sect 4) mathematically nonlinear and the result of top-down processing.

To investigate cue interaction beyond the 30 degree gravitational tilt angles used by Witkin, Howard and colleagues constructed a small cubic room mounted on a horizontal axle that could be fully tumbled with a human subject inside. The gravitational orientation of the subject could be independently manipulated. The room was furnished with a table, chair, door, and other everyday objects. Howard and Hu (2001) showed that the normal effect of visual cues on gravitationally erect subjects is dramatically enhanced if the subject's body axis is tilted away from the normal gravitationally upright position. For example, if both the subject and the room are both tilted 90 degrees from the gravitational vertical, two thirds of adult subjects will judge they are gravitationally upright.

What properties of the visual scene influence the strength of the visual vector? Howard and colleagues (Howard and Childerson, 1994; Howard and Hu, 2001; Howard et al., 2005; Jenkin et al., 2006) argue that at least five scene properties contribute to perceived tilt:

1. "Intrinsic polarity" cues: Many familiar objects, such as desks, trees, and people are almost always seen in a consistent orientation with respect to gravity. Intrinsically polarized visual objects can strongly influence perceived orientation. Intrinsically polarized objects have identifiable principal axis with one end perceptually the "top" and the other end perceptually the "bottom". Large, readily recognizable environmental surfaces presumably also fall into this class, such as a lawn, water surface - or the earth viewed from orbit. The orientation of the human body –

either another person, or a downward glance at one's own torso and legs - also provides a significant intrinsic polarity cue.

2. "Extrinsic polarity cues": Objects that lack intrinsic polarity can acquire polarity by their placement relative to other supporting objects in the scene. Examples include objects hanging or lying on shelves, tapered objects that would fall over unless they were large end "down".
3. Environmental symmetry cues: The walls, ceilings, floors, and large stationary objects present in most scenes define axes of symmetry. Howard (1982) referred to these as "frame" cues to emphasize the correspondence to the luminous frame used in Witkin's experiments.
4. Background location: Polarized objects are more effective when placed in the background rather than the foreground of the visual scene.
5. Field of view: the more polarized objects and surfaces the observer is able to see, the stronger the effect.

Although these five factors are known to influence the magnitude of the visual effect, it is important to note that frame and polarity cues are perceptual, not physical quantities. They depend on the visual attention, expectations and the prior experience of the observer. Ultimately the model for cue interaction is empirical: For a given subject and visual scene, if one manipulates the orientation of the visual scene, body axis, and gravity, it is possible to fit a mathematical model to data and estimate the component visual, body axis and gravitational vectors. However one cannot physically measure scene polarity or symmetry cues, and use it to predict the magnitude of the visual vector.

4.3 Top-down processing and surface identities

In many situations involving ambiguous sensory cues, the resulting perceptions show evidence of "top down" processing. Prior assumptions or equivalently an internal mental model determines what is perceived. For example there are typically multiple axes of symmetry in a visual scene, and which one provides the dominant "frame" cue depends on where the subject expects the vertical to be. Howard and Childerson (1994) placed subjects gravitationally upright inside an unfurnished cubic chamber, and then rolled the chamber about the subject's visual axis. Presumably the chamber's surfaces provided only visual symmetry cues. Significantly, these subjects reported a sensation of oscillating tilt, not full rotation as in a fully furnished room. We (Oman and Skwersky, 1997) have repeated these experiments, and noted that subject reports of "oscillating tilt" are linked to a change in the perceived identity of the chamber surfaces. Apparently the subjects assume that the surface nearest their feet and most closely aligned with gravity is a "floor", the opposite surface is a "ceiling" and the intermediate surfaces are "walls". However as the "floor" surface rotates away from the horizontal, the wall on the opposite side becomes more horizontal. Eventually, the identity of the surfaces becomes ambiguous. As chamber rotation proceeds, the original "floor" suddenly switches subjective identity and becomes a "wall", and simultaneously wall on the opposite side becomes the new "floor". Since the new "floor" is oriented 90 degrees from the previous one, the subject suddenly reports feeling tilted in the opposite direction. As the rotation proceeds, tilt sensation oscillates. The "oscillations" have a paradoxical quality, since there is no concomitant change in vestibular cue. Apparently at the perceptual level the interior surfaces of the chamber are generic visual objects whose perceived wall/ceiling/floor attributes are determined not only by specific polarized objects on them, but also by top-down assumptions as to the expected orientation of the vertical (e.g. gravitationally down and beneath the feet). It is interesting that some subjects tested seated gravitationally erect in Howard's furnished, highly polarized tumbling room also experience oscillating tilt and not full rotation. Presumably gravity and body axis cues dominate over the rotating polarity cues. Subjects are aware of the paradoxical surface attributes and say "that surface which I can see is actually a ceiling now seems like a floor". Thus, surface identity seems linked in top-down fashion to an unseen allocentric reference frame, determined by gravitational, body axis, and object polarity cues. The correspondence between their reports and the VRI descriptions of Skylab and Spacelab astronauts is obvious.

The top down linkage between allocentric orientation, perceived tilt, and perceived surface identity has been explored in other experiments. For example, Mast and Oman (2004) showed that if subjects view an ambiguously polarized room (Fig. 13-5) tilted at 45 degees, top down processing determines perceived object and surface identity and the direction of the perceived vertical in the scene, and this in turn influences even low level visual processing, such as the horizontal-vertical line-length illusion.

4.4 Human visual orientation experiments in orbit

On several missions early in the Shuttle era, our laboratory (Young et al., 1986) studied how the absence of gravity and footward force applied with bungee cords influenced illusory rolling sensations induced with a rotating dot display. In weightlessness, crews gave greater weight to visual flow and haptic cues. However the response varied between subjects, suggesting crew differed in "perceptual styles".

Figure 13-5. Room with ambiguous frame and polarity cues (Mast and Oman, 2004). View the figure upright and rotated 90 deg. clockwise. Which surface seems to be the floor ?

Parker and Harm (1993) summarized comments from several other Shuttle astronauts, and concluded that some astronauts apparently increased the weight given to static visual cues, while others apparently became more idiotropic. During the 1998 "Neurolab" Shuttle mission, (Oman et al., 2003) we studied VRI and motion illusion susceptibility, visual vs. idiotropic tendencies and the interdependency of self-orientation and visual shape perception among four astronauts, who wore a head- mounted display (Fig. 13-6). We tested the crew on several occasions preflight and postflight and on the third or fourth day of the mission. None of our subjects reported persistent inversion illusions during testing in flight.

In one experiment, our subjects indicated the direction of subjective "down" while viewing a virtual spacecraft interior tilted with respect to their body by an angle that varied randomly over successive trials. Responses were classified as aligned with scene architectural visual axes, body (idiotropic) axes, or other.

Most all the inflight responses were closely aligned to either the visual scene or idiotropic axes. Comparing an average measure of visual vs. idiotropic dependency across mission phases (Fig. 13-7), we saw clear differences between subjects consistent with the notion of individual perceptual styles. Those astronauts who were strongly visually dependent or independent prior to flight remained so in orbit. Three of the subjects (A,B &C) were visually independent preflight. One (A) became more visually dependent inflight, showing greater orienting response to scene polarity, and then reverted postflight indicating an adaptive response. However the other two remained idiotropic when tested inflight – consistent with the high incidence of VRIs under operational conditions in orbit.

Figure 13-6. 1998 Neurolab Shuttle mission experiments on individual differences in visual orientation and shape perception. (Oman et al., 2003). The head mounted display provided controlled visual stimuli in the otherwise cluttered and busy laboratory.

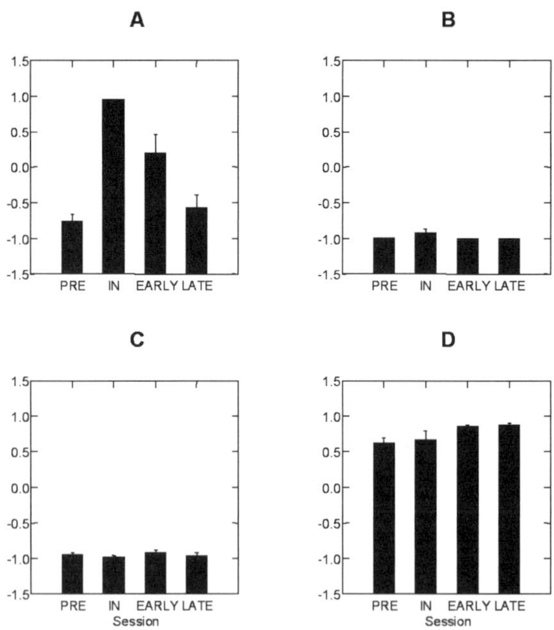

Figure 13-7. Visual-idiotropic dependency coefficient for Neurolab subjects A-D by mission phase (pre=preflight, in=idays 3-4 in orbit, early=first 3 postflight days, late=postflight days 4-5. A value of +1 indicates strong visual dependence, and -1 indicates strong idiotropic dependence. See Oman et al. (2003) for details.

For practical reasons we could not measure VRI susceptibility to real scenes under operational conditions to compare with our data, nor was the mission long enough to determine whether VRI susceptibility (real or virtual) decreases in orbit eventually. Perhaps one day these answers can be obtained aboard ISS.

In a second experiment, three of four subjects who viewed rotating polarized or dotted scenes while free floating experienced stronger roll motion illusion than on the ground, confirming Young et al's earlier finding. When the scene motion corresponded to virtual motion down a long hallway, perception of linear self motion increased dramatically. Other experiments (Young et al., 1996; Liu et al., 2002) in parabolic flight have shown that the linear and angular motion illusion enhancement happens immediately upon entry into 0-G. It may be that 0-G more immediately and consistently enhances the perceptual weight given to visual flow cues as opposed to static frame and polarity cues. This phenomenon could also explain the enhanced sensation of orbital motion occasionally described by EVA astronauts, and the persistence of VRI reports on long duration missions.

In a third experiment, when our subjects viewed a physically flat but gradient shaded disk, three out of four experienced a change in illusory disk convexity after cognitively initiating a VRI so perceived self orientation changed from floating perpendicular to parallel to the deck. That such a change in perceived object convexity occurred after a VRI would be expected, since as every art student knows, perceived convexity/concavity of surfaces is known to be based on a "light comes from above" assumption. The result demonstrated the interdependency of shading interpretation and self-orientation perception, even in weightlessness.

4.5 Visual reorientation, mental rotation and perspective taking

Visual reorientation mechanisms allow people to recover their sense of location and direction after becoming momentarily disoriented, both in normal terrestrial environment (Wang and Spelke, 2002), in orbital flight, and in desktop virtual reality experiments and games, where vestibular cues confirming visual motion are missing. In order to reorient or remain oriented while free floating within a spacecraft cabin, astronauts must be able to recognize visual landmarks from an arbitrary relative orientation. Hence spatial orientation in 0-G likely depends on individual ability to recognize individual 3D objects after rotation (Shepard and Metzler, 1971) and to correctly mentally visualize the appearance of an object array after an imagined change in location or viewing direction (Huttenlocher and Presson, 1979). Individual 3D mental rotation and imaginary perspective taking abilities are experimentally distinguishable personal characteristics (Kozhevnikov and Hegarty, 2001). In mental rotation tests, error rates and response times increase with visual rotation angle. Mental rotation abilities of the genders overlap, but on average men perform better. Among women, spatial abilities vary across the menstrual cycle (Hausmann et al., 2000). There is no public data on individual differences among astronauts, but among MIT graduate students, we routinely see large inter-individual variability in these skills. Leone et al. (1995) tested the 3D mental rotation performance of five Mir cosmonauts, and showed that individual abilities are unchanged in weightlessness as compared to on the ground. Imaginary perspective taking ability has not yet been tested in orbit. However in ground experiments, Creem et al. (2001) showed that self-rotations are more easily imagined about the body axis, perhaps because in our upright terrestrial lives, most imagined rotations take place about that axis. They also demonstrated that the subject's orientation to gravity has little effect on imaginary perspective taking.

We tested the abilities of several large subject groups to visualize the direction to objects inside a simulated space station node after large changes in relative viewing angle (Oman et al., 2002; Richards et al., 2003; Shebilske et al., 2006). We consistently found that performance correlated with several well known tests of 2D and 3D mental rotation abilities. Most subjects said that they memorized the environment from a prototypical orientation. Many invented rules to help them mentally reconstruct the space, such as memorizing opposite or adjacent pairs of objects. As with many spatial tasks, performance improved with practice. Most – but not all – eventually performed adequately. Manipulation of the subject's orientation to gravity had little effect on performance, nor did it in Creem's experiments Most subjects described the mental rotation/visualization task as "something done in your head". Collectively these findings suggest that 3D orientation ability in weightlessness probably varies between subjects, even among the highly select astronaut population, but should improve with experience and training, particularly if people are taught strategies for choosing and remembering appropriate landmarks. Validated tests of 3D mental rotation and perspective taking abilities may be helpful in identifying particularly vulnerable individuals, and in customizing their training.

4.6 3D Navigation

Wang and Spelke (2002) argue that both humans and many animals navigate - keep track of their orientation and position - via similar fundamental neural mechanisms supporting reorientation, place recognition, path integration, and cognitive map formation. Most experimental studies have focused on terrestrial navigation in a 2D horizontal plane. Path integration involves continuous updating of position and orientation relative to a starting point using vestibular and motoric cues, and without reference to fixed environmental landmarks. When people encounter a novel environment, they first identify landmarks and associate individual landmarks with specific actions, such as turning left or right, and eventually learn a sequence of landmarks and actions as a route (Siegel and White, 1975). Route knowledge consists of declarative topologic rules that becomes automatic with practice. Most older children and adults recognize common landmarks on interconnected routes and develop an ability to take shortcuts, to point to unseen landmarks, and even do so from a different, imagined location. This kind of ability requires configurational environmental knowledge and is frequently described as a "cognitive map" (e.g. Tolman, 1948), though this not meant to imply a person actually has a mental image of a cartographic map. The physiological basis of cognitive maps and how they are acquired is the subject of debate (e.g. Wang and Spelke, 2000, 2002). There is evidence (e.g. Sadalla et al., 1980; Colle and Reid, 1998) that configurational knowledge is hierarchical. Local objects are coded relative to room landmarks, which in turn are coded relative to buildings, and so on up to larger geographic scales. Even local room scale spatial mental models are based on conceptions rather than perceptions, and people imagine local object locations using a spatial framework employing both salient environmental axes and their body axes to establish referent categorical directions. Most adults employ a mix of landmark, route, and cognitive map based navigation strategies, often resorting to landmark and route techniques when unsure of their orientation, or simply out of convenience. Particularly when disoriented, astronauts apparently do the same.

The ability of astronauts to physically perform actual physical three dimensional wayfinding/navigation tasks has not yet been tested in orbit. However performance in simulated navigation tasks in 3D mazes has been tested in several non-immersive virtual reality experiments conducted on the ground (Aoki et al., 2003, 2005; Vidal et al., 2004) and also in 3 cosmonauts aboard the International Space Station (Vidal et al., 2003). Though the Aoki and Vidal maze architectures and methods differed in details, all routes required a succession of 90 degree turns in various directions through a virtual maze. Both sets of experiments showed that subjects generally had difficulty building a correct mental representation of their path whenever the path required a body rotation other than in yaw (azimuth). There was no major difference between Vidal's ground and orbital flight results. Practice generally improved performance, particularly with the complex configurations. Vidal et al. (2004) concluded that "although humans can memorize 3D-structured environments, their innate neurocognitive functions appear to be specialized for natural 2D navigation about a gravitationally upright body axis." Aoki et al. (2003) explained their results by assuming that whenever their subjects made a pitch rotation, they "did not recognize the rotation of their frame of reference." Although Vidal and Aoki did not specifically ask their subjects about changes in subjective surface identity, or explain their results in terms of VRIs, one can account for both by assuming that whenever subjects made a 90 degree turn in pitch or roll, and entered the next maze segment, they experienced a VRI, and as result failed to correctly rotate their unseen allocentric navigational reference frame as a result of scene movement. Subjects may be able to reconstruct their orientation and position relative to a global allocentric frame by remembering the direction of successive turns, but this requires a series of mental rotations that likely becomes increasingly prone to error as the number of turns increases. Unfortunately, (Sect 2.1) it is usually impractical to design spacecraft with globally congruent visual verticals, or hatches large enough so astronauts can avoid pitches resulting VRIs when transiting through them.

In virtual reality based 3D orientation training experiments (Richards et al., 2003; Benveniste, 2004; Oman et al., 2006) subjects responded fastest when module interiors were presented in a visually upright orientation and looking in a specific direction. This suggests that subjects remember each module's landmark arrangement from a canonical viewpoint that establishes a local reference frame. When modules were attached to each other with local reference frames incongruently oriented, and the subjects had to make spatial judgments between them, they required several seconds longer, suggesting subjects mentally interrelated the two modules though some kind of 3D mental rotation process. If Mir and ISS crews had to perform complex mental rotations to interrelate module interiors, this may explain why they found it so difficult to maintain their allocentric orientation relative to the entire station. When first learning the actual flight configuration in orbit, if they experienced an unrecognized VRI when transiting between modules, their sense of direction would be mis-oriented relative to the larger coordinate frame of the station. Their mental cognitive map of the station interior would then be incomplete or erroneous, as in the case of the Mir astronaut who felt he was living in a single story house (Sect. 3). In terrestrial situations, miscoding of the orientation of a local cognitive map with respect to a larger scale one can create "wrong door" disorientation in room scale environments (Lackner and DiZio, 1998), and "direction vertigo" on building and city scales. Once learned, such miscoding can be difficult to unlearn (Jonsson, 2002). Therefore it may be important to teach astronauts the actual flight configuration of their spacecraft interiors very early in the ground training process. Preflight virtual reality training where astronauts learn the allocentric relationships between visually incongruent spacecraft modules – for example using "see through" walls or miniature 3D models of the station interior and exterior (e.g. Marquez et al., 2002) - and where they learn rules relating specific adjacent/opposite landmark pairs both within and between modules should be a useful 0-G disorientation countermeasure.

4.7 Animal experiments in 0-G

Does the CNS actually maintain an internal allocentric coordinate frame in weightlessness that establishes a "floor"-like navigation plane? Over the past two decades, the neural basis of spatial memory in humans and animals has become better understood based on electrophysiological studies in animals, and functional neuroimaging in humans. Portions of the limbic system, including the hippocampus, post-subiculum, thalamic nuclei, and entorhinal cortex function together to interrelate various external (e.g. visual) and internal (e.g. vestibular and haptic) sensory cues and determine place and direction relative to the environment. Wiener and Taube (2005) provide a comprehensive review. One type, "head direction" cells (Taube, 1998), are found in several limbic areas and consistently discharge as a function of a rat's head direction in the spatial plane the animal is walking in, independent of place or head pitch or roll up to 90 degrees. The direction of maximum response ("preferred direction") varies from cell to cell. The range of firing is typically about 90 degrees. The preferred directions of the entire ensemble of cells reorient in unison when distant visual landmarks in the room are rotated about the animal. Comparable cells have also been found in primate. Head direction cells in turn provide the essential azimuthal reference input to at least two other classes of limbic cells: "grid cells" (Hafting et al., 2005) and "place cells" (Best et al., 2001) that ensemble code various attributes of the rat's location- also in the two dimensional plane of the animal's locomotion. (It is important to note that though these particular cell classes respond in a 2D plane, the animals show 3D orienting behavior. Presumably there are other as-yet-undiscovered limbic cell classes that code other orientation or place attributes in third dimension defined by the orientation of this 2D locomotion plane - e.g. height, elevation angle or roll angle).

A critical question is the extent to which gravity anchors the orientation of the response plane of these cell classes. In 1-G laboratory experiments, head direction cells usually maintain directional tuning when the animal climbs a vertical wall, but if the rat crawls inverted across a gridded ceiling, many cells show reduced directional tuning, or lose it entirely (Calton and Taube, 2005). In parabolic flight experiments, we monitored rat head direction cell responses while animals in a visually up-down symmetrical cage successively experienced 1G, 0G and 1.8G (Taube et al., 2004). Allocentric directional tuning was maintained in 0-G while the animal crawled on the familiar floor of the cage, despite the absence of gravity.

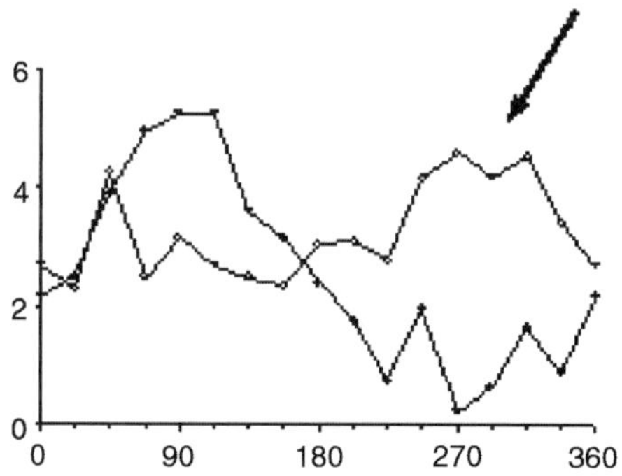

Figure 13-8. Rat head direction cell directional tuning curves on cage ceiling and floor during 0-G parabolic flight. Data recorded on ceiling indicated with arrow (Taube et al., 2004).

When we manually transferred the rat to the ceiling in 0-G, most cells lost directional tuning, and statistically showed an increase in overall background firing level, which could reflect an instability in orientation perception. We predicted that if the rat occasionally experienced a VRI and adopted the ceiling rather than the floor as the navigation reference plane, but continued to use a primary visual landmark to determine azimuth, the preferred firing direction should flip across the visual axis of symmetry of the cage. Bursts of firing in other than the original preferred direction occurred on the ceiling in several animals, and in some animals were 2-3 times more frequent in the expected ceiling-preferred directions than in the original floor-preferred directions. Fig. 13-8 shows the ceiling and floor tuning curves for one such cell, which shifted through about 180 deg in azimuth. Such shifts in azimuth may correspond to the common human perception during a 180 degree VRI that one is in a familiar but somehow mirror-reversed place, since objects remembered on the left are now to be found on the right.

In a related experiment conducted in on the Neurolab Shuttle mission, Knierim et al. (2000, 2003) recorded place cell activity as trained rats walked across three surfaces defining the corner of a cage. Their path required a yawing 90 degree turn while on each surface, followed by a pitching 90 degree turn to move onto the next surface. After a total of 3 yaws and 3 pitches, they returned to the original starting point. The investigators' original hypothesis was that in 0-G only the yaw rotations would be taken into account, and the animal would have to yaw 360 deg. and traverse four successive surfaces to do it before the same place cell would fire again. However, when tested on the fourth flight day, one animal's place cells responded in only a single area of the 6 turn track, suggesting this animal had incorporated the pitch rotations, and was maintaining a 3D allocentric sense of place within the cage. In the other two animals, place cell fields were abnormal, with one of them exhibiting symmetric firing fields on each successive surface. We suggested that this would fully be expected if the animal experienced the equivalent of human VRIs: After each pitch back, the view of the track ahead was virtually identical on each surface, so they might have the illusion of traversing the same one turn segment of the track three times in succession. The third animal did not exhibit consistent place fields – which might be expected if it was disoriented, and simply following the track using a route strategy. However when tested after five more days in weightlessness, the place fields of the second and third animals appeared unimodal, suggesting they had learned to orient to the entire cage, rather than successive locomotion surfaces.

Taken together, these experiments show that even in the physical absence of gravity, limbic head direction and place cells in animals responses define a two dimensional navigation plane parallel to the "floor" of the animal's environment. In 0-G if the animals crawl or are placed on adjacent or opposite surfaces, direction and place tuning can disappear or change in ways suggesting the navigation plane has reoriented into alignment with the adjacent or opposite surface. Note that humans and animals not only spatially "re-orient", but also "re-position". We cannot ask animals their perceptions of surface identities, but the neural behavior of their limbic navigation plane in 0-G does correspond to that posited for humans, based on the character of 0-G disorientation and VRIs.

So far head direction and place cell responses have been characterized only in terrestrial animals. It is interesting to speculate about what we will ultimately find in other vertebrate species. Birds, marine mammals and cartilaginous fish rely on dynamic lift to oppose gravity, and usually fly/swim upright. Most bony fish have gas bladders, which ballast them upright. Certain species - notably the marine mammals - apparently have the ability to remain allocentrically oriented while performing multiple graceful rotations about axes perpendicular to gravity, yet it is ecologically important for them to remain allocentrically oriented with respect to the ocean surface or bottom. Do marine mammals have a more robust ability than rodents and humans apparently do to maintain allocentric orientation when gravitationally inverted or in weightlessness? To what extent can vertebrate limbic neural networks reorganize during life to respond to new environmental challenges?

4.8 Sensory integration in weightlessness

The theories and experiments reviewed in Sects 4.1-4.7 account for many of the perceptual phenomena described Sections 1-3. As detailed in Oman

(2003), one can formally combine Mittelstaedt's original notions of gravireceptor bias and body axis (idiotropic) cues with Howard's concepts for visual frame and polarity cues into a model for sensory cue interaction. However, several new assumptions are required. One is that the net gravireceptor bias may be different than that measured in 1-G. The second - and more important - assumption is that though sensory cues can be represented by vectors, their resultant is not simply a mathematical vector sum. Rather, they are interpreted in nonlinear, top down fashion based on visual attention and the assumed orientation of an internal 3D coordinate frame that codes the remembered location of local cabin landmarks, and that assigns corresponding surface identities. When an astronaut floats visually upright in a familiar cabin, the internal mental coordinate frame is properly anchored, surface identities are correctly perceived, and objects are in remembered locations. However, if the visual scene has multiple axes of visual symmetry and/or polarized objects have inconsistent visual orientations, as shown in the Figure 13-9 example, the perceived orientation and surface identity is multistable and depends on body orientation and visual attention. The internal mental coordinate frame can alternate between a veridical orientation, and one that does not correspond with reality. When it does, the astronaut notes a change in perceived cabin surface identity, the hallmark of a VRI. Frequently VRIs are triggered when the astronaut looks at his own legs, since their intrinsic visual polarity is aligned with the body axis rather than environmental polarities.

To the extent that object polarities result from prior terrestrial experience in an upright body orientation, experience viewing an environment from multiple body orientations may eventually reduce polarity effects. One of the goals of preflight virtual reality based training is to accelerate this process. However the continuing occasional susceptibility of long duration astronauts to VRIs suggests that certain types of polarity are innate, and that the disorienting body axis orientation effect does not disappear entirely.

If asked to indicate the direction of "up" or "down", most astronauts will point perpendicular to the perceived floor. "I take my down with me, and it attaches to whatever surface seems beneath me". A few with very strong idiotropic tendencies may report "down" seems aligned with their head to foot axis, even though paradoxically if they change their body orientation, they never feel that they are stationary and the spacecraft is rotating around them. When VRIs occur, the internal mental coordinate frame aligns with local axes of the spacecraft cabin interior. In contrast, during inversion illusions or an episode of EVA height vertigo, the internal mental reference frame jumps beyond the spacecraft. The latter two are the only situations in which astronauts describe strong "gravitational vertical" perceptions.

Figure 13-9. VRI in the ISS US Laboratory module. The equipment and labeling on the true floor and ceiling are oriented 90 degrees counterclockwise from those on the true walls and floor. The square cross section of the module means the major physical axes of symmetry are also 90 degrees apart. When a crewmember floats in this body orientation, which surface is perceived as a floor depends on visual attention. VRIs due to such ambiguities can only be prevented by attending to learned landmarks.

EVA height vertigo occurs when crew have a wide view of the Earth in their lower visual field, and extrinsic visual polarity and haptic cues are consistent with supported by/hanging from the spacecraft. The external reference frame jumps "down" to the Earth, and suddenly there is a strong perception of height. Inversion illusions are likely when strong head ward gravireceptor bias cues (perhaps from fluid shift during the first several days of flight) are strong enough to overcome environmental visual polarity and foot ward body axis cues regardless of relative body orientation. In this situation, the only sensory interpretation possible is that there must be an unseen gravitational coordinate frame, far beyond spacecraft cabin and aligned so "down" is always in the foot-to-head direction. Hence when floating upside down in the cabin, they feel gravitationally inverted. When visually upright, they feel upright but the entire spacecraft seems gravitationally upside down.

How people mentally represent physical space is often determined by conceptions, not direct perceptions (Tversky, 2003). On Earth our spatial knowledge of the layout a familiar building is generally derived by concatenation of our spatial knowledge of the layout of the individual rooms, and we cannot see through the walls. This process is relatively effortless, since all the "floors" lie in the same plane. Most people can point in the direction of the front door of a building regardless of what room they are in. Unfortunately, for engineering reasons, all spacecraft from Apollo to ISS have required crews to work in areas with incongruently aligned coordinate frames. Anecdotal reports from astronauts and evidence from virtual reality simulations (e.g. Aoki, Vidal, Benveniste) suggests that crews have great difficulty concatenating their knowledge of incongruently aligned local coordinate frames, and often cannot correctly point in the direction of unseen landmarks in distant modules, such as the emergency exit. Crews probably have difficulty maintaining a spacecraft-fixed rather than local-module fixed internal coordinate frame, since the latter are more useful when working in individual modules. This way, labels appear upright, and objects are in remembered places. When transiting between modules, crews usually deliberately initiate a VRI and work in the local coordinate frame. Another goal of virtual reality based preflight orientation training is to teach crewmembers the relationships between important landmarks in different modules relative to a single spacecraft-fixed allocentric navigation frame.

Will astronauts who live in weightlessness for years eventually lose their susceptibility to VRIs, inversion illusions and height vertigo, and be able to interrelate the reference frames of adjacent modules or work areas, regardless of orientation ? When the first human children ultimately are born and mature in weightlessness, will their spatial abilities and neural coding be fundamentally different than ours ? Or will they – like today's astronauts – still show evidence of their terrestrial evolutionary heritage? Perhaps one day future astronaut-researchers will discover the answers.

ACKNOWLEDGEMENTS

My thanks to the Shuttle and ISS crews cited herein for sharing their experiences and insights, and to Drs. H. Aoki, A. Natapoff, and J. Taube for their useful comments. Research results of Calton and Taube (2005), Howard and Hu (2001), Howard (2003), Mast and Oman (2004), Oman (2003), Oman et al. (2002, 2003, 2006), Richards et al. (2003), and Shebilske et al. (2006) cited herein, and the writing of this chapter were supported by a grant to the author through NASA Cooperative Research

Agreement NCC 9-58 from the National Space Biomedical Research Institute.

REFERENCES

Aoki, H., Ohno, R., and Yamaguchi, T., 2003, A study of spatial orientation in a virtual weightless environment. Part 2 Causes of spatial cognition errors, *Journal of Architecture, Planning, and Environmental Engineering* **563**:85-92.

Aoki, H., Ohno, R., Yamaguchi, T., 2005, The effect of the configuration and the interior design of a virtual weightless space station on numan spatial orientation, *Acta Astronaut* **56**:1005-1016.

Aubert, H., 1861, Eine scheinbare Drehung von Objekten bei Neigung des Kopfes nach rechts oder links, *Virchows Arch* **20**:381-393.

Benveniste, D., 2004, Cognitive conflict in learning three-dimenaional space station structures, SM Thesis, Department of Aeronautics and Astronautics, Massachusetts Institute of Technology, Cambridge, MA, September, 2004.

Best, P. J., White, A. M., and Minai, A, 2001, Spatial processing in the brain: the activity of hippocampal place cells, *Annu Rev Neurosci* **24**:459-86.

Brandt, T., Arnold F, Bles, W. and Kapteyn T.S., 1980, The mechanism of physiological height vertigo I: Theoretical approach and psychophysics. *Acta Otolaryngol* **89**:534-40.

Burrough, B., 1998, *Dragonfly: NASA and the crisis aboard the MIR*, Harper Collins, New York

Calton, J. L., and Taube, J.S., 2005, Degradation of head direction cell activity during inverted locomotion, *J Neurosci* **25**:2420-2428.

Cooper, H.S.F. Jr. 1976, *A House In Space*, Holt, Rinehhart and Winston, Austin, TX.

Coss, R. G., Clearwater, Y. A., Barbour, C. G., and Towers S. R., 1989, Functional décor in the International Space Station: body orientation and picture perception. *NASA TM 102242*, November, 1989.

Cheung, B., 2004, Nonvisual illusions in flight. In: *Spatial Disorientation in Aviation*, F. H. Previc and W. R. Ercoline, eds, Volume 203, Progress in Astronautics and Aeronautics, AIAA, Reston, VA, pp. 243-281.

Colle H. A., and Reid G. B., 1998, The Room Effect: metric spatial knowledge of local and separated regions, *Presence* **7**:116-128.

Creem, S. H., Wraga, M.-J., Proffitt, D. R., 2001, Imagining physically impossible self-rotations: geometry is more important than gravity, *Cognition* **81**:41-64.

Dyde, R., Jenkin, M. R., and Harris L. R., 2006, The subjective visual vertical and the perceptual upright, *Exp Brain Res*, in press.

Gazenko, O., 1964, *Medical studies on the cosmic spacecraft Vostok and Voskhod*, NASA TTF-9207.

Glasauer, S., and Mittelstaedt, H., 1992, Determinants of orientation in microgravity, *Acta Astronaut* **27**:1-9.

Graybiel, A., and R. S. Kellogg, 1967, Inversion illusion in parabolic flight: its probable dependence on otolith function, *Aviat Space Envir Md* **38**(11):1099-1013.

Gurovskiy, N. N., Kosmolinskiy, F. P., and Melnikov, L. N, 1980, *Designing the Living and Working Conditions of Cosmonauts*, Moscow, Mashinstroyeniye.

Hafting, T., Fyhn, M., Molden, S., Moser, M. B., and Moser E. I., 2005, Microstructure of a spatial map in the entrohinal cortex, *Nature* **436**:801-806.

Hausman, M., Slabbekoorn, D., Van Goozen, S., Cohen-Kettenis, P., and Gunturkun, O., 2000, Sex horomones affect spatial abilities during the menstrual cycle, *Behav Neurosci* **114**:1245-1250.

Homan, D., 2001, NASA Johnson Space Center VR Laboratory DOUG software, personal communication.

Howard, I. P., 1982, *Human Visual Orientation*, Wiley, Toronto.

Howard, I. P., and Childerson, L., 1994, The contribution of motion, the visual frame and visual polarity to sensations of body tilt, *Perception* **23**:753-762.

Howard, I. P., and Hu, G., 2001, Visually induced reorientation illusions, *Perception* **30**: 583-600.

Howard, I. P., Hu, G., Saxe, R., and Zacher J. E., 2005, Visual orientation in a mirror world tilted 90 degrees, *Perception* **34**:7-15.

Huttenlocher, J., and Presson, C., 1979, The coding and transformation of spatial information. *Cognitive Psychol* **11**:375-394.

Jenkin, H. L., Zacher, J. E., Oman, C. M., and Harris L. R., 2006, Effect of field of view on a visual reorientation illusion: does the levitation illusion depend on the view seen or the scene viewed? Abstract presented at the 7th Symposium on the Role of the Vestibular Organs in Space Exploration, Noordwijk, the Netherlands, June 8, 2006 (available online: www.congrex.nl/06a07).

Johnston, R. S., and Deitlein, L. F., 1977, *Biomedical Results from Skylab*, NASA SP-377, US Government Printing Office, Washington DC.

Jonsson, E., 2002, *Inner Navigation: Why we get lost and how we find our way*, Scribner, New York.

Kitmacher, G. H., 2002, Design of the Space Station habitable modules, Paper presented at the 53rd International Astronautical Congress, 10-19 Oct, 2002, Houston, TX. Paper IAC-02-IAA.8.2.04. http://www.spacearchitect.org/pubs/IAC-02-IAA.8.2.04.pdf

Knierim, J. J., McNaughton, B. L., and Poe, G. R., 2000, Three –dimensional spatial selectivity of hippocampal neurons during space flight, *Nature Neurosci* **3**:209-210.

Knierim, J. J., Poe, G. R., and McNaughton, B. L., 2003, Ensemble neural coding of place in zero-g, in: *The Neurolab Spacelab Mission: Neuroscience Research in Space: Results from the STS-90 Neurolab Spacelab Mission*. J. C. Buckey, Jr., and J. L. Homick, eds., NASA SP-2003-535, pp.63-68.

Kornilova, L. N., 1997, Orientation illusions in spaceflight, *J Vestib Res* **7**:429-439.

Kornilova, L. N., Muller, K. H., and Chernobylskiy, L. M., 1995, Phenomenology of Illusory Reactions in Weightlessness, *Fiziologii Cheloveka* **21**:50-62.

Kozhevnikov, M., and Hegarty, M., 2001, A dissociation between object manipulation spatial ability and spatial orientation ability, *Mem Cognition* **29**(5):745-756.

Lackner, J. R., 1992, Spatial orientation in weightlessness, *Perception* **21**:803-812.

Lackner, J. R., and Graybiel, A., 1983, Perceived orientation in free-fall depends on visual, postural, and architectural factors, *Aviat Space Envir Md* **54**(1):47-51.

Lackner J. R., and DiZio P., 1998, Spatial Orientation as a Component of Presence: Insights Gained from Nonterrestrial Environments, *Presence*, **7**:108-115.

Leone, G., Lipshits, M., Gurfinkel, V., and Berthoz, A., 1995, Is there an effect of weightlessness on mental rotation of three-dimensional objects? *Cognitive Brain Res* **2**:255-267.

Linenger, J., 2000, *Off the Planet*, McGraw Hill, New York.

Liu, A. M., Duda, K., Oman, C. M., Natapoff, A., 2002, Effects of parabolic flight zero-gravity on looming linear vection, *J Vestib Res* **11**(3-5):325.

Lu, E., 2005, Ed Lu's Journal Entry #12: Which Way is Up? NASA Johnson Space Center press release, Sept. 2003. http://www.edlu.com/whichWay.pdf

Marquez, J. J., Oman, C. M., Liu, A. M., Beall, A. C., 2002, Spacecraft in miniature: a tool for the acquisition of mental representations of large environments. *J Vestib Res* **11**:338.

Mast, F. W., and Oman, C. M., 2004, Top-down processing and visual reorientation illusions in a virtual reality environment, *Swiss J Psychol* **63**(3):143-149.

Mittelstaedt, H., 1983, A new solution to the problem of the subjective vertical. *Naturwissenschaften* **70**:272-281.

Mittelstaedt, H., 1987, Inflight and postflight results on the causation of inversion illusions and space sickness, in: *Proceedings of the Norderney Symposium on the Scientific Results of the German Spacelab Mission D-1*. Norderney, Germany, 27-29 August, 1986, P. R. Sahm, R. Jansen, and H. M. Keller, eds., German Ministry of Research and Technology, Bonn, Germany.

Mittelstaedt, H., 1989, Interactions of form and orientation, in: *Spatial Displays and Spatial Instruments*, , S. R. Ellis and M. K. Kaiser, eds., NASA Conference Publication 10032. pp. 42.1-42.14.

Mittelstaedt, H., 1996, Somatic gravireception, *Biol Psychol* **42**:53-74.

Mukai, C., and Igarashi, M., 1995, Microgravity and space motion sickness. *J Otolaryngol Head Neck Surg* **11**(7):967-972.

NASA, 1995, Man-Systems Integration Standards (MSIS), NASA Johnson Space Center Bioastronautics/Habitability and Human Factors Office.

NASA Space Station Program Office, 1999, Flight crew integration design requirements SSP 50005, NASA Johnson Space Center, Houston TX.

Oman, C. M., 2003, Human visual orientation in weightlessness, in: *Levels of Perception*, L. Harris and M. Jenkin, eds, Springer-Verlag, New York, pp. 375 398.

Oman, C. M., 1987, The role of static visual orientation cues in the etiology of space motion sickness, in: *Proceedings of the Symposium on Vestibular Organs and Altered Force Environment.*, Houston, TX, October, 1987, honoring A. Graybiel, M. Igarashi, ed., NASA/Space Biomedical Research Institute, pp.25-37.

Oman, C. M, Benveniste, D., Buckland, D. A., Aoki, H., Liu, A., and Natapoff, A., 2006, Spacecraft module visual verticals and training affect spatial task performance, *Habitation* **10**(3/4):202 203.

Oman, C. M., and Skwersky, A., 1997, Effect of scene polarity and head orientation on illusions in a tumbling virtual environment, *Aviat Space Envir Md* **68**:649.

Oman, C. M., Howard, I. P., Smith T., Beall, A. C., Natapoff, A., Zacher, J. E., and Jenkin H. L., 2003, The role of visual cues in microgravity spatial orientation, in: *The Neurolab Spacelab Mission: Neuroscience Research in Space*: Results from the STS-90 Neurolab Spacelab Mission, J.C. Buckey, Jr. and J.L. Homick, eds, NASA SP-2003-535, pp. 69-82.

Oman, C. M., Lichtenberg, B. K., and Money, K .E., 1984, Symptoms and signs of space motion sickness on Spacelab-1. NATO-AGARD Aerospace Medical Panel Symposium on Motion Sickness: Mechanisms, Prediction, Prevention and Treatment, Williamsburg, VA, NATO AGARD CP-372. Later published as:

Oman, C. M.,Lichtenberg, B. K., and Money, K. E., 1990, Symptoms and signs of space motion sickness on Spacelab-1, in: *Motion and Space Sickness*, G. H. Crampton, ed., Boca Raton, FL, CRC Press, pp.217-246.

Oman, C. M, Lichtenberg, B. K, Money, K. E, and McCoy, R. K., 1986, M.I.T./Canadian vestibular experiments on the Spacelab-1 mission: 4. Space motion sickness: symptoms, stimuli, and predictability, *Exp Brain Res* **64**: 316-334.

Oman, C. M., Shebilske, W. L., Richards, J. T., Tubre, T. C., Beall, A. C., and Natapoff, A., 2002, Three dimensional spatial memory and learning in real and virtual environments, *Spatial Cognit Comput* **2**:355-372.

Parker, D. E., and Harm, D. L., 1993, Perceived self-orientation and self-motion in microgravity, after landing and during preflight adaptation training, *J Vest Res* **3**:297-305.

Reschke, M. F., Bloomberg, J. J., Paloski, W. H., Harm, D.L., and Parker D.E., 1994a, Physiologic adaptation to space flight - neurophysiologic aspects: sensory and motor function, in: *Space Physiology and Medicine*, A. E. Niccogossian, C. L. Huntoon, and S. L. Pool, eds., Lea and Febiger, pp. 261-285.

Reschke, M. F., Harm, D. L., Parker, D.E., Sandoz, G.R., Homick, J.L. and Vanderploeg, J.M., 1994b, Physiologic adaptation to space flight - neurophysiologic aspects: Space Motion Sickness, in: *Space Physiology and Medicine*, A. E. Niccogossian, C.L. Huntoon, and S.L. Pool, eds., Lea and Febiger, pp. 228-255.

Richards, J. T., Oman, C. M., Shebilske, W. L., Beall, A. C., Liu, A. M., and Natapoff, A., 2003, Training, transfer, and retention of three-dimensional spatial memory in virtual environments, *J Vestib Res* **12**:223-238.

Schmitt, H. H. and Reid D. J., 1985, Anecdotal information on space adaptation syndrome. University Space Research Association Division of Space Biomedicine, Houston, TX Report, July 1985.

Tversky, B., 2003, Structures of mental spaces: how people think about space. *Environ Behav* **35**:66-80.

Walz, C., 2002, Expedition Four: Spacewalking, ISS Style, available online http://www.spaceflight.nasa.gov/station/crew/exp4/eva/spacewalking.html

Sadalla, E. K., Burroughs, W. J., and Staplin L. J, 1980, Reference points in spatial cognition. *J Exp Psychol-Hum L* **6**:516-525.

Shebilske, W. L., Tubre, T, Tubre A. H., Oman, C. M., Richards, J. T., 2006, Three-dimensional spatial skill training in a simulated space station: random vs. blocked designs. *Aviat Space Envir Md* **77**(4):404-409.

Shepard, S. and Metzler, D., 1971, Mental rotation of three dimensional objects, *Science* **171**:701-703.

Siegel, A. W. and White, S. H., 1975, The development of spatial representations of large-scale environments, in: *Advances in child development and behaviour*, H. W. Reese, ed., Academic Press, New York, pp. 9-55.

Smart, K., Bowen, C., Carr D., and Novak J., 2001, ISS emergency egress: a system for crewmember pathway indication, paper presented at the 31st International Conference on Environmental Systems, Orlando, FL, July 9, 2000, SAE Technical Paper 2001-01-2184.

Taube J. S., 1998, Head direction cells and the neurophysiological basis for a sense of direction. *Prog Neurobiol* **55**:225-256.

Taube, J. S., Stackman, R. W., Calton, J. L., and Oman, C. M., 2004, Rat head direction cell responses in zero-gravity parabolic flight, *J Neurophysiol* **92**:2887-2997.

Tolman, E. C., 1948, Cognitive Maps in Rats and Men, *Psychol Rev* **55**:189-208.

Titov, G. and Caidin M., 1962, *I am Eagle !* Bobs-Merrill.

Viguier, C., 1882, Le sense de l'orientation et ses organs chez les animaux et chez l'homme, *Rev Philos* **14**:1-36.

Wang, R. F. and Spelke, E. S., 2002, Human spatial representation: insights from animals, *Trends Cogn Sci* **6**:376-382.

Wiener, S. I. and Taube, J. S., 2005, *Head direction cells and the neural mechanisms of spatial orientation*, MIT Press, Cambridge.

Witkin, H. A. and Asch, S. E., 1948, Studies in space orientation: IV. Further experiments on perception of the upright with displaced visual fields, *J Exp Psychol* **38**:762-782.

Young, L. R., Oman, C. M., Watt, D. G., Money, K. E., Lichtenberg, B. K., Kenyon, R. V., and Arrott, A. P., 1986, MIT/Canadian vestibular experiments on the Spacelab-1 mission:

1. Sensory adaptation to weightlessness and readaptation to one-g: an overview, *Exp Brain Res* **64**:291-298.

Young, L. R., Mendoza, J. C., Groleau, N., and Wojcik, P. W., 1996, Tactile influences on astronaut visual spatial orientation: Human neurovestibular experiments on Spacelab Life Sciences 2, *J Appl Physiol* **81**:44-49.

Yuganov, E. M., Gorshkov, A. I., Kasyan, I. I., Bryanov, I.I., Kolosov, I. A., Kopanev, V. I., Lebedev, V. I., Popov, N. I., and Solodovnik, F. A., 1966, Vestibular reactions of cosmonauts during the flight in the Voskhod spaceship, *Aerospace Med* **37**: 691.

Chapter 14

SPATIAL REPRESENTATIONS IN THE RAT: CASE STUDY OR PERSPECTIVE ON EPISODIC MEMORY?

Françoise Schenk, Delphine Preissmann and Chiara Sautter
Department of Physiology and Institute of Psychology, University of Lausanne, Switzerland

Abstract: Spatial orientation in animals or in men requires memory as an essential feature and may be considered as a complex manifestation emerging from multiple brain structures with the hippocampus at "the crossroad". In this chapter, we present the underlying biological mechanisms of spatial behavior along a contextual and historical dimension, in an ethological perspective. We propose that study of spatial memory in mammals, and more precisely in laboratory rats, sheds some light on the development and evolution of episodic memory.

Key words: Multiple memory systems; spatial navigation; hippocampus; LTP; place units; exploration.

1. FROM STEREOTYPED BEHAVIOR TO SPATIAL MEMORY: A HISTORICAL DETOUR

Episodic memory has very specific properties leading to the capacity to *"behave in a particular way because we remember a prior event and not* (only) *because this event has happened"* (Morris, 2001). As discussed by Morris (see also Martin and Morris, 2002), this condition of an explicit memory is hardly met in animal work, even in the elegant experiment by Clayton and Dickinson (1998) in which jays decide where to orient their search for food depending on a memory of what they have eaten, where and when.

However, the demonstration of a conscious recollection must be considered as an ultimate condition for an experimental approach of episodic memory in animals, but not as a prerequisite. Indeed, as has been abundantly discussed in the case of the "theory of the mind", the explicit dimension of episodic memory might be an additional property, particularly developed in humans. The phylogenetic basis of episodic memory does not have to be an explicit process. We consider that a fundamental property of this memory is the organization of events in spatio-temporally tagged episodes. This suggests that episodic memory emerges from the evolution of the mechanisms mediating the automatic encoding of life events in a spatio-temporal framework. In this perspective, the eventual explicit statement "*Je me souviens*" comes as an additional property related to higher level conscious processing.

In fact, no biological mechanism can be fully understood in an "arrêt sur image" but requires both a historical Darwinian and a contextual dimension (Morange, 2005). We propose an ethological like perspective, following the rules listed by Tinbergen that to *understand a behavior*, one must take into account not only its *physiological mechanism*, but also its *evolutionary history*, its *current ecological significance* and its *development* (Tinbergen, 1963).

To address this issue, we will first describe briefly how the actual view of animal spatial cognition emerges from the convergence of psychological, ethological and neurophysiological lines of research.

1.1 Animal cognition in the field and in the laboratory

During the first part of the XXth century, ethology was grown up an important research field, under the impulsion of Max Von Frisch, Konrad Lorenz, Robert Hinde or Niko Tinbergen, to cite just a few of the most famous ethologists. Two important European reviews, *"Behaviour"* and *"Zeitschrift für Tierpsychologie"* provided repertoires of animal adaptive strategies in natural conditions. These observations were mostly described in terms of inherited capacities to react to a given stimulus, thus promoting the view that adaptive behavior was mainly due to preprogrammed sequences triggered by appropriate stimuli or released by an internal innate mechanism. This exclusive alternative of internal innate representations of goal oriented behavior or of automatic responses to relevant external stimuli ruled out the possibility that animals might memorize representations of the environment.

In this perspective, the performance of squirrels retrieving buried nuts was mainly attributed to their insistence in digging at the basis of a vertical wall or cylinder, such as tree trunks, both for caching and retrieving food. If all the animals of the same species were sharing the same rule, it was supposed, an individual did not need to memorize where it had buried nuts. The probability to dig out a nut in such a standard context was relatively high, whether finding its own nuts or those cached by another squirrel. Thus, animals did not need projects or plans, just a perseverance in expressing innate stereotyped motor responses. This type of explanation was commonly accepted, even reinforced by the expected fundamental difference between animals and men.

Meanwhile, experimental psychologists such as Edward Tolman were working in the laboratory, observing Norwegian rats in mazes which they considered as the best substitute for their natural environment (tunnels or rubbish accumulations). In a famous paper dated 1948, Tolman described rats' orientation in mazes as implemented by a "cognitive map". He based his major argument on the rats' capacity to make a detour or a short cut in a slightly modified familiar environment. His main opponent was Clark Hull, who tried to analyze rats' sequence of choices in similar situations as emerging from the progressive binding of chains of stimulus responses. Eventually, the design used by Tolman was regarded as having obvious flaws, which helped rejecting the idea of a rat *"lost in thoughts"*.

Thus, until the end of the seventies, it was hardly possible to find the word "memory" or "learning", even less the adjective "cognitive" in the titles of the papers reporting animal behavior experiments in neuroscience journals. However, two books published in 1978 made an outbreak. *"The hippocampus as a cognitive map"* by John O'Keefe and Lynn Nadel and *"Cognitive processes in animal behavior"* by Steward Hulse, Harry Fowler and Werner Honig. The first book was a real "*pavé dans la mare*" as it proposed that the hippocampus, a structure of the limbic system supposed to be man's hidden and despised heritage of animal brain, was implementing spatial cognitive maps "à la Tolman" in both animals and humans. It was supported by the development of electrophysiological recording in the hippocampus of freely moving rats revealing the existence of brain networks engaged in spatial representations (O'Keefe, 1976). The second book analyzed animal behavior as based on intentions and plans, thus providing a firm basis for the researchers interested in cognitive processing in animals.

One major weakness of this research line was that laboratory rats had been bred and selected for hundreds of generations and were moving in artificial laboratory environments, driven by artificially modified motivations. To some scientists, the results obtained from such experiments were as far from what could be expected from wild animals in their environment as were in vitro from in vivo experiments.

John Krebs, in Cambridge, contributed to bridging the gap during the early 80ies. He analyzed memory capacities from marsh tits, in the field and in aviaries in which the birds were allowed to cache food for later retrieval, depending on their motivation (Shettleworth and Krebs, 1982; Krebs, 1983). To make a long story short, this line of research illustrated the remarkable spatial memory capacities of different bird species and revealed that these capacities were positively correlated with the development and expansion of the hippocampus in food storing birds (Clayton and Krebs, 1995).

This last step provided a strong impulse to a new field of neuro-eco-ethology, in which scientists collected experimental laboratory and field observations linking the development and function of the hippocampus with a specialization in food caching and retrieval, both in birds and in mammals (Jacobs and Liman, 1991, Jacobs, 1995).

1.2 The hippocampus: a memory system like any other?

To approach episodic memory as an emergent capacity, while specifying the role played by the hippocampal formation in this form of memory, it is preferable to concentrate on *the memory of systems* rather than on *systems of memory*, as proposed by Fuster (1995). The idea of a system of memory requires an a priori definition of its specificity and also ignores the memory properties of other brain structures. Indeed, all brain systems have a certain degree of plasticity and consequently they mediate some kind of memory, depending on how they process and encode information.

A comprehensive theory of the memory of systems has been presented by White (see White and McDonald, 2002; McDonald et al., 2004). It implies parallel information processing by different brain structures with different styles, i.e. the processing of 1) *stimulus-stimulus relation* in the hippocampus, of 2) *stimulus – response* in the striatum and of 3) *stimulus-reinforcement* association in the amygdala. This work brings an important clarification in the field of memory. It emphasizes also that if different memory systems are functioning in parallel, the important question is how these different systems interact, when they can stand for each other and when they complement each other. One key behavioral test is based on the old "place vs. response" paradigm, in short, "reaching a goal vs turning right or left" (Restle, 1957), now associated with a balance between a dominant hippocampus or striatum based activity, depending on how familiar the task is (see Packard and McGaugh, 1996; Packard, 1999).

However, the hippocampus itself is a modular structure. Anatomical, electrophysiological and behavioral approaches revealed three major functional properties of the hippocampus: the plasticity of the three synaptic pathway, the fact that the dentate gyrus is the seat of active neurogenesis in adult rats and the behavioral correlates of the place units.

1.2.1 A critical position for plasticity

In 1949, the psychologist Donald Hebb was the first to propose a rule for learning: "*When an axon of cell A is near enough to excite B and repeatedly or persistently takes part in firing it, some growth process or metabolic change takes place in one or both cells such that A's efficiency, as one of the cells firing B, is increased*" (*The Organization of Behavior*, p. 62).

No such modification was observed in neurons, until Bliss and Lomo discovered the mechanism of *Long Term Potentiation* (LTP) in the hippocampus (1973). More precisely, they provided evidence that a tetanic stimulation of the perforant path, the main input from the entorhinal cortex to the dentate gyrus, had long term effects on the reactivity of this circuit to a calibrated stimulus, due to enhanced synapse efficacy. The effect was long lasting and corresponded to the Hebbian rule.

Because the LTP discovery was the first demonstration of plasticity in the brain, in a structure that was already known to be involved in memory, a common oversimplification led to the proposition that the LTP was critical to any form of spatial memory in rats.

In a first series of experiments, Barnes (1979) showed that the decreased facility to induce LTP in some aged rats was correlated with a marked impairment in spatial discrimination. In a complementary approach, the pharmacological blockade of the N-methyl-D-aspartate (NMDA) glutamate receptors by a chronic intraventricular infusion of aminophosphonovaleric acid (AP5) blocked LTP. This allowed to the first experimental demonstration that the controlled disruption of LTP impaired place learning and memory (Morris et al., 1986). The major criticism of this result was that the rats might have suffered other non specific motor impairments interfering with a normal acquisition (Cain et al., 1997).

However, it was already evident from the data presented by Morris and colleagues that even the rats with NMDA blockade had some memory of the trained position since their escape performance was affected by finding the platform in a new position in the pool. Obviously, the treatment induced a dissociation between subcomponents of place learning and preserved some learning capacities, particularly when the rats had been previously trained to some aspects of the task (Bannermann et al., 1995; Saucier and Cain, 1995; Good and Bannerman, 1997). However, there was no theoretical proposition to account for a modular function of the hippocampus and the question was whether LTP in the hippocampus had anything to do with spatial learning without reference to a modular involvement of the hippocampus in spatial memory. There was also no discussion as how spatial memory would emerge from a series of complementary processes.

Recent research papers have emphasized the "selective" implication of LTP in some memory processes. For example, the activation of NMDA receptors appears involved in "episodic like" memory since infusion of NMDA receptor antagonist into the hippocampus results in a delay dependent impairment in a one trial learning task in rats (Steele and Morris, 1999). Moreover, the retrieval of previously acquired information is not affected by the blockade of NMDA receptors, an observation that was confirmed in another one-trial learning task (Morris and Day, 2003)

Taken together, these results suggest that although LTP is critically involved in spatial memory, its disturbance reveals non-NMDA dependent subcomponents of spatial memory.

Interestingly, LTP is associated not only with a functional facilitation but also provokes structural changes such as new dendritic spines, requiring protein synthesis. Thus, two "distinct" phases of LTP have been identified. The first one, or early LTP, does not involve new protein synthesis and is relatively short lasting (2-3h). A second phase, called the late LTP, requires the synthesis of new proteins and lasts longer (until weeks in vivo) (see for a review Lynch, 2004).

These phases are characterised by successive activation of different genes (Alberini, 1999) and open the important question of memory consolidation and reconsolidation (see Nader, 2003). Actually, a "pure" consolidation process, in the absence of any reactivation phase, seems very unlikely, since most life events are likely to reactivate past memories and even post training sleep phases appear to reactivate the very circuits activated during the original experience (Wilson and McNaughton, 1994). This means also that previously stored information is recalled during or after particular events and combined for (re)consolidation.

For our purpose, we will keep in mind that since different events can happen in the same place, but not at the same time, the process of reconsolidation must contribute to the organization of episodic memory, allowing future recollection of different interconnected episodes. In addition, the controversies as to the role played by LTP in spatial memory add to the fact that spatial memory is not a unitary process and emerges from the integrated activity of several hippocampal subregions.

1.2.2 A locus for neurogenesis

Until recently, a "dogma" about the absence of neurogenesis in the adult brain was dominant (see for a review Gross, 2000). It was in line with the citation of Santiago Ramon y Cajal: *"In adult centres the nerved paths are something fixed, ended, immutable. Everything may die, nothing may be regenerated. It is for science of the future to change, if possible, this harsh decree"* (cited by Fuchs and Gould, 2000).

In the middle of the sixties, Joseph Altman and Shirley Bayer reported the existence of newly generated neurons in the dentate gyrus and in the olfactory bulb of adult rats (Altman and Das, 1965; Bayer et al., 1982; Bayer, 1985) but this systematic work was however not able to question the dogma. Neurogenesis had also been demonstrated in canaries (Goldman and Nottebohm, 1983) and this production of new neurons was seasonal, at the time when birds have to learn different songs (Alvarez-Buylla et al., 1990). However, there was a solid resistance to admit that it could also be the case in mammalian brains. This resistance was supported by several results showing an absence of neurogenesis in adult brains (Rakic, 1985(a); Rakic, 1985(b); Eckenhoff and Rakic, 1988).

During the late eighties, a new technique using bromodeoxyuridine (BrdU) incorporated in the DNA of stem or progenitor cells during the S phase of mitosis allowed to count the labelled (hence new) cells with immunocytochemichal techniques (Miller and Nowakowski, 1988). Finally, the existence of neurogenesis in the dentate gyrus in adult brain of mammalian was established in mice (Kempermann et al., 1997), in rats (Kuhn et al., 1996), even in primates (Gould et al., 1999a; Kornack and Rakic, 1999) and humans (Eriksson et al., 1998). This neurogenesis is not a minor phenomenon, since about 9'000 new cells are produced daily in adult rats (Cameron and McKay, 2001).

In the dentate gyrus, different phases are identified, such as proliferation, differentiation and survival. It is now admitted that the majority of new cells would become neurons and half of them are likely to die within a month (Dayer et al., 2003). New neurons are integrated in the hippocampal circuitry (extending axonal projections to CA3) (Hastings and Gould, 1999), with electrophysiological proprieties similar to those of older granule cells.

The functional significance of this production is still unclear. In rodents, several factors modulating the neurogenesis in the adult brain have been identified, among which the voluntary exercise (running in a wheel for mice) (van Praag et al., 2005), living in an enriched environment (Kempermann et al., 1997; Kempermann et al., 2002; Brown et al., 2003) and hippocampus-dependent learning (Leuner et al., 2004). Thus, it was shown that hippocampus-dependent, but not hippocampus-independent learning, enhanced neurogenesis in the dentate gyrus (Gould et al., 1999b). Moreover, the induction of LTP is facilitated in young neurons, as compared to mature neurons, supporting the implication of neurogenesis in the memory formation (Schmidt-Hieber et al., 2004).

A comprehensive description of neurogenesis changes associated with acquisition and overtraining in a spatial task reveals three successive phases

(Abrous et al. 2005): 1) increased new cell survival rate (initial phase of learning), 2) increased cell death (performance stabilization) and 3) increased proliferation surge (further consolidation and integration of stored information with more recent events). This complex dynamic evolution certainly accounts for the difficulty to characterize the role of neurogenesis in memory consolidation.

As for the LTP, the neurogenesis decreases with aging (Kuhn et al., 1996; Seki and Arai, 1995), and this reduction is correlated with reduced spatial memory (Drapeau et al., 2003; Wati et al., 2006). However, the exact involvement of neurogenesis in learning and memory is still in debate as there have been contradictory results. Some studies found no correlation between neurogenesis level and memory in hippocampus-dependent tasks (see for example Merrill et al., 2003; Bizon and Gallagher, 2003).

Inhibiting neurogenesis in the hippocampus with an anti-mitotic agent, the methylazoxymethanol (MAM), or by ionizing irradiation that stops cellular proliferation (Peissner et al., 1999), confirmed that neurogenesis plays a role in memory. This role seems to be selective, as blocking neurogenesis affects learning and memory in certain conditions, while sparing it in others (Shors et al., 2002; Winocur et al., 2006; Madsen et al., 2003). A recent finding suggests a role of the newly generated neurons in long term retention. Blocking neurogenesis impaired place retention in the Morris water maze after a 14-day, but not after a 7-day retention delay (Snyder et al., 2005).

Together with the results obtained from blocking NMDA dependent LTP, the blockade of neurogenesis has selective effects on spatial memory components, which confirms the hypothesis that the hippocampal function is not a homogenous one. In the same way, one should consider the modularity of spatial representations and the multiple stage memory consolidation by which episodic memory is organized as a network of related ensembles of events.

1.2.3 Place units: the link between hippocampal activity and spatial location

Place units are hippocampal pyramidal neurons whose complex-spike activity is correlated with the location of the freely moving animal. In particular, the firing rate increases when the subject moves from the periphery to the center of a precisely located firing field (about the size of a rat from head to tail), showing therefore a strong spatial specificity (O'Keefe and Dostrovsky 1971).

The first recordings took place on a plus maze in a cue controlled environment (O'Keefe and Conway, 1978; O'Keefe and Speakman, 1987) or in a circular arena (Muller and Kubie, 1987) in which the rats were trained to collect randomly distributed pellets. Different controlled cues were placed around the cross maze, a salient white card was fixed on the arena wall. A more recent set-up (Rossier et al., 2000) allowed to the recording in an open arena in which a specific place was defined, where the rat had to return and wait between pellet chasing bouts. This procedure allowed to compare place unit properties in different parts of the arena, and in relation with this target.

The rotation of the controlled cues induced a corresponding displacement of the recorded firing field, as if determined by cue position. Removing the cues was not necessarily accompanied by the disappearance of the firing field, and, in the plus maze, the specific spatial firing persisted on an arm that was in the usual position relative to that where the rat expected food (O'Keefe and Speakman, 1987). When a second cue card was placed in the circular arena, Sharp and colleagues (1990) observed that the position of the place field remained constant relative to the usual starting position, as if the second card had been ignored, as if the hippocampal unit had fired where the rat "*thought it was*".

For many years, different experimental designs addressed the question of the control of spatial distribution of firing field by sensory input (for a review see Poucet et al., 2000) and on identifying the pathways by which internal or external sensory input are transmitted to hippocampus (Poucet et al., 2003).

An important condition for the development of stable spatial correlates of unit firing is the continuity and the coherence of visual input and motion related cues. In fact, when the animal was systematically disoriented before being introduced into the curtained arena, the place fields often showed a "remapping" (i.e a complete modification of the firing field distribution) in spite of the presence of the salient cue card (Knierim et al., 1995).

In the absence of visual input and if the animal was not disoriented before entering in the curtained environment, stable firing fields could be obtained only when the olfactory traces were available (Save et al., 2000). Since motion-related cues alone can maintain the stability of place units distribution only for short time (Knierim et al., 1995), the place units neural network appears to integrate an internal sense of direction (normally updated by motion related cues) with external sensorial inputs.

Proximal and distal landmarks appeared to have a differential control on the firing of place units. Distal landmarks are most often visual cues located beyond the animal's "locomotor space", and proximal cues can be defined as objects that can be directly approached during exploration. Cressant and colleagues (1997; 1999) showed that a configuration of distributed objects controlled the activity of place units if they were situated at the periphery of the arena, but not when grouped in the center. This suggested that proximal cues alone cannot control place firing fields. Correspondingly, when only a group of local cues was available, spatial orientation was also less accurate (Gothard et al., 1996; Benhamou and Poucet, 1998). The strong control exerted by distal landmarks on navigation might be related to the fact that they provide a stable spatial frame of reference: in contrast to proximal landmarks, they do not appear to change their relative position according to animal locomotion.

1.2.4 Place units and spatial learning

The exact role of synaptic plasticity in the development of place units system is not clear. Anomalies in place units (firing field reduction or spatial instability) caused by lesioning non-hippocampal systems (Leutgeb and Mizumori, 1999; Poucet et al., 2000), or by normal physiological process such as aging (Barnes et al., 1983; Tanila et al., 1997a, 1997b; Ikonen et al., 2002), are linked with impaired spatial accuracy, supporting the hypothesis of a link between synaptic plasticity and the spatial encoding mediated by the network of place units.

Mutant mice expressing continuously activated CaMKII lack low frequency LTP and show impaired learning capacities. They have a reduced number of hippocampal pyramidal neurons showing spatial correlates. They had also smaller and less stable place fields than normal subjects (Rotenberg et al., 1996). An other mutation, decreasing the expression of protein Kinase A and inducing an instability in LTP, was accompanied by place field instability and decreased memory performance, 24 hours - but not 1 hour - after contextual fear conditioning (Rotenberg et al., 2000).

Similar results were obtained with pharmacological inhibition of LTP. The injection of NMDA antagonist given before the animal was introduced in a novel or familiar environment did not affect the firing field "map" in the novel environment (Kentros et al., 1998). However, 24 hours after the injection, the neural representation previously observed in the novel environment appeared completely modified.

In a very elegant work, Nakazawa and colleagues (2002) found that genetically manipulated mice, in which NMDA receptor gene was selectively ablated in CA3, showed normal place learning and place units activity in a cue controlled environment. However, both their behavioral spatial accuracy and the accuracy of CA1 place fields were significantly reduced in comparison with that of control mice when some of the cues had been removed. This was attributed to the impaired function of the CA3 field in pattern completion, or more objectively, in confirming the absolute spatial position in a slightly modified context.

This short and limited review of some properties of hippocampal place units emphasizes the abstract properties of spatial memory. It reveals also that sensory information are likely to be of lower status when the subjects explore a familiar area in which some confirmation of the environment is sufficient to keep track of the position. It is also clear that the different pyramidal fields of the hippocampus play a complementary role in the maintenance of a stable spatial representation, as we will briefly discuss in the last section of this chapter. The activity of the CA3 neurons seems to contribute with a more purely spatial invariant signal, whereas the activity of the CA1 cells would be more dependent on details of the local context.

2. SPATIAL MEMORY IN THE LABORATORY

The functional properties of the hippocampus briefly reviewed above suggest that spatial representations are abstract and synthetic, what is largely unexpected from animals such as rats. Thus we will provide some evidence of this abstraction, based on the detailed analysis of orientation strategies in controlled laboratory conditions.

Two different lines of research, often combined, have contributed to the study of spatial memory in laboratory rats. The ethological approach, using controlled experimental set ups, provides a description of basic information processing from the analysis of behavioral orientation strategies throughout the entire life span. The neurophysiological approach relates these behavioral capacities to brain functions or dysfunctions, based on electrophysiological recording (place or direction units), specific brain lesions, or focal pharmacological treatments.

Spatial learning has its own rules and can be best understood from experimental situations with an ecological relevance, in which free choice is allowed. They reveal which strategies develop spontaneously, depending upon individual characteristics such as age, rearing conditions and sex. They are most likely to inform on the nature of spatial representations, which, as stated by Neisser (1976), "*...are not pictures in the head but plans for obtaining information from potential environments*".

2.1 Basic orientation mechanisms

Path integration is an ancient component of spatial orientation, since it is found throughout invertebrate and vertebrate taxa (Maaswinkel and Whishaw, 1999; Wehner et al., 1996). As an animal explores an unfamiliar environment, it uses internal and external cues to relate its current position to the start point of this journey. Internal cues such as *self-generated movement cues* inform how far and in which direction it has moved from a start position. During their journey, the animals estimate the angular and linear components of their locomotion through motion cues from the vestibular system (inertial signals due to angular and linear acceleration), from proprioceptive reafferences from joints, tendons and muscles and from the efference copy (see B.J.M. Hess in this volume). These informations are integrated as a permanently updated vector coding the actual distance and direction of the start of the journey (Etienne et al., 1990; Etienne et al., 1996; Etienne, 2003; Etienne et al., 2004).

An *internal compass* keeps track of head direction (that is supposed to coincide with the subject's x-axis and therefore with the direction of motion) by assessing angular head movement only. This internal compass contributes to the orientation of the head direction cells detected in the thalamic nuclei anterior thalamic nucleus and prosubiculum (see Wiener and Taube, 2005).

The question is thus whether merely extra-hippocampal signal processing accounts for homing based on pure path integration, in the field or in laboratory experiments. Indeed, one can expect that the information provided by this vector can be updated and integrated in a more complex representation of related positions implemented by the hippocampus. Indeed, there are increasing evidence that vestibular signals are processed in the hippocampus (see a review in Smith et al., 2005) and contribute to detailed spatial representation beyond a mere path integration vector.

2.1.1 Integration of path integration information

Gallistel (1990) has formalized a widely accepted model of spatial navigation. He considers two main sources of information for navigation, *idiothetic* and *allothetic* cues. The former are generated by the animal's movement and are supposed to be processed for path integration. The latter cues are sources of sensory stimuli from stable external cues or landmarks associated to objects or positions. They can be coded on an egocentric reference frame (referred to the subject's body, possibly depending on a specific position in space) or on an allocentric reference frame (position of the different cues relative to one another). It is admitted that idiothetic and allothetic cues must be integrated to form a spatial representation.

However, these apparently simple and clear dichotomies cover uncertainties as to the information provided by allothetic cues in path integration. Even the definition of idiothetic cues provided above and used by Whishaw (1998), introduced sources of dynamic external information such as "sensory flow". Olfactory or visual flows play a critical role in these experiments (Maaswinkel and Whishaw, 1999). Distant visual cues constitute an important complement to purely internal cues. It was thus critical to assess whether path integration was implemented by extrahippocampal structures and would be maintained following hippocampal lesion (Whishaw and Gorny, 1999), but the results from behavioral experiments were controversial (see Alyan and McNaughton, 1999). It is thus not clear whether pure path integration, i.e. the coding of a return vector, is sufficient information for navigating in certain conditions. How is this vector associated with other sources of information in a familiar environment? Why and when is it reset?

2.1.2 Maze patrolling as an organized exploratory strategy

In rats, learning about the spatial relations between places is based on spontaneous exploratory bouts. The structure of these bouts determines the sequence of perceived local views. Movement decisions rely on some basic win shift rules (for a review, see Dember and Richman, 1989), but it is also necessary that a fixed spatial reference such as a "home" base or start base constitutes a temporary reference position to relate to other visited places (Eilam and Golani, 1989). However, unlike bees (Collett and Baron, 1994), rats can rapidly free themselves from any fixed reference. One can thus hypothesize that basic rules of movements are predominant during the early exploration phase. Later on, the spatial representation takes a progressively more dominant role in movement selection, based on the relative novelty of yet unvisited areas.

Elimination tasks in regular 8-arm mazes have been widely used to assess spatial memory. A small amount of food is placed at the end of radially distributed alleys, so that the best strategy is to enter all the alleys only once. Already in the first experiment, the most impressive aspect of rats' performance in avoiding re-entries was that it was neither based on simple algorithms such as visiting adjacent arms in sequence, nor on the identification of olfactory marks deposited in the already visited alleys (Olton and Samuelson, 1976). Moreover these movements are not addressed to discrete objects associated with each arms' end, but they are related to the objects' spatial configuration (Suzuki et al., 1980). From this, it appears that exploratory movements are far from being randomly distributed, but are not stereotyped either. To the contrary, they appear constrained by very simple rules, which are evident from experiments conducted in different maze structures and visual environments.

We have worked with simple transparent Plexiglas tunnels (12 x 12 x 60 cm) arranged in different maze configurations. When the regular radial structure (45 degrees between two adjacent arms for an 8-arm maze) was distorted by alternating larger and smaller angles, the foraging performance was very poor and the rats re-entered into some of the arms before finding the baits placed in each arm (Schenk and Grobéty, 1992). Interestingly, their most frequent visits were addressed to a selection of 5 of the 8 arms composing the most regular structure possible in this maze. When two arms were made parallel in a regular 45-degree maze, the rats did neglect one of these arms, as if patrolling a 7-arm maze (Schenk et al., 1990). However, if the two parallel arms were separated by an opaque wall, as if placed in two different subspaces, the parallel neglect was not observed (Schenk and Grobéty, 1992).

From this series of experiments, one can conclude that exploratory movements are highly constrained to satisfy two apparently opposite principles, which facilitate both an optimal exploration (i.e., a collection of information about the global structure of the environment) and optimal foraging for food (i.e., finding the food sources in an efficient manner). The first principle, best known as "spontaneous alternation", reduces the frequency of successive visits in nearby places in a vast environment, and encourages the selection of widely distributed reference axes. In our experiments, this capacity is based on an integration of proprioceptive and visual cues, since parallel arms leading to visually separate compartments were not neglected. Thus, the rats were not reluctant to make a U-turn, rather they seemed to avoid revisiting the "same" place. The second principle, more difficult to understand, implies that rats have a bias towards regularly distributed movements (the regular 5 arm selection in the distorted maze). This suggests that the learning of spatial relations is based on a regular reference grid of radially distributed movements, which might facilitate the calibration of the environment during exploration, as suggested by recent demonstration of the existence of so called "grid maps" in the entorhinal cortex (Hafting et al., 2005).

The fact that obtaining information about the environment appears to be of a higher priority than collecting food is also evident in a particular training condition in which optimal patrolling (i.e. visiting all the 8 arms) persists in young adults, in spite of the presence of food in only 4 easily discriminated arms (Grandchamp and Schenk, 2006). In comparison, 24 month-old rats select the baited arms first, as if age reduced the priority for patrolling in the same time as it reduced the efficacy to avoid re-entries.

2.1.3 Place learning: the default option

The most important and new concept developed by O'Keefe and Nadel (1978) was that animals possess the capacity to identify a given position in space, on the basis of its relations with other positions. This abstract capacity was exemplified by adult rats in the well known Morris navigation task (1984), in which rats must reach an invisible submerged platform at a fixed position in a circular pool of opaque water. To meet the requirement for place learning in this situation, they must reach the escape position with help of distant information only, with little training, following a direct path from any starting position in the pool. In addition, the normal rats spend a significant amount of time searching on its exact position when the escape platform is absent. They can also memorize a new escape position in the same environment following one demonstration trial only. Rats with hippocampal lesions are unable to satisfy the third criterion and require a special training procedure to meet the two first conditions.

Interestingly, the very concept of place learning by rats appears to be counterintuitive, probably because anyone expects animals to be attached to local objects and not to behave on the basis of an abstract representation. However, it appears to be the default option of rats in any spatial task. If briefly trained to find a visible platform at a fixed position in space, they will mainly remember its spatial position for further escape trials, ignoring the visible cue if the latter was placed at another position in the pool (see McDonald et al., 2004, for a recent review).

This hippocampus dependent strategy appears to be the primary one in normal subjects, when they are not submitted to intense stress. During an exploratory phase in a stable environment, selecting simple features for later *stimulus response* associations could be regarded as an economic and simple strategy. However, such process would take time and require repeated exposures in the same environment. Thus, an early spatial representation is more likely to emerge from a global encoding based on distributed movement. The association of simple responses in a cumulative manner is more likely to occur during overtraining and in subjects with a dysfunctional hippocampus. Indeed, rats with hippocampal lesions which cannot process *stimulus-stimulus* relations from general "panoramas", are also more rapid than normal subjects in learning discrete stimulus response associations (Packard et al., 1989; McDonald and White, 1993; Packard, 1999). One can thus hypothesize that the two systems are antagonist in a new situation, where the dominant hippocampus promotes "spatial optimal foraging" strategies, disregarding the food reinforcement rule for which stimulus response strategies might be more efficient.

This suggests a hierarchy of orientation strategies depending on the stability of the environment and on the regularity with which an animal has been allowed to follow a given trajectory. For the squirrels retrieving nuts, the place where they had buried a nut is of higher relevance than the search for nut odor or other stereotyped behavior (Jacobs, 1995). A similar bias toward a spatial position was described in food caching birds trained to find nuts in salient feeders in an aviary (Brodbeck, 1994). In a more general perspective, this suggests that the organization of spatial orientation aims at overcoming local variability in an environment in which local cues are likely to change over time.

2.2 Sensory information for spatial orientation in rats: vision rather than olfaction

Considering that in rats, the *place* is the default hypothesis to remember in which circumstances a particular event happened is not very intuitive, as we discussed above. Moreover, we will show that the nature of the sensory information used in spatial orientation by rats is also somewhat counterintuitive and mainly based on vision in adults. As discussed by Kosslyn et al. in the first chapter of this volume, vision provides a unique ensemble of spatial information. As we will see, the role played by olfaction in infant rats is fundamental in calibrating other sensory references such as those provided by path integration and vision, but mature rats rely mainly on visual landmarks, whenever available.

2.2.1 An implicit memory for odor cues when the light is on

The fact that the environment of radial mazes or water mazes is mainly visual seems to be a mere result of our anthropomorphism, based on the predominance of vision for orientation in man. Anyone would suggest that olfactory cues would be much more efficient to be used by rats, but would agree that they are much more difficult to control.

Of course, they are difficult to control in the Morris watermaze, and it is commonly admitted that since water is not a stable substrate for odor trails, these would be of little importance for locating the platform. Indeed, the first experiments in which rats were trained to reach the platform in the dark or in a dark region of the pool (for a review, see Schenk, 1998) revealed that they were unable to discriminate the training position in such conditions.

One would agree that the discrimination of the training position in the watermaze must rely primarily on visual cues, since even proprioception must be of little efficacy in this condition, due to inertia and poor accuracy of the swim movements. As a matter of fact, several experiments have shown that rats are more disoriented by vestibular perturbation in walking maze tasks than they are in the watermaze (Martin et al., 1997; Dudchenko et al., 1997).

When rats patrol open fields or mazes placed on solid ground, they seem to pay an utmost attention to olfactory marks. However, in an extensive series of experiments, we found that controlled olfactory cues were of little importance in comparison with the visuospatial frame of reference. When adult rats had to decide if a Petri dish should contain food, which hole should be connected to an external issue, or which alley they had not yet explored, the dominant decision rule was based on visuospatial cues (Lavenex and Schenk, 1995; Schenk et al., 1995; Lavenex and Schenk, 1996).

The relevance of food spatial position in an open field was analyzed in a Petri dish arena with 18 small Petri dishes placed on a hexagonal grid. During training, food was available at a fixed spatial position, in a small box with a salient black cover bearing a specific olfactory mark. In a special probe trial with no baited box, the significant search position was the reinforced spatial location, in spite of the displacement of the salient olfactory cue and black cover on two other boxes (Lavenex and Schenk, 1995). However, it appeared soon that two reference systems were intermingled: the rats tended to explore and sit on the usually reinforced spot, while uncovering the boxes marked with the training odor or visual cue when passing by. This suggests that the spatial reference framework guides locomotion and displacements, whereas more "local" cues would trigger motor acts such as opening of a cover.

When rats were trained to collect food in a radial maze made of transparent tunnels in the dark (under infra-red light which they cannot see), their efficacy was much lower than in the same situation with the room light on (Lavenex and Schenk, 1996). This emphasizes the primary role of vision in such patrolling and indicates that the mere uncontrolled olfactory cues in the maze (whether left by the test rat itself, or the other rats) are not sufficient to allow efficient patrolling. This was also the case in interruption trials, when the rat was reinforced for entering 4 arms not accessible during a previous phase. Adding "controlled" olfactory cues marking the ceiling of each tunnel with a trail of drops of an alimentary aroma solution increased the performance of the rats in the dark to an optimal level. In this condition, they were also able to recognize unvisited arms based on its specific olfactory cue in spite of the maze's rearrangement disturbing the spatial and olfactory configuration.

In comparison, when the light was on, any conflict between the visible room environment and the internal olfactory configuration due to maze rotation or arm perturbation lead to arms' choices based on their spatial position, in spite of arms' specific olfactory cues possibly remembered from a previous visit (Lavenex and Schenk, 1996).

The prevalence of the spatial reference was related to maze illumination even in the absence of relevant visible cues. Indeed, the rats performed poorly in white translucent tunnel mazes as if unable to rely on the salient olfactory cues for arm choice. This contrasts with the excellent performance in the dark with the same cues and suggests that the darkness context releases more primitive association rules not depending on the hippocampus.

2.2.2 Juvenile rats rely on olfactory cues to calibrate and develop an abstract spatial framework

To assess when and how spatial abilities based on distant visuo-spatial cues emerge during ontogeny, we used a homing task in which the rats were trained to reach their home cage via an escape hole at a fixed spatial position in a large circular open field (diameter 160 cm). Like the Morris watermaze this is a place learning task, but the rats are free to walk, not forced to swim. Eighteen or fifteen identical holes, equally covered by a small plastic disk easily removed by rats, are regularly distributed on the table surface. All the holes are blocked by a tightly fitted foam rubber stopping masked by the plastic cover. Only the hole at the training position is connected under the table with the home cage (Schenk, 1989).

Intertrial rotations of the walled arena prevents associations between the escape position and local olfactory traces. Location of the escape position is related to distant room cues only. In this task, we observed a slow development of place discrimination until the fifth week postnatal (see Wiener and Schenk, 2005). The development of spatial abilities was synchronous in the Morris watermaze.

If the table remained fixed between trials and the uncontrolled olfactory traces around the holes were undisturbed, rats aged 18 days expressed a marked discrimination of the training position during probe trials with no hole connected. Rotating the table 120 degrees before the test placed the marked hole at a new spatial position and the 18-day old rats were only concentrated on the hole surrounded by the familiar traces. However, in the same experimental design, rats aged 20 days concentrated their movements both on the marked hole and on the spatially defined training position, although the latter was bereft of relevant olfactory traces (unpublished data).

Although they paid attention to the olfactory traces left on the table, rats were also able to integrate distant room cues from the age of 20 days. However, rats of the same age, trained with the local olfactory cues made irrelevant by intertrial table rotation, did not discriminate the training position in a significant manner, as if the absence of olfactory confirmation during training trials had prevented the acquisition of place memory.

This potentiation of place memory by local olfactory marking was still evident in experiments conducted in adult rats with an additional olfactory cue marking the spatially defined training position (Lavenex and Schenk, 1998). Place learning was more rapid in this condition and place discrimination was enhanced during probe trials in spite of the removal of the scented cue. Familiar olfactory cues can thus facilitate the encoding of significant places, but are not necessary for later place recognition.

2.2.3 Place learning with limited number of visual or olfactory cues

Standard laboratory environments are highly variable and it is not at all clear which cues are used for place learning. We worked in the dark with three light emitting diodes as visual cues forming an irregular triangle around the table (Rossier et al., 2000) or controlled olfactory cues (five distant olfactory cues made of scented Velcro tapes placed on the table, as in Rossier and Schenk, 2003). Juvenile rats (24 or 48 days) and adult rats (above 4 months) were trained in the visual or olfactory condition or in combined visual and olfactory condition.

These comparisons revealed that the 24-day rats were unable to show place learning with only the configuration of visual cues, although they reached the hole if it was directly cued by one of these diodes. As expected, they were efficient in using the olfactory cues. Rats of the second group, aged 48 days, were able to discriminate the training position with help of visual cues, only if they had been allowed a prior training phase of mixed olfactory and visual cues. The 4 month-old subjects showed very efficient place discrimination with the visual cue configuration only.

Thus, spatial navigation abilities of rats developed slowly during the first two months of life, and, although the subjects could see the visual cues from the fourth week of life, they were not able to discriminate a position based on the spatial relation between this limited number of pinpoint cues. The interesting observation was that, during the second month of life, the rats benefited from the help of olfactory cues for calibrating space and learning, in a second step, to rely on the associated visual cues.

2.2.4 A supramodal representation

These experiments show that optimal spatial learning relies on organized exploratory movements. This organization is based on movement algorithms mediating regular patrolling in a new environment. It relies also on the systematic approach of salient cues, a phenomenon particularly evident at certain stages of the development (Wiener and Schenk, 2005). This sort of affordance is likely to play an important role in the calibration of successive reference frames during the ontogeny in order to elaborate a supramodal spatial representation (Swanson, 1983). The very slow development of these capacities suggests that, in rats, the end of the second month of life might reveal some neurophysiological changes corresponding to those occurring around 16-20 years in man.

3. A DUAL MAP TO LINK EPISODES ON A SPATIO-TEMPORAL FRAMEWORK

Along this chapter, we have proposed several evidence of a modular function of the hippocampus. The hippocampal formation is composed by different substructures, each one with a different functional organization: the pyramidal fields CA1 and CA3, the dentate gyrus and the subiculum. It has been known for many years that selective lesions of a substructure of the hippocampal formation affect specific components of spatial memory. Impaired local searching with preserved approach accuracy followed lesions of the entorhinal cortex, a main source of afferences to the dentate gyrus and hippocampal fields (Schenk and Morris, 1985). Preserved short term but not long term memory was related to lesions of the subiculum (Morris et al., 1990). Following selective hippocampal lesions, place learning was preserved in conditions of massed training, but not following single trial sessions (Morris et al., 1990; Whishaw et al., 1995).

3.1 The two parallel components of spatial memory

We will briefly propose how this modularity can be interpreted in terms of spatial memory function, in which places are organized on a large spatial reference framework, based on the integrated activity of two separate hippocampal modules. We will then discuss how this 2-D map system might incorporate time as an additional dimension to mediate the temporal discrimination of successive episodes.

3.1.1 Two categories of cues for two types of maps

Two main categories of cues provide spatial information, *directional cues* and *positional landmarks,* as described in Jacobs and Schenk (2003).

Path integration is a primary *directional cue*. It is the key information for continuity and coherence in the processing of spatial information. Besides, the physical properties of the environment are often distributed in gradients, providing other sources of directional information, based on how the animal perceives intensity changes during locomotion (Schöne, 1984). Only few gradients are evenly distributed (one example is the magnetic field, in spite of local anomalies), but different cues (trails) may provide constantly changing intensity as the animal moves on a short distance (towards a perfumed tree, along a slope). Evidence for gradient maps have been found in reptilians (Lohmann and Lohmann, 1996) and in pigeons (Papi, 1992; Wallraff, 1996). Distant cues, such as trees or mountains, provide also directional information since their angular position relative to that of the subject remains stable during linear accelerations. We propose that the integration of – at least two – gradients or sources of directional information can support a 2-D *bearing map*.

Positional landmarks are local objects, usually disposed in an array. In this case, they can be either processed separately, as discrete elements, or combined in a topographical map, i.e. a *sketch map* (Jacobs and Schenk, 2003). If the configuration is asymmetrical, its geometrical shape can also provide directional information and mediate place learning in the Morris task (Benhamou and Poucet, 1998). But it is not clear how geometrical cues from the environment shape are processed for orientation (Hayward et al., 2004; Graham et al., 2006). This requires a comprehensive theory as to how the relative salience of different cues stems from their simultaneous availability.

It is significant that normal rats take much time to learn a position relative to a configuration of three objects in a curtained environment (Parron et al, 2004), as if their information content was not sufficient *per se* to allow accurate place learning (see also Cressant et al., 1997, 1999). Although rats express a memory of the platform position based on its relation with controlled intra-curtain cues, the distribution of their search trajectories indicate that they react to the mismatch induced by rotating the intra curtain cues relative to the invisible room environment (McGauran et al., 2005). These experiments reveal that, when an animal enters a compartment, it is expecting a continuity – or a coherence - with its previous spatial position. In the procedure used by Parron et al., the platform position was related to the cue configuration, but its orientation relative to the environment outside the curtain was unpredictable. Sketch maps might thus have no real orientation properties unless they are – or have been in the past - associated with a directional reference, such as provided by a bearing map.

We propose that the coactivation of *bearing* and *sketch maps* constitutes an integrated map in which the local configurations are assembled on a spatial framework. This can mediate the capacities expected from a "cognitive map". Moreover, this coactivation during the learning phase is a prerequisite to the spatial dimension of sketches.

These parallel maps are produced by the activation of two separate intrahippocampal channels (for a detailed description, see Jacobs and Schenk, 2003). The bearing map is mediated by subcortical and entorhinal afferent paths projecting to the dentate gyrus and CA3 subfield. The sketch map is mediated by the CA1 subfield and its cortical connections (directly or through the subiculum). The integrated map emerges from the synchronous activation of these two channels.

3.1.2 Specific properties of each channel

This theory predicts that optimal spatial abilities are provided by the integrated map. Moreover, the balance between the two channels might be adjusted depending on the category of cues provided by the environment, on the species specific adaptation strategies, on the subject's emotional level, or sex hormone impregnation. This is supported by the selective target of sex or stress hormones in the hippocampus (Kavushansky et al., 2006) and by the species specific development of the hippocampus in relation with food caching habits or home range (Jacobs, 1995).

Two lines of experiments might support this dual channel system. First, the selective blockade of one channel – by lesion or conditional gene deletion - must reveal the processing capacities mediated by the preserved structure. Second, unit firing in the CA3 and CA1 fields must reveal different environmental correlates, i.e. more large scale spatial correlates in the CA3 and more local properties in CA1.

A double dissociation has been induced by selective lesions of the CA3 or CA1 fields (Gilbert et al., 2001). In this work, the CA3 field appeared critical for spatial pattern separation (i.e., fine direction discrimination), whereas the field CA1 was more implicated in temporal pattern separation. Selective lesions of the CA3 region impair spatial separation and working memory (Gilbert and Kesner, 2006; Kesner et al., 2004; Lee et al., 2005).

Selective reductions of the NMDA dependent LTP, due to local treatment or to genetic engineering of the receptors in CA3, provide a support for a complementary role of these two channels in mediating spatial memory. Often, the spatial accuracy is reduced in an aspecific manner. However, the results obtained by Nakazawa et al. (2002) confirm our hypothesis that a pure sketch map is very sensitive to fine changes of the cue configuration and that intact CA3 function is required when mice have to find the platform within an incomplete configuration of cues.

A similar dissociation has been observed in the adaptation of place unit firing in the CA1 or in the CA3 fields (Leutgeb et al., 2004, 2005). In spite of similar firing characteristics of individual place cells in CA3 and CA1, the reaction of the place fields revealed independent population codes for what might be considered as a large scale location reference framework and more local cue configurations, respectively. In CA3, the sparser firing activity contribution to representations was nearly independent in room A and B, as if informing in which room the rat was. In contrast, in CA1, the overlap was more important, suggesting that common local features were processed in a similar way in the two rooms. A change of the box location in a specific room induced a more important remapping in the CA3 field (position on the bearing map) whereas the CA1 unit firing developed more rapidly and coded for individual landmark configurations or local sketch feature. Vazdarjanova and Guzowsky (2004), using a novel cellular imaging method, compared the reaction of CA3 and CA1 field activity during successive placements in the same partially modified environment or in two different environments. Detailed changes of local configurations in a given environment was characterized by a more stable activity in CA3 than in CA1 fields. As in Leutgeb et al (2004, 2005) changing the large scale environment had more dramatic effects on CA3 field activity.

These observations support the proposition of two independent "map circuits", as proposed in our parallel map theory. They support the view that the CA3 dependent bearing map provides the large scale spatial reference framework, whereas the CA1 field is more sensitive to the local configuration. These results do also emphasize the fact that spatial discrimination requires CA3 activation as a primary feature. Interestingly, this channel appeared earlier in phylogeny, as discussed elsewhere (Jacobs and Schenk, 2003; Jacobs, 2003).

3.2 Beyond metaphors?

Memory has a long tradition of spatial metaphors, illustrated by the method of the loci, i.e. to remember a sequence of items by associating each one to a different house in a street. These metaphors propose tempting shortcuts to understand episodic memory, whereas a scientific approach requires what might appear as a circuitous approach in comparison.

Using path integration as a conducting wire, we proposed that the parallel map theory of hippocampal function organizes memory as the product of spatially related sketches or episodes. Indeed, the continuity between episodes is a prerequisite to their integration in a relational framework provided by one or several bearing maps.

A main problem when comparing spatial and episodic memory is the absence of temporal dimension in the spatial representation. The later is time invariant, by definition. Thus, to discriminate among two similar episodes with the same spatial address, one should consider a secondary classification by which successive episodes are linked to each other on a time base (daily meals in the same restaurant), or tagged to enter into recognizable categories depending for instance on a tradition, or linked with a particular event.

The question of the temporal separation (see for instance Lee et al., 2005 or Leutgeb et al., 2004) depends on at least two different mechanisms. The first is based on a direct continuity rule. In this way, any episode has a privileged relation with the immediately preceding one, in the case of spatial memory it depends on where the subjects come from (which bearing map axis was followed, which sketch was previously visited). This provides sketch maps with an internal spatial coherence and a position in a sequence. Second, any episode might reactivate similar episodes in the same or other places, indicating that (re)consolidation *per se* provides a main temporal dimension, together with a secondary network of between sketches relations.

Thus, place stability requires a combined input from the CA3 field, providing information about the large scale reference frame of the bearing map. This stability provides the spatial invariant property of a sketch map, disregarding temporal differences and minor variability in composition (over seasons, for example). Whereas the reactivity of the CA1 field to subtle local changes might allow to a temporal discrimination between episodes. Included in these changes is the access path into a local configuration, in other words, where the subject comes from. In this way, different episodes might be discriminated in a spatial manner, whereas the sequential coding of successive visits into a given sketch would mediate a temporal differentiation.

In any situation in which the integrated mapping process would be prevented, two consequences are expected from our model. The sketch itself loses its inner (spatial) coherence or configuration. Different sketches will not be interrelated, which might break the continuity rule. As matter of fact, anyone having experienced a stressful event might understand what is meant by this disruption.

ACKNOWLEDGEMENT

This work is supported by the Swiss National Science Foundation (grant 100A0-105765/1).

REFERENCES

Abrous, D. N., Koehl, M., and Le Moal, M., 2005, Adult neurogenesis: from precursors to network and physiology, *Physiol Rev* **85**: 523-569.

Alberini, C. M., 1999, Genes to remember, *J Exp Biol* **202**, 2887-2891.

Altman, J. and Das, G. D., 1965, Autoradiographic and histological evidence of postnatal hippocampal neurogenesis in rats, *J Comp Neurol* **124**: 319-335.

Alvarez-Buylla, A., Kirn, J. R., and Nottebohm, F., 1990, Birth of projection neurons in adult avian brain may be related to perceptual or motor learning, *Science* **249**: 1444-1446.

Alyan, S., and McNaughton, B. L., 1999, Hippocampectomized rats are capable of homing by path integration, *Behav Neurosci* **113**: 19-31.

Bannerman, D. M., Good, M. A., Butcher, S. P., Ramsay, M., and Morris, R. G. M., 1995, Distinct components of spatial learning revealed by prior training and NMDA receptor blockade, *Nature* **378**: 182-186.

Barnes, C. A., 1979, Memory deficits associated with senescence: a neurophysiological and behavioral study in the rat, *J Comp Physiol Psych* **93**: 74-104.

Barnes, C. A., McNaughton, B. L., and O'Keefe, J., 1983, Loss of place specifity in hippocampal complex spike cells of senescent rat, *Neurobiol Aging* **4**: 113-119.

Bayer, S. A., Yackel, J. W., and Puri, P. S., 1982, Neurons in the rat dentate gyrus granular layer substantially increase during juvenile and adult life, *Science* **216**: 890–892.

Bayer, S. A., 1985, Neuron production in the hippocampus and olfactory bulb of the adult rat brain: Addition or replacement?, *Ann NY Acad Sci* **457**: 163–172.

Benhamou, S., and Poucet, B., 1998, Landmark use by navigating rats (Rattus norvegicus): Contrasting geometric and featural information, *J Comp Psychol* **112**: 317–322.

Bizon, J. L., and Gallagher, M., 2003, Production of new cells in the rat dentate gyrus over the lifespan: relation to cognitive decline, *Eur J Neurosci* **18**: 215-219.

Bliss, T. V., and Lomo, T., 1973, Long-lasting potentiation of synaptic transmission in the dentate area of the anaesthetized rabbit following stimulation of the perforant path, *J Physiol* **232**: 331-356.

Brodbeck, D. R., 1994, Memory for spatial and local cues : a comparison of a storing and a non storing species, *Anim Learn Behav* **22**: 119-133.

Brown, J., Cooper-Kuhn, C. M., Kempermann, G., Van Praag, H., Winkler, J., Gage, F. H., and Kuhn, H. G., 2003, Enriched environment and physical activity stimulate hippocampal but not olfactory bulb neurogenesis, *Eur J Neurosci* **17**: 2042-2046.

Cain, D. P., Saucier, D., and Boon F., 1997, Testing hypotheses of spatial learning: the role of NMDA receptors and NMDA-mediated long-term potentiation, *Behav Brain Res* **84**: 179-93.

Cameron, H. A., and McKay, R. D., 2001, Adult neurogenesis produces a large pool of new granule cells in the dentate gyrus, *J Comp Neurol* **435**: 406-417.

Clayton, N. S., and Krebs, J. R., 1995, Memory in food-storing birds: from behaviour to brain, *Curr Opin Neurobiol* **5**: 149-154.

Clayton, N. S., and Dickinson, A., 1998, Episodic-like memory during cache recovery by scrub jays, *Nature* **395**: 272-274.

Collett, T. S., and Baron, J., 1994, Biological compasses and the coordinate frame of landmark memories in honeybees, *Nature* **368**: 137-140.

Cressant, A., Muller, R. U., and Poucet, B., 1997, Failure of centrally placed objects to control the firing fields of hippocampal place cells, *J Neurosci* **17**: 2531-2542.

Cressant, A., Muller, R. U., and Poucet, B., 1999, Further study of the control of place cell firing by intra-apparatus objects, *Hippocampus* **9**: 423-431.

Dayer, A. G., Ford, A. A., Cleaver, K. M., Yassaee, M., and Cameron, H. A., 2003, Short-term and long-term survival of new neurons in the rat dentate gyrus, *Journal of Comparative Neurology* **460**: 563-572.

Dember, W. N. and Richman, C. L., 1989, *Spontaneous alternation behavior*, Springer Verlag, New York.

Drapeau, E., Mayo, W., Aurousseau, C., Le Moal, M., Piazza, P.-V., and Abrous, D. N., 2003, Spatial memory performances of aged rats in the water maze predict levels of hippocampal neurogenesis, *P Natl Acad Sci USA* **100**: 14385-14390.

Dudchenko, P. A., Goodridge, J. P., Seiterle, D. A., and Taube, J. S., 1997, Effects of repeated disorientation on the acquisition of spatial tasks in rats: dissociation between the appetitive radial arm maze and aversive water maze, *J Exp Psychol Anim B* **23**: 194-210.

Eckenhoff, M. F., and Rakic, P., 1988, Nature and fate of proliferative cells in the hippocampal dentate gyrus during the life span of the rhesus monkey, *J Neurosci* **8**: 2729-2747.

Eilam, D., and Golani, I., 1988, The ontogeny of exploratory behavior in the house rat (Rattus rattus): The mobility gradient, *Dev Psychobiol* **21**, 679–710.

Eriksson, P. S., Perfilieva, E., Bjork-Eriksson, T., Alborn, A. M., Nordborg, C., Peterson, D. A., and Gage, F. H., 1998, Neurogenesis in the adult human hippocampus, *Nat Med* **4**: 1313-1317.

Etienne, A. S., Joris, S., Maurer, R., and Teroni, E., 1990, Enhancing the impact of visual extra-maze cues in a spatial orientation task, *Behav Brain Res* **38**: 199-210.

Etienne, A. S, Maurer, R., and Seguinot, V., 1996, Path integration in mammals and its interaction with visual landmarks, *J Exp Biol* **199**: 201-209.

Etienne, A. S., 2003, How does path integration interact with olfaction, vision and the representation of space ?, in: *The neurobiology of spatial behaviour*, K. Jeffery, ed., Oxford University Press, Oxford.

Etienne, A. S, Maurer, R., Boulens, V., Levy, A., and Rowe T., 2004, Resetting the path integrator: a basic condition for route-based navigation, *J Exp Biol* **207**: 1491-1508.

Fuchs, E., and Gould, E., 2000, Mini-review: in vivo neurogenesis in the adult brain: regulation and functional implications, *Eur J Neurosci* **12**: 2211-2214.

Fuster, J. M., 1995, *Memory in the cerebral cortex: an empirical approach to neural networks in the human and nonhuman primate*, MA: MIT Press, Cambridge.

Gallistel, C. R., 1990, *The organization of learning*, 1st ed., MA: MIT Press, Cambridge.

Gilbert, P. E., Kesner, R. P., and Lee I., 2001, Dissociating hippocampal subregions: double dissociation between dentate gyrus and CA1, *Hippocampus* **11**: 626-636.

Gilbert, P. E., and Kesner, R. P., 2006, The role of the dorsal CA3 hippocampal subregion in spatial working memory and pattern separation, *Behav Brain Res* **169**: 142-149.

Goldman, S. A., and Nottebohm, F., 1983, Neuronal production, migration, and differentiation in a vocal control nucleus of the adult female canary brain, *P Natl Acad Sci USA* **80**: 2390-2394.

Good, M. and Bannerman, D., 1997, Differential effects of ibotenic acid lesions of the hippocampus and blockade of N-methy-D-aspartate receptor-dependent long-term potentiation on contextual processing in rats, *Behav Neurosci* **111**: 1171-1183.

Gothard, K. M., Skaggs, W. E., Moore, K. M., and McNaughton, B. L., 1996, Binding of hippocampal CA1 neural activity to multiple reference frames in a landmark-based navigation task, *J Neurosci* **16**: 823-835.

Gould, E., Reeves, A. J., Fallah, M., Tanapat, P., Gross, C. G., and Fuchs, E., 1999a, Hippocampal neurogenesis in adult Old World primates, *P Natl Acad Sci USA* **96**: 5263-5267.

Gould, E., Beylin, A., Tanapat, P., Reeves, A., and Shors, T. J., 1999b, Learning enhances neurogenesis in the hippocampal formation, *Nat Neurosci* **2**: 203-205.

Granchamp, N., and Schenk, F., 2006, Adaptive changes in a radial maze task: Efficient selection of baited arms with reduced foraging in senescent hooded rats, *Behav Brain Res*, **168**: 161-166.

Gross, C. G., 2000, Neurogenesis in the adult brain: death of a dogma, *Nat Rev Neurosci* **1**: 67-73.

Hafting, T., Fyhn, M., Molden, S., Moser, M. B., and Moser, E. I., 2005, Microstructure of a spatial map in the entorhinal cortex, *Nature* **436**: 801-806.

Hastings, N. B. and Gould, E., 1999, Rapid extension of axons into the CA3 region by adult-generated granule cells, *J Comp Neurol* **413**: 146-154.

Hebb, D., 1949, *The Organization of Behaviour*, Wiley, New York.

Hulse, S. H., Fowler, H., and Honig, W. K., 1978, *Cognitive processes in animal behavior*, Lawrence Erlbaum Associates, Hillsdale.

Ikonen S., McMahan R., Gallagher, M. Eichenbaum, H., and Tanila, H., 2002, Cholinergic system regulation of spatial representation by hippocampus, *Hippocampus* **12**: 386-397.

Jacobs, L. F., and Liman, E. R., 1991, Grey squirrels remember the locations of buried nuts, *Anim Behav* **41**: 103–110.

Jacobs, L. F., 1995, The ecology of spatial cognition: Adaptive patterns of hippocampal size and space use in wild rodents, in: *Studies of the brain in naturalistic settings*, E. Alleva, A. Fasolo, H.-P. Lipp and L. Nadel, eds., Kluwer Academic, Dordrecht, Germany, pp. 301–322.

Jacobs, L., 2003, The evolution of the Cognitive Map, *Brain Behav Evolut* **62**,128-139.

Jacobs, L. F. and Schenk, F., 2003, Unpacking the cognitive map : the parallel map theory of hippocampal function, *Psychol Rev* **110**: 285-315.

Kavushansky, A., Vouimba, R. M., Cohen, H., and Richter-Levin, G., 2006, Activity and plasticity in the CA1, the dentate gyrus, and the amygdala following controllable vs. uncontrollable water stress, *Hippocampus* **16**: 35-42.

Kempermann, G., Kuhn, H. G., and Gage, F. H., 1997, More hippocampal neurons in adult mice living in an enriched environment, *Nature* **386**: 493-495.

Kempermann, G., Gast, D., and Gage, F. H., 2002, Neuroplasticity in old age: sustained fivefold induction of hippocampal neurogenesis by long-term environmental enrichment, *Ann Neurol* **52**: 135-143.

Kentros, C., Hargreaves, E., Hawkins, R. D., Kandel, E. R., Shapiro, M., and Muller, R. V., 1998, Abolition of long-term stability of new hippocampal place cell maps by NMDA receptor blockade, *Science* **280**: 2121-2126.

Kesner, R. P., Lee I., and Gilbert, P., 2004, A behavioral assessment of hippocampal function based on a subregional analysis, *Rev Neurosci* **15**: 333-351.

Knierim, J. J., Kudrimoti, H. S., and McNaughton, B. L., 1995, Place cells, Head direction cells, and the learning of landmark stability, *J Neurosci* **15**: 1648-1659.

Kornack, D. R. and Rakic, P., 1999, Continuation of neurogenesis in the hippocampus of the adult macaque monkey, *P Natl Acad Sci USA* **96**: 5768-5773.

Krebs, J. R., 1983, Animal behaviour. From Skinner box to the field, *Nature* **304**: 117.

Krugers, H. J., Alfarez, D.N., Karst, H., Parashkouhi, K., van Gemert, N., and Joels M., 2005, Corticosterone shifts different forms of synaptic potentiation in opposite directions, *Hippocampus* **15**: 697-703.

Kuhn, H. G., Dickinson-Anson, H., and Gage, F. H., 1996, Neurogenesis in the dentate gyrus of the adult rat: age-related decrease of neuronal progenitor proliferation, *J Neurosci* **16**: 2027-2033.

Lavenex, P. and Schenk, F., 1995, Influence of local environmental olfactory cues on place learning in rats, *Physiol Behav* **58**: 1059-1066.

Lavenex, P. and Schenk, F., 1996, Integration of olfactory information in a spatial representation enabling accurate arm choice in the radial arm maze, *Learn Mem* **2**: 299-319.

Lavenex, P., and Schenk, F., 1998, Olfactory cues potentiate learning of distant visuospatial information, *Neurobiol Learn Mem* **68**: 140-153.

Lee, I., Jerman, T. S., and Kesner, R. P., 2005, Disruption of delayed memory for a sequence of spatial locations following CA1- or CA3-lesions of the dorsal hippocampus, *Neurobiol Learn Mem* **84**: 138-147.

Leuner, B., Mendolia-Loffredo, S., Kozorovitskiy, Y., Samburg, D., Gould, E., and Shors, T. J., 2004, Learning enhances the survival of new neurons beyond the time when the hippocampus is required for memory, *J Neurosci* **24**: 7477-7481.

Leutgeb, S., and Mizumori, S. J., 1999, Excitotoxic septal lesions result in spatial memory deficits and altered flexibility of hippocampal single-unit representations, *J Neurosci* **19**: 6661-6672.

Leutgeb, S., Leutgeb, J. K., Treves, A., Moser, M. B., and Moser, E. I., 2004, Distinct ensemble codes in hippocampal areas CA3 and CA1, *Science* **305**: 1295-1298.

Leutgeb, S. Leutgeb, J.K., Barnes, C.A., Moser, E.I., McNaughton, B. L., and Moser, M. B., 2005, Independent codes for spatial and episodic memory in hippocampal neuronal ensembles, *Science* **309**: 619-623.

Lohmann, K. J., and Lohmann, C. M. F., 1996, Orientation and open-sea navigation in sea turtles, *J Exp Biol* **199**: 73–81.

Lynch, M. A., 2004, Long-term potentiation and memory, *Physiol Rev* **84**: 87-136.

Maaswinkel, H., and Whishaw, I. Q., 1999, Homing with locale, taxon, and dead reckoning strategies by foraging rats: sensory hierarchy in spatial navigation, *Behav Brain Res* **99**: 143-152.

Madsen, T. M., Kristjansen, P. E. G., Bolwig, T. G., and Wörtwein, G., 2003, Arrested neuronal proliferation and impaired hippocampal function following fractionated brain irradiation in the adult rat, *Neuroscience* **119**: 635-642.

Martin, G. M., Harley, C. W., Smith, A. R., Hoyles, E. S, and Hynes, C. A., 1997, Spatial disorientation blocks reliable goal location on a plus maze but does not prevent goal location in the Morris maze, *J Exp Psychol Anim B* **23**:183-193.

Martin, S. J., and Morris, R. G., 2002, New life in an old idea: the synaptic plasticity and memory hypothesis revisited, *Hippocampus* **12**: 609-636.

McDonald, R.J., and White, N. M., 1993, A triple dissociation of memory systems: hippocampus, amygdala, and dorsal striatum, *Behav Neurosci* **107** : 3-22.

McDonald, R. J., Hong, N. S., and Devan, B. D., 2004, The challenges of understanding mammalian cognition and memory-based behaviours: an interactive learning and memory systems approach, *Neurosci Biobehav R* **28**: 719-745.

McGauran, A.M., O'Mara, S. M., and Commins S., 2005, Vestibular influence on water maze retention: transient whole body rotations improve the accuracy of the cue-based retention strategy, *Behav Brain Res* **158**: 183-187.

Merrill, D. A., Karim, R., Darraq, M., Chiba, A. A., and Tuszynski, M. H., 2003, Hippocampal cell genesis does not correlate with spatial learning ability in aged rats, *J Comp Neurol* ***459***: 201-207.

Miller, M. W., and Nowakowski, R. S., 1988, Use of bromodeoxyuridine-immunohistochemestry to examine the proliferation, migration and the time of origine of cells in the central nervous system, *Brain Res* **457**: 44-52.

Morange, M., 2005, *Les secrets du vivant : contre la pensée unique en biologie*, Sciences et société, Editions La Découverte, Paris.

Morris, R. G. M., 1984, Development of a water-maze procedure for studying spatial learning in the rat, *J Neurosci Meth,* **11**: 47-60.

Morris. R. G., Anderson, E., Lynch, G. S., and Baudry, M., 1986, Selective impairment of learning and blockade of long-term potentiation by an N-methyl-D-aspartate receptor antagonist, AP5, *Nature* **319**: 774-776.

Morris, R. G. M., Schenk, F., Tweedie, F., and Jarrard, L., 1990, Ibotenic lesions of the hippocampus and/or subiculum: Dissociating components of allocentric spatial learning, *Eur J Neurosci* **2**: 1016–1029.

Morris, R. G., 2001, Episodic-like memory in animals: psychological criteria, neural mechanisms and the value of episodic-like tasks to investigate animal models of neurodegenerative disease, *Philos T Roy Soc B* **356**: 1453-1465.

Morris, R. G. M., and Day, M., 2003, Places and flavours: one-trial paired associate learning as a model of episodic-like memory in the rat, *International Congress Series* **1250**: 183-198.

Muller, R. U., and Kubie, J. L., 1987, The effects of changes in the environment on the spatial firing of hippocampal complex-spike cells, *J Neurosci* **7**: 1951-1968.

Nader, K., 2003, Memory traces unbound, *Trends Neurosci* **26**: 65-72.

Nakazawa, K., Quirk, M. C, Chitwood, R. A, Watanabe, M., Yeckel, M. F, Sun, L. D, Kato, A., Carr, C. A, Johnston, D., Wilson, M. A., and Tonegawa, S., 2002, Requirement for hippocampal CA3 NMDA receptors in associative memory recall, *Science* **297**: 211-218

Neisser, U., 1976, *Cognition and reality. Principles and implications of cognitive psychology*, Freeman and Compagny, San Francisco.

O'Keefe, J., and Dostrovsky, J., 1971, The hippocampus as a spatial map. Preliminary evidence from unit activity in the freely moving rat, *Brain Res* **34**: 171-175.

O'Keefe, J., 1976, Place units in the hippocampus of the freely moving rat, *Exp Neurol* **51**: 78–109.

O'Keefe, J., and Conway, D. H., 1978, Hippocampal place units in the freely moving rat: why they fire where they fire, *Exp Brain Res* **14**: 573-90.

O'Keefe, J., and Nadel, L., 1978, *The hippocampus as a cognitive map*, Oxford University Press, Oxford, England.

O'Keefe, J., and Speakman, A., 1987, A single unit activity in the rat hippocampus during a spatial memory task, *Exp Brain Res* **68**: 1-27.

Olton, D. S., and Samuelson, R. J., 1976, Remembrance of places passed: spatial memory in rats, *J Exp Psychol Anim B* **2**: 97-116.

Packard, M.G., Hirsh, R., and White, N.M., 1989, Differential effects of fornix and caudate nucleus lesions on two radial maze tasks: evidence for multiple memory systems, *J Neurosci* **9**: 1465-1472.

Packard, M. G., and McGaugh, J. L., 1996, Inactivation of hippocampus or caudate nucleus with lidocaine differentially affects expression of place and response learning, *Neurobiol Learn Mem* **65**: 65-72.

Packard, M. G., 1999, Glutamate infused posttraining into the hippocampus or caudate-putamen differentially strengthens place and response learning, *P Natl Acad Sci USA* **96**: 12881-12886.

Papi, F., 1992, *Animal Homing*, Chapman and Hall, London.

Parron, C., Poucet, B., and Save, E., 2004, Entorhinal cortex lesions impair the use of distal but not proximal landmarks during place navigation in the rat, *Behav Brain Res* **154**: 345-52.

Peissner, W., Kocher, M., Treuer, H., and Gillardon, F., 1999, Ionizing radiation-induced apoptosis of proliferating stem cells in the dentate gyrus of the adult rat hippocampus, *Mol Brain Res* **71**: 61-68.

Poucet, B., Save, E., and Lenck-Santini, P. P., 2000, Sensory and memory properties of hippocampal place cells, *Rev Neurosci* **11**: 95-111.

Poucet B., Lenck-Santini, P. P., Paz-Villagran V., and Save E., 2003, Place cells, neocortex and spatial navigation: a short review, *J Physiology-Paris* **97**: 537-546.

Rakic, P., 1985a, Limits of neurogenesis in primates, *Science* **227**: 1054-1056.

Rakic, P., 1985b, DNA synthesis and cell division in the adult primate brain, *Ann NY Acad Sci* **457**: 193-211.

Restle, F., 1957, Theory of selective learning with probable reinforcements, *Psychol Rev* **64**: 182-191.

Rossier, J., Grobety, M.-C., and Schenk, F., 2000, Spatial learning by rats across visually disconnected environments. *Anim Learn Behav* **28**: 16–27.

Rossier J., Kaminsky Yu., Schenk F. and Bures J., 2000, The place preference task, a new tool for studying the relation between behavior and place cell activity in rats, *Behav Neurosci* **114**: 273-284.

Rossier J. and Schenk F., 2003, Olfactory and/or Visual Cues for spatial navigation through ontogeny : olfactory cues enable the use of visual cues, *Behav Neurosci* **117**: 412-425.

Rotenberg, A., Mayford M., Hawkins, R. D., Kandel, E. R., and Muller, R. U., 1996, Mice expressing activated CaMKII lack low frequency LTP and do not form stable place cells in the CA1 region of the hippocampus, *Cell* **87**: 1351-1361.

Rotenberg A., Abel, T., Hawkins, R. D., Kandel, E. R., and Muller, R. U., 2000, Parallel instabilities of long-term potentiation, place cells, and learning caused by decreased protein kinase A activity, *J Neurosci* **20**: 8096-8102.

Saucier, D., and Cain, D. P., 1995, Spatial learning without NMDA receptor-dependent long term potentiation, *Nature* **378**: 186-189.

Save, E., Nerad, L., and Poucet, B., 2000, Contribution of multiple sensory information to place field stability in the hippocampal place cells, *Hippocampus* **10**: 64-76.

Schenk, F., and Morris, R. G. M., 1985, Dissociation between components of spatial memory in rats after recovery from the effects of retrohippocampal lesions, *Exp Brain Res* **58**: 11–28.

Schenk, F., 1989, A homing procedure for studying spatial memory in immature and adult rodents, *J Neurosci Meth*, **26**: 249–258.

Schenk F., Contant B., and Werffeli, P., 1990, Intrahippocampal cholinergic grafts in aged rats compensate impairments in a radial maze and in a place learning task, *Exp Brain Res* **82**: 641-50.

Schenk, F., and Grobéty, M.-C., 1992, Interactions between directional and visual environmental cues in a radial maze task and in spontaneous alternation tests. *Learn Motiv* **23**: 80-98.

Schenk, F., Grobéty, M.-C., Lavenex, P., and Lipp, H.-P., 1995, Dissociation between basic components of spatial memory in rats, in: *Behavioral Brain Research In Naturalistic And Semi-Naturalistic Settings: Possibilities And Perspectives*, E. Alleva, A. Fasolo, H-P. Lipp, L. Nadel and L. Ricceri, eds., Kluwer Press, Dordrecht, The Netherlands, pp. 277-300.

Schenk, F., 1998, The Morris water maze (is not a maze), in: *Interacting with the environment: a handbook of spatial research paradigms and methodologies*, N. Foreman and R. Gillett, eds., Psychology Press, Hove.

Schmidt-Hieber, C., Jonas, P., and Bischofberger, J., 2004, Enhanced synaptic plasticity in newly generated granule cells of the adult hippocampus, *Nature* **429**: 184-187.

Schöne, H., 1984, *Spatial Orientation: The Spatial Control of Behavior in Animals and Man*, Princeton University Press, Princeton, NJ.

Seki, T., and Arai, Y., 1995, Age-related production of new granule cells in the adult dentate gyrus, *Neuroreport* **6**: 2479-2482.

Sharp, P. E, Kubie, J. L., and Muller, R. U.,1990, Firing properties of the hippocampal neurons in a visually symmetrical environment: Contributions of multiple sensory cues and mnemonic processes, *J Neurosci* **10**: 3093-3105

Shettleworth, S. J., and Krebs, J. R., 1982, How marsh tits find their hoards: the roles of site preference and spatial memory, *J Exp Psychol Anim* **8** 354-375.

Shors, T. J., Townsend, D. A., Zhao, M., Kozorovitskiy, Y., and Gould, E., 2002, Neurogenesis may relate to some but not all types of hippocampal-dependent learning, *Hippocampus* **12**: 578-584.

Smith, P. F., Horii, A., Russell, N., Bilkey, D. K., Zheng, Y., Liu, P., Kerr, D. S., and Darlington, C. L., 2005, The effects of vestibular lesions on hippocampal function in rats, *Prog Neurobiol* **75**: 391-405.

Snyder, J. S., Hong, N. S., McDonald, R. J., and Wojtowicz, J. M., 2005, A role for adult neurogenesis in spatial long-term memory, *Neuroscience* **130**: 843-852.

Steele, R. J., and Morris, R. G. M., 1999, Delay-dependent impairment of a matching-to-place task with chronic and intrahippocampal infusion of the NMDA-antagonist D-AP5, *Hippocampus* **9**: 118-136.

Suzuki, S. W., Augerinos, G., Black, A. H., 1980, Stimulus control of spatial behaviour on the eight-arm radial maze in rats, *Learn Motiv* **11**: 1-18.

Swanson, L. W., 1983, The hippocampus and the concept of the limbic system, in: *Neurobiology of the hippocampus*, W. Seifert, ed., Academic Press, New York, pp. 3-19.

Tanila, H., Sipila, P., Shapiro, M., and Eichenbaum, H., 1997a, Brain aging: impaired coding of novel environment cues, *J Neurosci* **17**: 5167-5174.

Tanila, H., Shapiro, M., Gallagher, M., and Eichenbaum, H., 1997b, Brain aging: changes in the nature of information coding by the hippocampus, *J Neurosci* **17**: 5156-5166.

Tinbergen, N., 1963, On the aims and methods of ethology, *Z Tierpsychol* **20**: 410–433.

Tolman, E. C., 1948, Cognitive maps in rats and men, *Psychol Rev* **55**: 189–208.

Van Praag, H., Shubert, T., Zhao, C., and Gage, F. H., 2005, Exercice enhances learning and hippocampal neurogenesis in aged mice, *J Neurosci* **25**: 8680-8685.

Vazdarjanova, A., and Guzowski, J. F., 2004, Differences in hippocampal neuronal population responses to modifications of an environmental context: evidence for distinct, yet complementary, functions of CA3 and CA1 ensembles, *J Neurosci* **24**: 6489-6496.

Wallraff, H. G., 1996, Seven theses on pigeon homing deduced from empirical findings, *J Exp Biol* **199**: 105–111.

Wati, H., Kudo, K., Oiao, C., Kuroki, T., and Kanba, S., 2006, A decreased survival of proliferated cells in the hippocampus is associated with a decline in spatial memory in aged rats, *Neurosci Lett* **399**: 171-174..

Wehner, R., Michel, B., and Antonsen, P., 1996, Visual navigation in insects: coupling of egocentric and geocentric information, *J Exp Biol* **199**: 129-40.

Whishaw, I. Q., Cassel, J.-C., and Jarrard, L. E., 1995, Rats with fimbria-fornix lesions display a place response in a swimming pool: A dissociation between getting there and knowing where, *J Neurosci* **15**: 5779–5788.

Whishaw, I. Q., 1998, Place learning in hippocampal rats and the path integration hypothesis, *Neurosci Biobehav R* **22**: 209-220.

Whishaw, I. Q., and Gorny, B., 1999, Path integration absent in scent-tracking fimbria-fornix rats: evidence for hippocampal involvement in "sense of direction" and "sense of distance" using self-movement cues, *J Neurosci* **19**: 4662-4673.

White, N. M., and McDonald, R. J., 2002, Multiple parallel memory systems in the brain of the rat, *Neurobiol Learn Mem* **77**: 125-184.

Wiener, S., and Schenk F., 2005, Behavioral studies of directional orientation in developing and adult animals, in: *Head direction cells and the neural mechanisms underlying directional orientation*, J. Taube and S. Wiener, eds., MIT Press, Cambridge, pp. 247-274.

Wiener, S., and Taube, J. S., 2005, *Head direction cells and the neural mechanisms underlying directional orientation*, MIT Press, Cambridge.

Wilson, M. A., and McNaughton, B. L., 1994, Reactivation of hippocampal ensemble memories during sleep, *Science* **265**: 676-679.

Winocur, G., Wojtowicz, J. M., Sekeres, M., Snyder, J. S., and Wang, S., 2006, Inhibition of neurogenesis interferes with hippocampus-dependent memory function, *Hippocampus* **16**: 296-304.

Chapter 15

SENSORIMOTOR TRANSFORMATIONS IN SPATIAL ORIENTATION RELATIVE TO GRAVITY

Bernhard J. M. Hess
Department of Neurology, University Hospital Zurich, Switzerland

Abstract: Self-orientation in space is based on multisensory interactions of visual, vestibular and somatosensory-proprioceptive signals. In this article, we analyze vestibular signal processing in terms of its capacity to provide inertial cues for self-orientation in space. We show that vestibular signals from both the otolith organs and the semicircular canals must be processed in a bootstrap-operation like manner in order to obtain true inertial head-in-space orientation.

Key words: Inertial signal processing; semicircular canals; otoliths; sensori-motor transformations; inertial navigation.

1. INTRODUCTION

Self-orientation in space during locomotion involves stabilization of the head, which carries the sense organs for visual, vestibular and auditory orientation. Although proprioceptive signals from the somatosensory system are involved in the control of head and body posture, it is the vestibular system that is of primary importance in guiding the head as an inertial platform to allow free visual orientation during walking, running and jumping without risking to fall. In the following we are concerned with the performance of the vestibular system in terms of decoding inertial motion signals from the peripheral sense organs. Our focus is on the computational capacity of this system in providing a gravity-centered reference frame to enhance spatial orientation during complex locomotor behavior.

2. THE FORWARD MODEL OF VESTIBULAR INFORMATION PROCESSING

Under natural conditions, the vestibular organs of the inner ear, the utricular and saccular otoliths and the semicircular canals operate like inertial detectors of angular and linear head acceleration. Over a large range of frequencies, the output of the semicircular canals corresponds to that of an integrating angular accelerometer providing head angular velocity information whereas the otoliths convey information on linear acceleration of the head, including inertial and gravitational components. Since there is no way to distinguish gravitational from inertial accelerations on physical grounds (Einstein's equivalence principle of gravity and inertial acceleration), the problem of linear acceleration encoding that we will also call the *forward model* can be summarized in the following vector equation (vectors will systematically be denoted in bold face):

$$\mathbf{a}(t) = \mathbf{g}(t) + \mathbf{f}(t) \qquad (1)$$

In this equation **g**(t) is the gravitational acceleration and **f**(t) is the inertial acceleration acting on the otoliths in the inner ear as a function of time that primary vestibular afferents collectively signal to the brain.

Consider a forward step in walking, where the head undergoes a transient acceleration, as schematically illustrated in Fig. 15-1 A. At a given point in time, the otolith organs encode the net linear acceleration that had changed from an initial pure downward directed acceleration (denoted by **g** in Fig. 15-1 A) to an acceleration directed downward and backward (vector **a** in Fig. 15-1 A) whereas the angular acceleration of the head, encoded by the semicircular canals, is zero during a pure translation. Both signals are encoded in some head-fixed coordinates, symbolized by the x-, y-, and z-axis in Fig. 15-1 A. To distinguish these signals from forces acting on the head relative to a space-fixed frame we will call them *inertial signals*. The task of the brain is to ultimately interpret the inertial motion of the head encoded by the vestibular afferent signals in a space-fixed frame, in the following called gravity-centered reference frame in order to estimate self-orientation and appropriately control motor behavior. We will call this task the *inverse problem of vestibular signal processing*. It involves the following two interrelated information processing steps (Hess and Angelaki 1997):

(a) The brain needs to express head angular velocity in a gravity-centered reference frame. Notice that the angular velocity from the semicircular canals are coded in a head-fixed frame that has no a priori established relation to any kind of space-fixed frame.

(b) The brain needs to know head orientation relative to gravity. For this, it must parse the afferent information from the otolith signals into its gravitational and inertial components.

The two tasks are interrelated because head angular velocity cannot be interpreted in a spatial frame without prior knowledge of head orientation relative to gravity. If the normally continuous updating of self-orientation relative to a gravity-centered reference is interrupted or disturbed, the brain can no longer correctly reference the signals from the semicircular canals with respect to space. Conversely, the process of parsing the net linear head acceleration into gravitational and inertial acceleration components requires in turn information about head angular velocity in a spatial reference frame. If this process of referencing of the head angular velocity signals to the physical space is disturbed, the brain can no longer correctly parse linear head acceleration signals into its gravitational and inertial components. To solve this problem, which we refer to as the *inverse problem of vestibular signal processing*, it appears that the vestibular spatial orientation mechanisms must be bootstrapped by an a priori estimate of head-in-space orientation. This process involves proprioceptive signals from the somatosensory and visual system. In the following paragraphs we will address the essential computational steps that are involved in a solution of the inverse problem.

3. AMBIGUOUS STIMULUS CONFIGURATIONS

How can the brain solve the inverse problem of vestibular signal processing? Before dwelling on this issue, let us first define the term "head attitude" as "the head's orientation relative to gravity". Notice that specifying head attitude involves specifying only two of the three rotational degrees of freedom of head-in-space orientation. Thus, any given head attitude encompasses infinitively many orientations according to the infinitively many heading directions that the subject can assume in a plane parallel to earth horizontal. Having stated this, we ask what knowledge the brain does need to have at hand in order to obtain a reliable estimate of head attitude. Clearly, all it needs to know is an estimate of the current orientation of gravitational acceleration in some head-fixed coordinates (vector **g** expressed in some x, y, z frame as illustrated in Fig. 15-1 A). Since gravitational accelerations are not distinguishable from other inertial accelerations, it is clear that the inverse problem cannot be solved without some additional information. This ambiguity is illustrated in Fig. 15-1 B which shows that there are infinitively many combinations of head orientations relative to gravity that can give rise to exactly the same net acceleration during a translation:

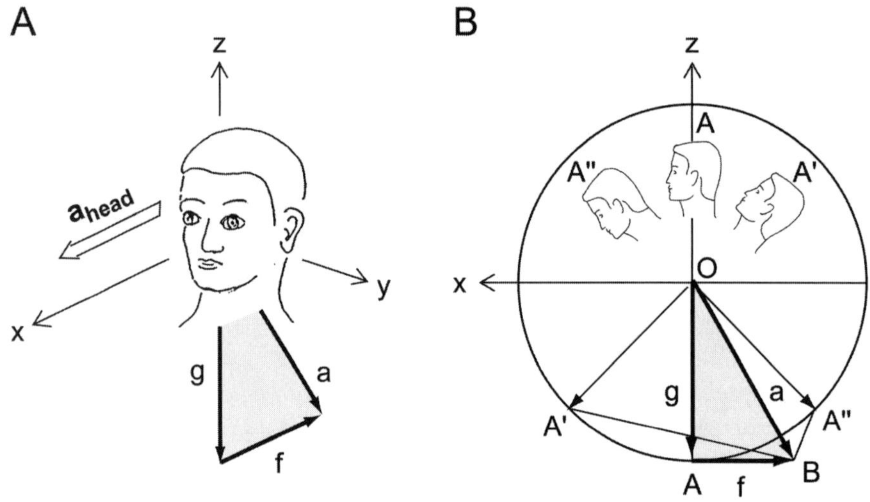

Figure 15-1. A. Forward model: Head attitude determines the action of gravity (**g**) and inertial acceleration (**f**) on the vestibular sensors in the inner ear (not shown). The brain analyses these inertial motion signals, shown here during a forward step (arrow $\mathbf{a}_{head}$ denotes the vector of translational acceleration of the head), in some intrinsic head-fixed coordinates, conceived here as a Cartesian frame with coordinates x, y, z. *B.* Inverse problem of vestibular information processing: The net acceleration (**a**) acting on the otolith organs within the x-z frame is compatible with infinitively many head attitudes, amongst them upright attitude (A), forward tilted attitudes (A') or backward tilted attitudes (A''). Note that each of these head attitudes corresponds to a different estimation of inertial acceleration **f** (= **a**-**g**). Note also that $\mathbf{a}_{head} \neq \mathbf{a}$.

Obviously, the vector **g** can point from O to any point A' on a circle of radius |**g**| around O. The translational acceleration **f** that has to be added to the vector **g** to complement the measured net head acceleration **a** is given by connecting point A' with B. In the following we will show that the solution to the inverse problem involves information about head-in-space rotation and rate of change of acceleration.

4. SOME REASONS WHY JERK SIGNALS ARE USEFUL

To separate components due to gravity from those due to acceleration, the processing of head acceleration signals must be based on the prior assumption that gravity is the constant portion and inertial acceleration is the transient portion (Mayne, 1974). This goal can be achieved by processing the rate of change of head acceleration, also called jerk, instead of the net acceleration in order to reject the constant value representing gravity. If the head attitude were constant over time, a simple integration of jerk signals would suffice to estimate the inertial linear acceleration vector of the head up to three free integration constants (linear acceleration = $\int \mathbf{da} = \mathbf{a} + \mathbf{c}$ where d**a**: increment in net acceleration over time interval dt, **c**: vector of integration constants). In general, however, motor control of head attitude is challenged by rotational disturbances as well such that the rate of change of net head acceleration (net jerk) includes a vector term due to the rate of change of gravity (gravitational jerk). One might ask, how there could be a finite rate of change of gravity although gravity is everywhere constant in magnitude. The only way this can happen is by a change of head orientation relative to gravity, i.e. a change in head attitude (see Fig. 15-1 B). A non-zero rate of change of gravity is thus indicative for a rotation of the head relative to the everywhere constant external field of gravity. The following equation captures mathematically the forward model for such rotation:

$$d\mathbf{g} / dt = -\omega_{\text{head}}(t) \times \mathbf{g}(t) = \omega(t) \times \mathbf{g}(t) \tag{2}$$

In this equation the vector ω(t) represents the angular velocity of **g** as a function of time, which is always directed opposite to the angular velocity of the head, ω_{head}, and "x" denotes the cross product. Thus, if the head rotates from upright towards left ear down, the gravitational jerk vector d**g**/dt points towards the left ear. Obviously, only head velocity components perpendicular to gravity will change head attitude. The solution to this ordinary linear differential equation is completely determined by the initial head attitude ($\mathbf{g}_0$= head attitude at time t=0, i.e. before motion onset) and the time evolution of head angular velocity. For small tilts:

$${}_0 + \varphi(t) \times \mathbf{g}_0 \text{ with } \varphi(t) = \int_0^t \omega(s)ds$$

For larger tilts, higher order terms in angular velocity (i.e., terms of the form $^1/_{k!}\Phi(t)^k\mathbf{g}_0$, k>1, with $\Phi(t)= \int\Omega(s)ds$ and $\Omega=[\omega_{ik}]$ etc.) must be included. The translational head acceleration is obtained by subtracting the current head attitude that is a solution to Eq. (2) from the net acceleration at each instant in time. Thus, Eq. (2) suggests in fact that updating of head attitude can be conceived as a feed forward process, starting from an initial head attitude $\mathbf{g}_0$, in contrast to what current inverse models of vestibular processing seem to imply by representing the respective computations in a feed back form: $\mathbf{g} = \int \boldsymbol{\omega} \times \mathbf{g}\, dt$. There are two important features that characterize a feed forward mechanism, one is minimal reaction times and the other is the requirement of calibration. To warrant appropriate reactions in motor control of head attitude a quick discrimination of whether the head is going to tilt or to translate or both can be crucial. This latter process is often summarized as tilt-translation discrimination or gravito-inertial force resolution, which in its most general form involves the two computational steps summarized above as inverse problem of vestibular processing. Reliable calibration of head attitude is probably a multi-sensory process, which includes proprioceptive information from the somatosensory system as well as from the visual system. None of these non-vestibular factors is, however, as directly linked to the inertial motion of the head as the vestibular signals.

Combining Eqs. (1)-(2) leads to the following forward model for the rate of change of net acceleration or jerk (with $\boldsymbol{\omega} = -\boldsymbol{\omega}_{head}$):

$$d\mathbf{a} / dt = d / dt(\mathbf{g} + \mathbf{f}) = \boldsymbol{\omega} \times \mathbf{g} + d\mathbf{f} / dt \qquad (3)$$

The first term on the right-hand side of Eq. (3) gives the change in net head acceleration due to the rate of change in head attitude while the second term describes the rate of change in translational acceleration. It has long been established that otolith afferent signals convey information about the net head acceleration as described by Eq. (1), yet many of these afferents carry a combination of both net head acceleration and the rate of change of net acceleration as described by Eq. (3) (Fernandez and Goldberg, 1976a, 1976b, 1976c). The reasons for this are not clear and have not attracted the due attention in vestibular research (see e.g., Hess 1992), although they are likely connected to an economic way of solving Eq. (3) in terms of head attitude and translation. Interestingly, the proportion of otolith afferents carrying jerk information differs depending on the species. Such differences in peripheral processing of net head acceleration signals likely reflect not only differences in locomotor behavior but also in central processing. Accordingly, some modeling work dealing with the inverse problem use central high-pass filters or rely implicitly on high-pass filter properties with the same effect of eliminating the constant portion of gravity (see e.g., Green and Angelaki, 2004).

A key question in vestibular signal processing is how the brain does manage to efficiently extract head orientation relative to gravity, i.e. head attitude, in order to measure the inertial acceleration of the head from the net acceleration signals. At first sight, it appears that this can easily be achieved if only the brain has access to independent information about head angular velocity (symbolized by the vector $\boldsymbol{\omega}_{head}$ on the right hand side of Eqs. (2) - (3)).

The rationale behind this assumption is that the rate of change of head attitude in Eq. (3) can be computed from the initial head attitude by means of head angular information from the semicircular canals. Estimation of the inertial acceleration of the head thus simply appears to require integration of the net jerk signals after having subtracted the rate of change of head attitude: $\int d\mathbf{a} - \int \omega \times \mathbf{g}\, dt = \mathbf{f} + \mathbf{c}$ (**c**: vector of integration constants).

To make this strategy work an accurate estimation of head attitude (second integral on the left) is crucial. A number of models have addressed this problem in different behavioral contexts (Merfeld et al., 1993; Merfeld, 1995; Mergner and Glasauer, 1999; Zupan et al., 2002; Green and Angelaki, 2003; Green and Angelaki, 2004; Green et al. 2005). Relatively little is known about its neuronal implementation. Only recently it has become possible to successfully tackle the intricate neuronal mechanisms using new stimulation techniques (see e.g. Angelaki et al 2004, Green et al 2005).

To what extent the brain masters inertial processing is demonstrated by locomotor behaviors that involve complex head and body movements like, for example, when monkeys jump and swing seemingly effortless from tree to tree while scrutinizing the visual surround. The problem thereby is not only that the semicircular canals encode angular head velocity faithfully only over a restricted frequency range. Rather the problem is that the semicircular canal velocity signals are intrinsically coded in coordinates that change their relation to gravity depending on current head attitude. Clearly if head attitude is not specified a priori, the orientation of head angular velocity with respect to gravity is likewise not specified.

5. SOLVING THE INVERSE PROBLEM OF VESTIBULAR INFORMATION PROCESSING

The formulation of the forward model in Eq. (3) is equivalent to the following equation (Viéville and Faugeras, 1990; Hess and Angelaki, 1997; Angelaki et al.,1999; with $\omega = -\omega_{head}$):

$$d\mathbf{f} / dt - \omega \times \mathbf{f} = d\mathbf{a} / dt - \omega \times \mathbf{a} \qquad (4)$$

This non-homogeneous ordinary differential equation can explicitly be solved for the translational acceleration (**f**) in terms of the net acceleration (**a**) and the net jerk (d**a**/dt), both singals being in principle collectively available to the brain from the otolith afferent inputs. At this stage we are not concerned with the mathematical details of the solution to Eq. (4). What is more important in this context is to emphasize that the spatial properties of this solution are linked to the coordinates, in which head angular velocity (represented by the intrinsic vector signal $\omega = -\omega_{head}$) is expressed. Since the translational acceleration is necessarily expressed in inertial, i.e. gravity-centered coordinates, it follows that the head angular velocity must also be expressed in such coordinates. Remember, however, that the intrinsic coordinates of head angular velocity from the semicircular canals are head-fixed coordinates. The crucial question therefore is how the vestibular system deals with these different coordinate systems. Although an equivalent set of equations have been used to postulate that head acceleration signals are centrally resolved into estimates of gravitational and translational accelerations (Mergner and Glasauer, 1999; Merfeld and Zupan, 2002; Green and Angelaki, 2004), we believe that the key to a deeper understanding of how to solve the inverse problem lies in the geometric link between properties of centrally integrated head acceleration and velocity signals (see inverse problem, point (a) and (b) above). This approach has only begun to be successfully exploited (Green and Angelaki 2004, Green and Angelaki 2005)

A proposal of the geometric link between centrally processed linear acceleration (otolith) and angular velocity (semicircular canals) signals in a common inertial reference frame is outlined in the flow diagram in Fig. 15-2. It suggests the existence of a tight geometric connection between the integrator network (for short called G-integrator, upper part of Fig. 15-2) that integrates net acceleration and jerk signals from the otolith organs and a complementary integrator (for short called Ω-integrator, lower part of Fig. 15-2) that integrates semicircular canal signals. Experimental evidence for the existence of such interconnected central integration processes comes from studies of the orientation properties of the vestibulo-ocular reflex (VOR). First, it has long been observed that time constants of the eye velocity decline of optokinetic and vestibulo-ocular responses following a step in constant velocity rotation were considerably longer than the activity generated by the semicircular canals. Since the central vestibular system appears to store the semicircular canal afferent velocity signals it has been termed "velocity storage system" (Ter Braak, 1936; Mowrer, 1937; Cohen et al., 1977; Raphan et al., 1977, 1979, Robinson 1977). Second, it has been found that the VOR not only stabilizes gaze in space against rotational or translational disturbances of the head, both usually occurring at high frequencies, but that it has also important orienting functions during head tilts that exhibit slower dynamics.

Concurrently, it was found that the time constants of the VOR depend on head orientation relative to gravity (Matsuo and Cohen, 1984; Raphan and Cohen, 1988; Raphan and Schnabolk, 1988; Angelaki et al., 1991; Dai et al., 1991; Raphan and Sturm, 1991; Wearne et al., 1999).

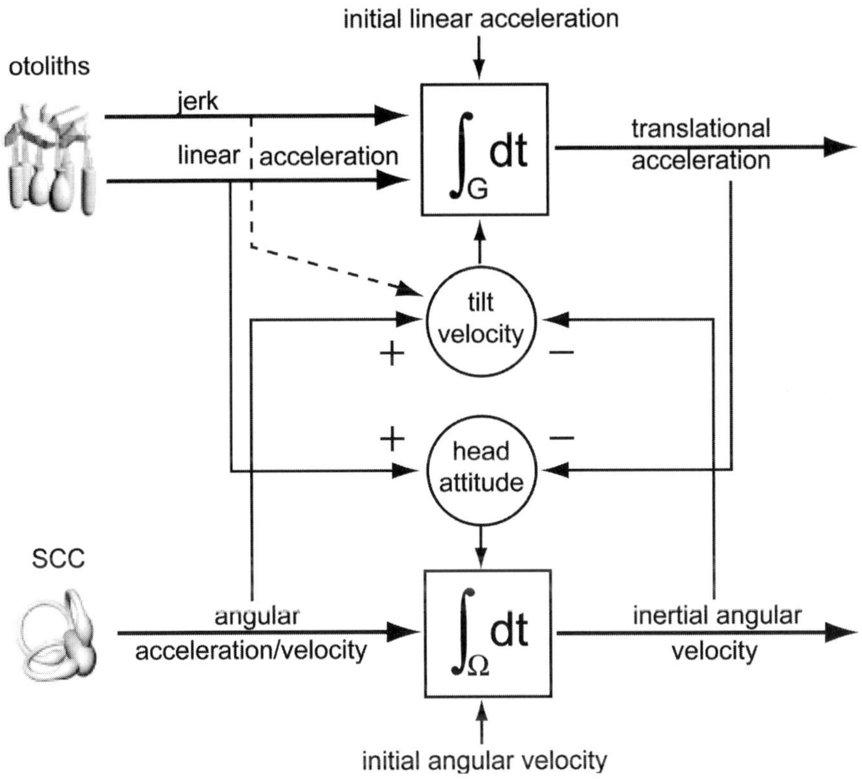

Figure 15-2. Inertial vestibular signal processing. To solve the inverse problem of vestibular signal processing, the otolith and semicircular canal signals feed into a network of two interconnected integrators, each with different spatiotemporal properties. The *otolith signal integrator* (upper part) processes linear acceleration and jerk signals to compute an estimate of translational acceleration. For this, it requires tilt velocity signals in gravity-centered coordinates that are computed either from semicircular canal signals and/or by collectively processing linear acceleration and jerk signals (dashed line). The *semicircular canal signal integrator* (lower part) transforms angular acceleration and velocity signals from head-fixed into gravity-centered coordinates. For this, it requires head attitude signals that are computed from tilt velocity and linear acceleration/jerk signals. Its output is an angular velocity signal that is aligned with gravity and, in turn, serves to compute tilt velocity in gravity-centered coordinates. The outputs of this network drive the translational VOR and/or can be observed in the postrotatory VOR.

And finally, it has been proposed that all these phenomena are manifestations of a central vestibular network that computes head velocity in gravity-centered coordinates (Angelaki and Hess 1994; Hess and Angelaki 1997).

Although it has long been proposed that the brain integrates information from both the otolith and the semicircular canal afferents to differentiate tilt from translation (Guedry, 1974; Mayne, 1974; Young, 1984), the relevance of such integration process has only recently experimentally been demonstrated in a series of elegant studies exploring inertial motion estimation during translation (Angelaki et al., 1999; Green and Angelaki, 2003; Angelaki et al., 2004; Green and Angelaki, 2004).

How does this parallel integrator network operate? A key feature of the here proposed organization of this central network is that each integrator receives crossed parametric inputs from the other integrator in order to guarantee operation in the same gravity-centered coordinate system. The sequence of operations can be conceived as follows: First, the G-integrator (upper part of Fig. 15-2) transforms acceleration and jerk signals into translational acceleration based on initial linear acceleration and head tilt velocity, which is coded in gravity-centered coordinates. This transformation implements the formal solution to Eq. (3). The required head tilt velocity in gravity-centered coordinates (see box "tilt velocity" in Fig. 15-2) comes essentially from the semicircular canals after a spatial transformation by the Ω-integrator. The output of the G-integrator is translational acceleration. The head attitude signal results from subtracting translational acceleration from the net acceleration input. Second, the Ω-integrator transforms semicircular canal signals into inertial angular velocity, based on initial angular velocity and head attitude information, which in turn depends on the G-integration network. This transformation step is needed to keep the coordinates of the Ω-integrator gravity-centered. A crucial point in this sequence of transformations is the question, how head tilt velocity in gravity-centered coordinates is ultimately generated. In essence, we propose that head tilt velocity results by vector subtraction of the inertial velocity signal from a dynamically adjusted semicircular canal signal. This adjustment could occur upstream or inside the G-integrator network (not explicitly shown in the scheme of Fig. 15-2). The tilt velocity signals originate from two different sensory sources: One source comprises velocity signals that come directly from the anterior and posterior semicircular canals and indirectly through the Ω-integrator. The other source results from otolith processing and will be considered in more detail in the paragraph below.

Why can the vestibular system not simply use the anterior and posterior semicircular canal signals to estimate head tilt velocity? Why should these signals be modified by a signal that passes through the Ω-integrator as proposed in this flow diagram? The reason for this transformation process is that the value of the semicircular canal velocity signals in terms of indicating a change in head attitude ("the tilt value") itself depends on head attitude (Green and Angelaki 2004; Green et al 2005). For example, if the head rotates in the roll plane (y-z plane, Fig. 15-1 A), the tilt value of the roll velocity signals, sent to the brain from the anterior and posterior vertical canals, would be maximal in upright position yet minimal in supine position. Therefore, we propose that the Ω-integrator signal serves as an error signal to correct the anterior and posterior canal velocity signals as a function of current head tilt.

From studies of the postrotatory VOR in the roll or pitch plane in monkeys, there is evidence that the Ω-integrator network transforms head angular velocity through a projection mechanism (Angelaki and Hess, 1994; Angelaki et al., 1995; Hess et al., 2005). Such a mechanism is perfectly suited to correct the tilt value of these signals as a function of head position because it filters out velocity components orthogonal to gravity. The vector difference of the input to and the output of the Ω-integrator thus signals tilt velocity in gravity-centered coordinates (Fig. 15-2). Notice that in head upright position this error signal is relatively small because velocity signals from the anterior and posterior semicircular canals will be close to orthogonal to gravity. On the other hand, signals from the lateral semicircular canal are unlikely to contribute to the estimation of tilt velocity. Studies in rhesus monkeys have suggested that the underlying spatial transformation mechanisms for these signals correspond to a rotation rather than a projection (Angelaki and Hess, 1994; Angelaki et al., 1995; Jaggi-Schwarz et al., 2000). If tilt velocity would be estimated by calculating the difference between the direct and the rotated semicircular canal signals, updating of head attitude would be incomplete, resulting in considerable underestimation of the head tilt at large tilt angles.

So far most evidence for the proposed transformation schema comes from studies of the VOR in animals. Some studies of the 3D spatio-temporal properties of postrotatory VOR and optokinetic nystagmus/afternystagmus in humans have suggested similar properties as those observed in rhesus monkeys (Harris and Barnes, 1987; Gizzi et al., 1994). Others studies have questioned the existence or effectiveness of these orientation responses in humans (Fetter et al., 1992, 1996).

6. SUBSTITUTES OF HEAD ANGULAR VELOCITY INFORMATION

The reliability of the forward model (Eq. 3) and the solution to the inverse model (Eq. 4) critically depends on the quality of head angular velocity detection. If the brain does not appropriately estimate head angular velocity (represented by the vector ω) from sensory inputs then the inertial acceleration of the head cannot be correctly computed from the model

Eq. (4). As a consequence, it will affect and limit one's motion repertoire because it is vital to obtain appropriate information about changes in head attitude during complex locomotor behavior. Studies of the dynamics of the peripheral vestibular system (Fernandez and Goldberg, 1971) have shown that the assumption of faithful head velocity coding is appropriate only in a limited frequency range. To estimate head velocity over a larger range, there exists adaptive dynamic compensation mechanisms based on neural representations of inverse models of the peripheral sensors. These mechanisms, however, are bound to be vulnerable in the low frequency working range because of the drop in gain and the decreasing signal-to-noise ratios of afferent responses. It is in the low frequency range where the inverse problem becomes ill-posed due to missing sensory information. If external visual references become unavailable or unreliable spatial disorientation will result, often due to tilt-translation illusions (Guedry, 1974; Young ,1984; Bos and Bles, 2002).

From studies of the vestibulo-ocular reflex (VOR) it has long been known that the otolith afferent signals can also provide head angular velocity information (for a short review see Hess and Angelaki, 1993). Many models have been proposed to explain the underlying signal processing (Hain, 1986; Raphan and Schnabolk, 1988; Angelaki, 1992a, 1992b; Hess, 1992; Mergner and Glasauer, 1999). In general, these models address the question of how angular velocity can be extracted from the gravitational jerk and acceleration signals (see Eq. 2), without addressing the estimation problem of tilt-translation discrimination. However, if we assume that the brain does a priori not know that the net head acceleration is identical to the gravitational acceleration, Eq. (2) must be replaced by a more general model, which includes a rate of change of acceleration magnitude beside the usual orientation term:

$$d\mathbf{a} / dt = \tilde{\omega} \times \mathbf{a} + d\,|\mathbf{a}|/dt\ \hat{\mathbf{a}} \quad \text{where } \hat{\mathbf{a}} = \mathbf{a}/|\mathbf{a}| \tag{5}$$

This equation states that the rate of change of acceleration can be represented as a vector sum of a jerk orientation term ($\tilde{\omega} \times \mathbf{a}$) and a term, which we call for short jerk magnitude term that describes the rate of change in magnitude in the current direction of acceleration. This latter term is conveniently expressed with the help of a vector ($\hat{\mathbf{a}}$) that points along the same direction as the net acceleration (**a**), yet has unit length ($|\hat{\mathbf{a}}|=1$). The physical interpretation of the jerk orientation term is ambiguous in the sense that in general it represents a combination of a real and an apparent tilt velocity with respect to gravity: $\tilde{\omega} = \omega_{real} + \omega_{apparent}$. We will refer to it as *general tilt velocity* because it describes the tilt velocity of the head with respect to the net acceleration (Fig. 15-1 A). This general tilt velocity ($\tilde{\omega}$) can be computed from the net acceleration (**a**) and the net jerk signals (d**a**/dt). It is thus a unique function of pieces of information that are available to the brain based on the otolith afferent input signals. Although this is the case, it is not possible to parse it into its true and apparent tilt velocity component without additional information. Since robust independent cues about head angular velocity are hard to obtain in the very low frequency domain, the vestibular system by itself is prone to confuse real and apparent head tilt in this domain.

Studies of the VOR during constant velocity rotation about an off-vertical axis suggest that the brain makes use of the general tilt velocity signal as rotational cue, which is, as mentioned earlier, computationally accessible on the basis of net otolith inputs alone. In this context, a reliable signature for a pure rotation would be the disappearance of the second term on the right-hand side of Eq. (5) during the entire movement. Geometrically this property indicates that the jerk vector is continuously perpendicular to the net acceleration vector (see Eq. 2). Estimation of this cue involves a computation of the sum of the products of the net jerk and net acceleration components (an algebraic operation called scalar product). If this product sums up to zero then the underlying head movement must be purely rotational. Otherwise, it could be a rotation mixed with translation or a pure translation. Notably, even if the brain would rely on such computational strategy for tilt-translation discrimination, it still would not allow it to differentiate between a combined rotation-translation and a pure translation since in either case the general tilt velocity does not disappear (see examples illustrated in Fig. 15-3 A, B). Another limitation of this strategy is its limited sensitivity in distinguishing small signals corrupted with noise from a noisy null signal.

In the monkey, it is found that otolith-born velocity signals can drive compensatory slow phase eye velocities of the VOR with high gain (~ 0.6-0.8) during off-vertical axis rotations for as long as the head rotates. The generation of these slow phase velocities shows low-pass filter characteristics that matches the high-pass characteristics of the semicircular canal born velocity signals (Angelaki and Hess, 1996; Jaggi-Schwarz et al., 2000). We assume that these low pass filter characteristics are probably associated with the computation of tilt velocity from the acceleration and jerk signals (Angelaki and Hess, 1996), a process that it is not explicitly outlined in the flow diagram of Figure 15-2. The well-developed head velocity detection mechanism using otolith inputs alone that is manifest in the monkey VOR during off-vertical axis rotation is likely correlated with the astounding balance and acrobatic skills of these animals, which requires perfect inertial vestibular signal processing capacity.

Whether the brain makes use of other, non-vestibular cues to parse the ambiguous orientation signals in the forward model Eq. (5) is not known. Studies in animals with a bilateral inactivation of the semicircular canals suggest that other sensory cues cannot substitute the loss of semicircular canal signals, which are instrumental in the tilt-translation discrimination, in order to generate otolith-born angular velocity signals, at least not on a long term basis (Angelaki et al., 2000; Angelaki et al., 2002).

7. TRANSIENT HEAD ACCELERATIONS

The jerk magnitude signal in the above forward model (second term in Eq. 5), which is a potentially important signature indicating a translation or a combined rotation-translation, becomes smaller and smaller as the frequency content of the movement decreases. It decreases in magnitude by 20 decibel/decade as frequency decreases. As a consequence, discrimination of tilt from translation appears to become more and more difficult the lower the frequency content of the head movement. It is in the low frequency range of passive whole head and body motions where spatial disorientation occurs under the conditions of unreliable or unavailable external visual references. A major factor for such disorientations lies probably in the combination of the vanishing head velocity and jerk magnitude signals that is bound to disrupt reliable vestibular-based tilt-translation discrimination in the very low frequency range of movements.

In the common head movement range that covers frequencies between 0.1 and 5 Hz (Howard 1986), the forward model (Eq. 5) provides, however, significant cues for a robust tilt-translation discrimination in both the spatial and temporal signal domain. A graphical illustration of the signals based on this model is presented in Figure 15-3 for two different motion states: a forward translating subject that keeps head attitude constant (Fig. 15-3 A) and a subject that simultaneously tilts backward during a forward translation (Fig. 15-3 B).

In the case of a simple forward translation, the net jerk is identical to the inertial jerk sensed by the otoliths (gray backward pointing arrow in Fig. 15-3 A). It has two mutually perpendicular components, indicating the momentary rate of change in acceleration direction (jerk orientation term) and in magnitude (jerk magnitude term). Notice that the vector diagrams in Figure 15-3 A, B represent only a momentary picture of the jerk signals anywhere in time before the moment indicated by the vertical dashed lines in Figure 15-3 C.

Although at a given moment in time the net acceleration can be exactly the same as suggested by the black diagrams illustrating the vector equation

a = **g** + **f** in Fig. 15-3 A, B the time evolution of the general tilt velocity (indicated by $\tilde{\boldsymbol{\omega}}$ in Fig. 15-3), which is in fact an accessible quantity to the brain as mentioned earlier, can be quite different. This time evolution together with the evolution of the net acceleration and net jerk is illustrated in Figure 15-3 C for some of its spatial components.

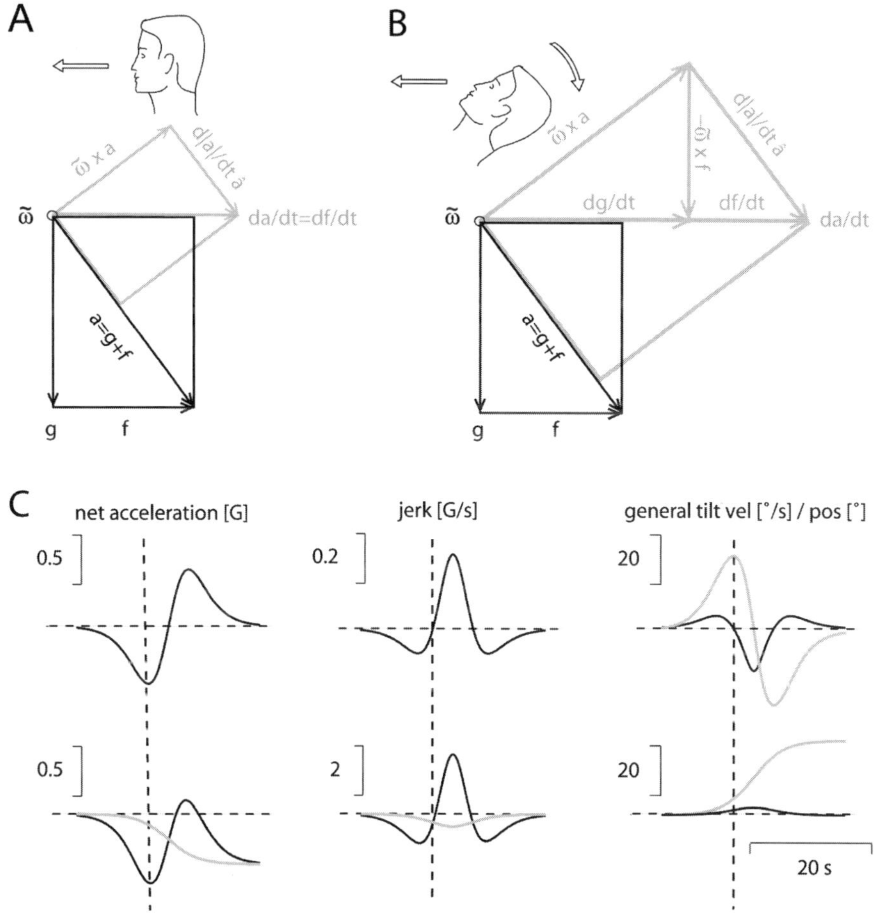

Figure 15-3. Geometric relations between the rate of change of net acceleration (d**a**/dt) and rate of change of real and/or apparent head tilt ($\tilde{\omega}$ x **a**) and rate of change of acceleration magnitude (d|**a**|/dt **â**). *A*. During forward translation, the rate of change of net acceleration (jerk) is associated with a unique apparent tilt velocity ($\tilde{\omega}$). The respective jerk signals are shown in gray. If the brain fails to predict that head attitude is constant (**g** = constant), the subject will perceive an apparent backward tilt (e.g., pitch up illusion on takeoff of the airplane). *B*. During forward translation while simultaneously backward tilting, the net acceleration (**a**=**f**+**g**) may have exactly the same geometric configuration as in *A*. Yet, the rate of change of net acceleration is in the general case associated with a general tilt velocity signal ($\tilde{\omega}$) that is the sum of a real and an apparent tilt velocity. *C*. Time course of net

acceleration and jerk along x-axis and of general tilt velocity about y-axis for forward translation as in *A* (upper row) and forward translation + backward tilt as in *B* (lower row). Gray lines show (from left to right) in the lower row gravity and rate of change of gravity, and in both rows change in tilt position. Open circles in *A*, *B*, indicate the angular velocity of general tilt (labeled $\tilde{\omega}$) perpendicular to plane of drawing. Vertical dashed lines in *C* indicate time of peak net accelerations. 1 G = 9.81 m/s^2.

In the following paragraph we will address this issue in more detail in the context of motor control of head attitude during locomotion.

Tilt-translation discrimination is the first step in motor control of head attitude. During locomotion it will often be important to maintain head attitude stable during periods of several seconds. The underlying inertial control mechanisms are therefore typically operating in a lower frequency band than do gaze stabilization mechanisms like the vestibulo-ocular reflexes. To appreciate the acceleration signals in the context of tilt-translation discrimination it is thus useful to analyze more closely the time course of a typical transient disturbance of head attitude that would last for several seconds if it were not counteracted. To enable fast discrimination we hypothesize that jerk orientation signals are of primary importance. They allow quick and efficient stabilization of head attitude against rotational disturbances through activation of vestibular neck reflexes since they lead the head acceleration signals in time and indicate the momentary direction, in which the acceleration of the head is going to change. A closer look at the time course of the acceleration and jerk signals in the exemplary situations of a forward translation or combined rotation-translation (Fig. 15-3 A, B) reveals that in the first 5-10s into the motion the dynamics of net acceleration and jerks signals is very similar. A major difference is the general tilt velocity signal: During the forward step it describes the change in orientation of the net acceleration. It must therefore have the same shape as the jerk signal but reversed signs. Since the motion is a pure forward step, the general tilt velocity signal describes the time evolution of the tilt of the net acceleration relative to the subject, which in fact represents an apparent rather than a real tilt of the subject with respect to gravity. The time integral of the general tilt velocity signal describes the angular orientation of the net acceleration relative to the head as a function of time. It shows no offset at the end of motion indicating that there is no net change in head orientation relative to the net acceleration or gravity at the end of the motion (see Fig. 15-3C gray line).

In contrast, during a forward motion combined with backward rotation, the general tilt velocity does not exhibit a jerk like profile despite the similarity in the general shape of the net acceleration profile with that of a pure forward step (Fig. 15-3 C). The time integral of the general tilt velocity signal indicates the net change of head attitude as a function of time. This change ends up in a backward tilt as expected (see Fig. 15-3 C gray line). The jerk signal along the x-axis during this combined rotation-translation motion has a peak excursion that is almost a magnitude larger than during a simple forward step although the net peak to peak acceleration is similar in the two situations. Not shown in Fig. 15-3 C is a smaller, but still significant jerk signal along the z-axis.

Despite the fact that the overall spatio-temporal pattern of net acceleration, jerk, and general tilt velocity signals are largely different in these two motion paradigms, the differences in the time course of these signals early in the motion are small. This illustrates the difficulty of a robust timely prediction of the nature of head attitude disturbances based otolith signals alone in the low frequency range. To efficiently stabilize the head attitude appropriate neck reflexes have to be activated in an early phase of the movement, which appears to require the assistance of additional sensory inputs.

ACKNOWLEDGEMENT

This work is supported by the Swiss National Sience Foundation (grant 3100A0-100802) and the Betty & David Koetser Foundation for Brain Research. I thank A. Green and D. Angelaki for valuable comments.

REFERENCES

Angelaki, D. E., Anderson, J. H., and Blakley, B. W., 1991, Changes in the dynamics of the vertical vestibulo-ocular reflex due to linear acceleration in the frontal plane of the cat, *Exp Brain Res* **86**:27-39.

Angelaki, D. E., 1992a, Two-dimensional coding of linear acceleration and the angular velocity sensitivity of the otolith system, *Biol Cybern* **67**:511-521.

Angelaki, D. E., 1992b, Detection of rotating gravity signals, *Biol Cybern* **67**:523-533.

Angelaki, D. E., and Hess, B. J., 1994, Inertial representation of angular motion in the vestibular system of rhesus monkeys. I. Vestibuloocular reflex, *J Neurophysiol* **71**: 1222-1249.

Angelaki, D. E., Hess, B. J., and Suzuki, J., 1995, Differential processing of semicircular canal signals in the vestibulo- ocular reflex, *J Neurosci* **15**:7201-7216.

Angelaki, D. E., and Hess, B. J., 1996, Organizational principles of otolith- and semicircular canal-ocular reflexes in rhesus monkeys, *Ann N Y Acad Sci* **781**:332-347.

Angelaki, D. E., McHenry, M. Q., Dickman, J. D., Newlands, S. D., and Hess, B. J., 1999, Computation of inertial motion: neural strategies to resolve ambiguous otolith information, *J Neurosci* **19**:316-327.

Angelaki, D. E., Merfeld, D. M., and Hess, B. J., 2000, Low-frequency otolith and semicircular canal interactions after canal inactivation, *Exp Brain Res* **132**:539-549.

Angelaki, D. E., Newlands, S. D., and Dickman, J. D., 2002, Inactivation of semicircular canals causes adaptive increases in otolith-driven tilt responses, *J Neurophysiol* **87**: 1635-1640.

Angelaki, D. E., Shaikh, A. G., Green, A. M., and Dickman, J. D., 2004, Neurons compute internal models of the physical laws of motion, *Nature* **430**:560-564.

Bos, J. E., and Bles, W., 2002, Theoretical considerations on canal-otolith interaction and an observer model, *Biol Cybern* **86**:191-207.

Cohen, B., Matsuo, V., and Raphan, T., 1977, Quantitative analysis of the velocity characteristics of optokinetic nystagmus and optokinetic after-nystagmus, *J Physiol (Lond)* **270** :321-344.

Dai, M. J., Raphan, T., and Cohen, B., 1991, Spatial orientation of the vestibular system: dependence of optokinetic after-nystagmus on gravity, *J Neurophysiol* **66**:1422-1439.

Fernandez, C., and Goldberg, J. M., 1971, Physiology of peripheral neurons innervating semicircular canals of the squirrel monkey. II. Response to sinusoidal stimulation and dynamics of peripheral vestibular system, *J Neurophysiol* **34** :661-675.

Fernandez, C., and Goldberg, J. M., 1976a, Physiology of peripheral neurons innervating otolith organs of the squirrel monkey. I. Response to static tilts and to long-duration centrifugal force, *J Neurophysiol* **39**:970-984.

Fernandez, C., and Goldberg, J. M., 1976b, Physiology of peripheral neurons innervating otolith organs of the squirrel monkey. II. Directional selectivity and force-response relations, *J Neurophysiol* **39**:985-995.

Fernandez, C., and Goldberg, J. M., 1976c, Physiology of peripheral neurons innervating otolith organs of the squirrel monkey. III. Response dynamics, *J Neurophysiol* **39**: 996-1008.

Fetter, M., Tweed, D., Hermann, W., Wohland-Braun, B., and Koenig, E., 1992, The influence of head position and head reorientation on the axis of eye rotation and the vestibular time constant during postrotatory nystagmus, *Exp Brain Res* **91**:121-128.

Fetter, M., Heimberger, J., Black, R., Hermann, W., Sievering, F., and Dichgans, J., 1996, Otolith-semicircular canal interaction during postrotatory nystagmus in humans, *Exp Brain Res* **108**:463-472.

Gizzi, M., Raphan, T., Rudolph, S., and Cohen, B., 1994, Orientation of human optokinetic nystagmus to gravity: a model-based approach, *Exp Brain Res* **99**:347-360.

Green, A. M., and Angelaki, D. E., 2003, Resolution of sensory ambiguities for gaze stabilization requires a second neural integrator, *J Neurosci* **23**:9265-9275.

Green, A. M., and Angelaki, D. E., 2004, An integrative neural network for detecting inertial motion and head orientation, *J Neurophysiol* **92**:905-925.

Green, A. M., Shaikh, A. G., and Angelaki, D. E., 2005, Sensory vestibular contributions to constructing internal models of self-motion, *J Neural Eng* **2**:S164-179

Guedry, F E, 1974, Phsychophysics of vestibular sensation. *in: Handbook of Sensory Physiology, vol. VI/2.* H. H. Kornhuber, (ed.), Springer-Verlag, Berlin, pp. 3-154.

Hain, T. C., 1986, A model of the nystagmus induced by off vertical axis rotation, *Biol Cybern* **54** :337-350.

Harris, L. R., and Barnes, G. R., 1987, Orientation of vestibular nystagmus is modified by head tilt. *in: The vestibular system: neurophysiological and clinical research,* M. D. Graham, and J. L. Kemink, eds., New York, pp. 539-548.

Hess, B. J., 1992, Three-dimensional head angular velocity detection from otolith afferent signals, *Biol Cybern* **67**:323-333.

Hess, B. J., and Angelaki, D. E., 1993, Modelling spatiotemporal properties of directionally sensitive multi-input single-output systems. *Biol Cybern* **69**:407-414.

Hess, B. J., and Angelaki, D. E., 1997, Inertial vestibular coding of motion: concepts and evidence, *Curr Opin Neurobiol* **7**:860-866.

Hess, B. J., Jaggi-Schwarz, K., and Misslisch, H., 2005, Canal-otolith interactions after off-vertical axis rotations. II. Spatiotemporal properties of roll and pitch postrotatory vestibuloocular reflexes, *J Neurophysiol* **93**:1633-1646.

Jaggi-Schwarz, K., Misslisch, H., and Hess, B. J., 2000, Canal-otolith interactions after off-vertical axis rotations I. Spatial reorientation of horizontal vestibuloocular reflex, *J Neurophysiol* **83**:1522-35.

Matsuo, V., and Cohen, B., 1984, Vertical optokinetic nystagmus and vestibular nystagmus in the monkey: up-down asymmetry and effects of gravity, *Exp Brain Res* **53**:197-216.

Mayne, R. A., 1974, A systems concept of the vestibular organs. *in: Handbook of sensory physiology vestibular system.* H. H. Kornhuber, ed., Springer-Verlag, New York, pp. 493-580.

Merfeld, D. M., Young, L. R., Oman, C. M., and Shelhamer, M. J., 1993, A multidimensional model of the effect of gravity on the spatial orientation of the monkey, *J Vestib Res* **3**: 141-161.

Merfeld, D. M., 1995, Modeling the vestibulo-ocular reflex of the squirrel monkey during eccentric rotation and roll tilt, *Exp Brain Res* **106**:123-134.

Merfeld, D. M., and Zupan, L. H., 2002, Neural processing of gravitoinertial cues in humans. III. Modeling tilt and translation responses. *J Neurophysiol* **87**:819-833.

Mergner, T., and Glasauer, S., 1999, A simple model of vestibular canal-otolith signal fusion, *Ann NY Acad Sci* **871**:430-434.

Mowrer, O. H., 1937, The influence of vision during bodily rotation upon the duration of post-rotational vestibular nystagmus, *Acta Otolaryngol* **25**:351-364.

Raphan, T., Cohen, B., and Matsuo, V., 1977, A velocity-storage mechanism responsible for optokinetic nystagmus (OKN), optokinetic after-nystagmus (OKAN) and vestibular nystagmus. *in: Control of gaze by brain stem neurons, developments in neuroscience.* R. Baker, and A. Berthoz, eds., Elsevier/North-Holland biomedical press, Amsterdam, pp. 37-47.

Raphan, T., Matsuo, V., and Cohen, B., 1979, Velocity storage in the vestibulo ocular reflex arc (VOR), *Exp Brain Res* **35**:229-248.

Raphan, T., and Cohen, B., 1988, Organizational principles of velocity storage in three dimensions. The effect of gravity on cross-coupling of optokinetic after-nystagmus, *Ann NY Acad Sci* **545**:74-92.

Raphan, T., and Schnabolk, C., 1988, Modeling slow phase velocity generation during off-vertical axis rotation, *Ann N Y Acad Sci* **545**:29-50.

Raphan, T., and Sturm, D., 1991, Modeling the spatiotemporal organization of velocity storage in the vestibuloocular reflex by optokinetic studies, *J Neurophysiol* **66**:1410-1421.

Robinson, D. A., 1977, Vestibular and optokinetic symbiosis:an example of explaining by modelling. *in: Control of gaze by brain stem neurons, developments in neuroscience.* H. Baker, and A. Berthoz, eds., Elsevier/North-Holland Biomedical Press, Amsterdam, pp. 49-58.

Ter Braak, J. W. G., 1936, Untersuchungen ueber optokinetischen Nystagmus, *Arch Neerland Physiol* **21**:309-376.

Viéville, T, and Faugeras, O. D., 1990, Cooperation of the inertial and visual systems. *in: Traditional and non-traditional robotic sensors.* T. C. Henderson, ed., Springer-Verlag, Berlin.

Wearne, S., Raphan, T., and Cohen, B., 1999, Effects of tilt of the gravito-inertial acceleration vector on the angular vestibuloocular reflex during centrifugation, *J Neurophysiol* **81**:2175-2190.

Young, L. R., 1984, Perception of the body in space: mechanisms. *in: Handbook of physiology, Section 1: The nervous system, Volume III. Sensory Processes, Part 1.* I. Darian-Smith, ed., American Physiological Society, Bethesda, Maryland, pp. 1023-1066.

Zupan, L. H., Merfeld, D. M., and Darlot, C., 2002, Using sensory weighting to model the influence of canal, otolith and visual cues on spatial orientation and eye movements, *Biol Cybern* **86**:209-230.

Chapter 16

SENSORIMOTOR CONTROL OF HUMAN DYNAMIC BEHAVIOR IN SPACE IMPLEMENTED INTO A HOMINOID ROBOT

Thomas Mergner, Christoph Maurer and Georg Schweigart
Neurology, Neurocenter, University of Freiburg, Germany

When a biologist and an engineer were once walking home together in the evening, they lost the key to their flat. The biologist starts to search the ground where they were last walking. In contrast, the engineer walks over to the other side of the street and starts to search there. Upon request, he justifies this: "But here we have the best chance to find the key, because here is the street light".

(Funny Rumor)

We like to add: We biologists have to bring both, the engineer and his systems approach (the light) over to our side of the street.

Abstract: To what extent can we claim nowadays that we understand sensorimotor control of human dynamic behavior in space? We try here to answer this question by exploring whether the available knowledge base suffices to build a hominoid robot such that its sensorimotor control functions mimic those of humans. It is, actually, our aim to build such a robot. We want to use it, in a systems approach, for simulations to better understand human sensorimotor control functions. We posit that a systems approach is necessary to deal with this complex non-linear control. We are especially interested in the sensory aspects of the control, the inter-sensory interactions ('multisensory integration' or sensor fusion) and the spatio-temporal coordination. Psychophysical work in our laboratory showed that the brain creates from sensory inputs internal estimates of the physical stimuli in the outside world (i.e., of the external constellation that caused a particular set of sensor stimuli). For example, the brain derives from vestibular and proprioceptive signals an estimate of body support surface motion. It then uses these estimates for sensorimotor feedback control (rather than the 'raw' sensory signals such as the vestibular signal). We hold that this internal reconstruction of the external physics is required for appropriate spatio-temporal coordination of the behavior. However, a problem arises from non-ideal sensors. An example is the vestibular sensor, which shows pronounced low-frequency noise. The solution of this problem involves sensory re-weighting mechanisms. Based on the discovered sensor fusion principles, we built a hominoid robot for control of upright stance (which we consider a simple prototype of sensorimotor control). It mimics human stance control even in complex behavioral situations. We aim to use it to better understand sensorimotor deficits in neurological patients and to develop new therapy designs.

Key words: model; multisensory integration; posture; robot; sensory re-weighting.

1. INTRODUCTION

Working in a neurological clinic, we tend to view the topic of human dynamic behavior in space from a clinical perspective. Thus, our ultimate aim is to better understand sensorimotor deficits in neurological patients and to design new and better therapies. We present here a vision of why and how we think dynamic models of sensorimotor functions can help us in this task. The essence of our approach is to build a hominoid robot that is able to perform the same sensorimotor tasks as healthy human subjects and in the same way. The robot's functions are derived from sensorimotor mechanisms which we have inferred in the past from systems analysis experiments in normal subjects and patients and then implemented in dynamic computer models. The physics parts of the model (body in the external world standing on a motion platform, etc.) as well as sensors and actuators are transformed into hardware, while the 'nervous functions' (sensory processing and control mechanisms) remain as a software simulation model. The hardware and software parts are combined in a novel approach called "hardware in the loop" simulation[5]. It allows us to modify internal control parameters of the robot online during task performance. The robot performs sensorimotor tests which we normally present to patients for diagnosis and validation of therapy. Based on parameter identifications in a given patient, a hypothesis of the patient's deficit is developed and implemented into the robot's control so as to mimic the patient's deficits. This will allow us to better understand the control functions and the deficit, to evaluate the effects of therapy, and to design new therapeutic concepts.

Why do we choose this complex approach? Looking back into the past, European medicine followed for about two thousand years the paradigm of

[5] Having the sensor fusion and sensory feedback mechanism as computer model, i.e. as software, in the simulations, one could call the approach actually a 'software in loop' simulation.

the Four Humors (blood, phlegm, yellow bile, and black bile; from about 400 yrs B.C. to the 19th century), after which 'modern medicine' with its orientation towards natural sciences took over. Now the Four Humors paradigm became replaced by a 'solid' paradigm, where diseases are explained mainly by destructive or disturbing effects through particles. This development started with the discovery of parasites, bacteria causing infectious diseases, followed by viruses, prions, etc. Yet, we still do not understand biological functions very well and medicine is still behind other fields of the natural sciences, with medical treatments today still being to a large extent empirical without a deeper understanding of the mechanisms that underlie the functions. A further change in paradigm is pending. We think the change is towards a 'systems' paradigm. This takes into account the mutual interactions between the various parts of a system ("the whole is more than the sum of its parts").

Why would we need a systems approach and why was it not possible in the past to infer from elementary mechanisms such as sensorimotor reflexes the behavior of an organism? As known already for some time and pointed out earlier in a previous work, the main reason is the high complexity of biological systems and their non-linear characteristics (see Mergner and Becker, 2003). Noticeably, because of these features it is essentially impossible to describe biological functions by means of mathematical formulas. But this is no longer a problem nowadays, because we now can use computer simulations instead. We already described in the previous work some sets of rules, which can be followed in the simulation approach, and biological constraints (such as modularity, flexibility, robustness). Here we are aiming at a possible application in medicine.

Using a behavioral scenario (Fig. 16-1) we try to explain the meaning of the term *systems neuroscience* and specify the nervous functions we will deal with in our approach. We limit the scope to brain control of human spatially oriented behavior. Figure 16-1 shows a situation where "The waiter on the ship brought me a glass of beer. Without looking at it, he balanced it across the swaying deck through a pushing crowd." This complex situation includes a specific meaningful task (not to spill my beer) as well as perception and action of a multi-segment organism and its interaction with the external world including compensation for external perturbations (force fields such as gravity, body support motion, contact forces). How can we dare to tackle this complex task although we have not fully understood the underlying basic simple mechanisms such as the postural reflexes or the interaction between voluntary action and these reflexes? We hold that we need not know all details, but have to focus on the behavioral framework and the rules that govern the interplay between this framework and the individual.

Figure 16-1. Scenario used to illustrate our approach: *"The Waiter on the Ship Brought Me A Glass of Beer. Without Looking at It, He Balanced It Across the Swaying Deck Through a Pushing Crowd."* The aim of our systems approach is to find simple formal descriptions of the waiter's sensorimotor control functions by which he is performing his job. Furthermore, the description should be such that we can test its predictive power concerning functions that are evaluated in the future.

Only then we will be able to decide which of the details are relevant for our enterprise and which are not.

In other words, when one is confronted with the task of describing a behavioral function with the systems approach, one first has to take a step back, in order to get an overview and to obtain a concept of the function as a whole. Then one can step closer, get into details and start to simplify, with the inferred concept of the system's function as guideline. Simplification, conceivably, is required when one tries to formalize the system and its function. Vice versa, formalization for its part is a prerequisite to obtaining a reasonable description of a complex system and to allow predictions of the outcome of future tests – thus the systems approach is a 'modeling' approach which moves back and forth between experiments and model simulations. In this approach, noticeably, the success criterion does not rely so much on an experimental "proof" or "falsification" of a (verbally described) hypothesis, but on whether a given model is superior to another in describing more behavioral aspects more accurately, and in showing more predictive power.

Another success criterion could be to demonstrate that a hardware realization of the model is able to perform the same task as human subjects and in a similar way (by building a machine as a 'demonstrator' of a model's validity). One can then go one step further by combining, in a hybrid approach, the hardware with a computer model. This would be advantageous, in that the simplifications and formalizations used for modeling, which tend to move us away from reality, can be restricted to the focus of our interest. In our scenario this focus would be a computer model of the waiter's sensorimotor control functions in the brain, while his body, the ship, the beer glass, etc. would be hardware. Thus we assume that the systems approach in the future often will include a hybrid approach in the sense of the "hardware in the loop" simulation. The background for this notion is that we have tried in vain over the past few years to combine in our simulations a detailed biomechanical software model (instead of the hardware) with our sensorimotor control model. The results were unsatisfactory, because the effort for accurate formalization of the system (the model design) was extremely high and simulation speed was very low.

Figure 16-2 shows where we locate the systems approach among the other approaches of brain research. In the figure it is located at the upper end of a hierarchical multi-layer structure called brain functions, where information processing occurs in exchange with the external world.

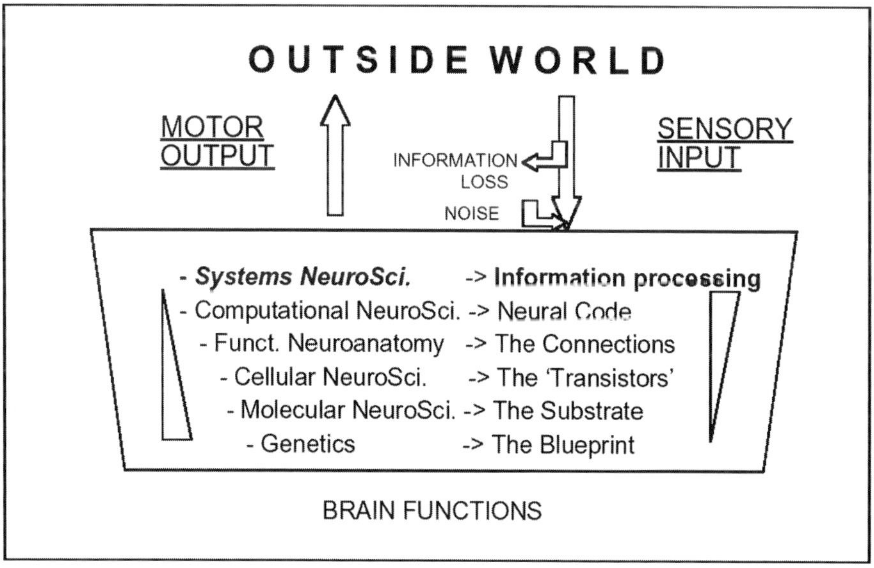

Figure 16-2. Systems neuroscience relates to the behavioral functions by which the brain interacts with the external world. We hold that this interaction is not ideal because of information loss and noise affecting the sensory signals. The brain functions are depicted here as a hierarchical multi-layer structure with systems neuroscience on top. The hierarchy reaches from this level, which is the level of organs (such as brain commanding muscles and receiving input from sense organs), down to molecular levels. The vertical inter-connections between the different layers and their functions are not well understood to date.

This upper level builds upon the 'neural code' level (dealing with the brain's "computer language" and modes of processing), on levels of the wiring connections, neural substrates, etc., and finally on genetics. Noticeably, the lower levels such as genetics relate much more indirectly to behavioral functions than the systems level. Therefore, we consider contemporary attempts to relate genetics directly to behavioral functions as premature. Even attempts to connect behavioral function with functional anatomy are still problematic (the reader may think of the problem of teaching students what the role of the cerebellum or the basal ganglia would be in the above scenario). The functional significance of the vertical interconnections in the figure are still to be established in the future. At each level, researchers apply specific scientific methods and formalisms to disclose and describe the relevant sets of rules. Noticeably, however, certain methods of the systems approach are recently applied also to molecular and genetic levels (in attempts to gain a more 'whole-istic' view of the functions performed on these levels). The formalisms used for the systems approach are often 'borrowed' from information theory and electronic engineering.

In our view the systems approach represents the successor of the cybernetics approach (which became stuck because of the biological complexities and non-linearities; see Mergner and Becker, 2003). In contrast, the computational neuroscientists deal with the rather recent issues of neural code, learning and optimization in neural nets (efforts which hopefully, in the future, will help us interpret the results of neuron recordings from trained animals, etc.). Both fields use tools from the neighboring disciplines in the engineering sciences. But they are distinct from those, because they deal with the biological rather than the technical implementation of signal processing. To explain this point: The control function of our envisaged hominoid robot are derived from experimental data in humans and are meant to mimic human brain functions (or their deficits in neurological patients), whereas robot construction in the engineering fields is free to choose any promising solution. Actually, most engineers use artificial neural nets for the multi-sensor control of their robots. However, these nets usually are learning and optimization automatons from which the sensor-fusion and control mechanism cannot be deduced. This is irrelevant when we have the robot vacuum cleaning our living room, for instance. But such a robot would not serve our aims here. *Our aim is to use the robot as a tool to describe and better understand sensorimotor control in humans, deficits of this control in patients, and possible treatments of the deficits - rather than to find some smart engineering solution for the control.*

What would we need to build the envisaged hominoid robot? In the following we describe where we have arrived on our way towards its construction. To this end, we first choose from the above scenario a prototype of a simple sensorimotor function and simplify and formalize it (section 2). On this basis we proceed by considering whether our knowledge of human sensors (3), sensor fusion (4), and sensorimotor control (5) suffices to serve as a blueprint for the envisaged robot control, although the brain tries to hide certain aspects (4.5). We then describe an envisaged clinical application (6), report on observations during the robot construction (7), and end with general conclusions on spatio-temporal coordination in sensorimotor functions (8).

2. A SIMPLE PROTOTYPE OF SENSORIMOTOR FUNCTION AND ITS FORMALIZATION (MODELING)

A sensorimotor system can be schematized as a circuit of information flow (see arrows in Fig. 16-2). Starting from the external physical world, information is flowing via sensors into an individual and its brain for information processing and back to the outside world via actuators. If performance of the individual were always absolutely accurate, we would have to assume that the brain is able to exactly represent internally the outside world (this would entail for us researchers a disadvantage: we would be unable to infer from the outside how the brain is doing this job). Phylogenesis has optimized over millions of years the information pickup via the sensors, the internal information processing, and the actuators. Yet, human performance in reality usually is not perfect. This is mainly due to information loss at the level of the sensors and to the noise that arises there and internally (Fig. 16-2). We consider this fact a great chance for our enterprise, because from comparisons between input and output (by means of a systems analysis) and some knowledge about sensor characteristics and internal signals we can derive a notion of how the brain proceeds in the internal reconstruction of the outside world. Before we proceed and specify this point, however, we take a snapshot from the 'waiter on the ship' scenario illustrated in Figure 16-1 in order to formalize and simplify it for our model.

We choose from the scenario a moment where the waiter produces a voluntary body lean in the presence of external perturbations (omitting locomotion and the beer). Thus the function considered in the following is stance control (task: make a 2° forward body lean in space despite external perturbations). The perturbations would be support-surface motion (swaying deck), external contact forces (pushing crowd), and gravity. Let us make the following assumptions (compare Fig. 16-3): (a) Only small body and deck angular motions are considered (i.e. rotations of a few degrees), so that gravity is continuously pressing the person's feet firmly onto the support (we then can suppose that the waiter is continuously in a working mode of stance control and can omit situations where a heel or forefoot lift-off in response to very strong stimuli would force him to switch to an emergency mode). Using small signals has furthermore the advantage that saturations and dead stops are avoided. (b) We restrict the motion to the anterior-posterior, a-p, rotational plane. (c) With the small rotations ('small signal approach'), we have to consider essentially body rotations about the ankle joints and can neglect very small hip and knee bending, i.e., we have a situation of an inverted pendulum where two segments (body consisting of head, trunk, and legs versus foot and its support) are inter-connected by the ankle joint. (d) An ideal actuator produces torque in the ankle joint to move the body.

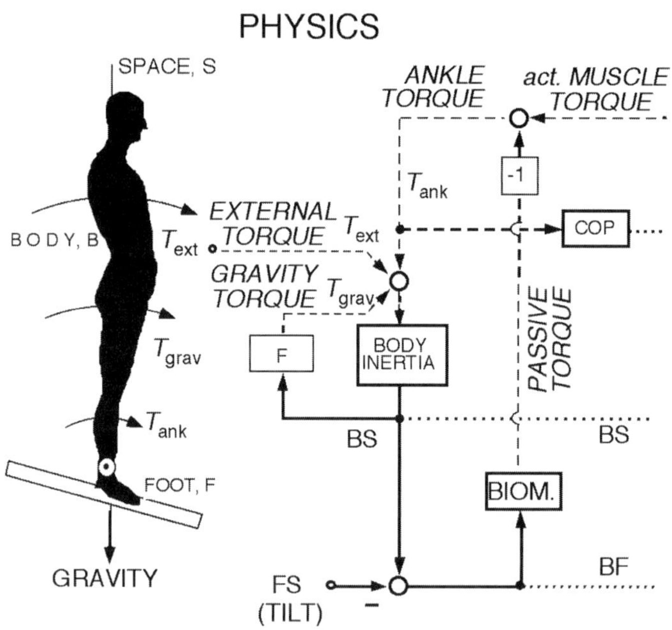

Figure 16-3. Inverted pendulum model of 'waiter on the ship' scenario. Body (head, trunk, leg) is connected to foot and its support by ankle joint, which allows anterior-posterior body rotations. Ankle torque (T_{ank}) results from summation of active muscle torque and passive torque (due to viscous-elastic muscle properties; box BIOM., for biomechanics). The torque accelerates the body and thus affects body-in-space position, BS. BS excursion leads to further body acceleration (via box F, and gravity torque, T_{grav}), as does a pull stimulus on the body ('external torque', T_{ext}). Body-to-foot depends on BS and foot-in-space (FS; tilt) in the form of BF= BS – FS. COP, Centre Of Pressure. Full lines: angular positions; dashed lines: torques; dotted lines: sensory inputs. (Modified from Maurer et al., 2006).

With these assumptions, we have the very simple situation depicted in Figure 16-3 in the form of a wiring diagram. When we now fill anthropometric data (body weight, height of centre of mass, COM, above ankle joint) of the subject as well as transfer functions into the boxes 'Body Inertia' and 'F' (for the calculation of 'gravity torque') and add the external perturbations 'support tilt' (Foot-in-Space, FS, TILT) and 'body pull' (External Torque), we could try to produce simulations that mimic the dynamics of the pendulum in response to the voluntary lean and the external perturbations. But this would always end with its 'fall'. The reason is that the pendulum is inherently instable and that gravity and other external stimuli tend to accelerate it. Therefore, we need a sensory feedback system that tries to compensate for gravity and to counteract the external perturbations. For this system we have to choose appropriate receptor systems.

3. CONCEPT OF SENSOR SYSTEMS

During dynamic behavior in space, many receptors in various sensor systems are activated in the body. Fortunately, the most relevant contributions come from a few sensors only: proprioceptive, vestibular, and visual (for details we refer to the text books). Their contributions are suggested by functional impairments following their loss (through experimental intervention or disease) or by specific behavioral responses upon their selective stimulation. There is furthermore evidence, although less complete, for a contribution from somatosensory receptors in the plantar soles (Maurer et al., 2000, 2001). From a theoretical viewpoint, one could argue that one does not need four sensors to stabilize the pendulum, at least in the absence of external contact forces, but that an ideal functioning vestibular sensor alone would be enough (Mergner, 2004). We will come back to this point further below, where we argue that it is for the brain not just a matter of information redundancy to use more sensors, but rather a necessity to avoid the vestibular signals whenever possible. Here we continue by describing briefly the main features of the four sensors and explain how they can be simplified and formalized in terms of technical equivalents.

3.1 Vestibular system

Anatomically, this system consists of two parts, the macular organs and the semicircular canals. Their functions are those of a 3D accelerometer and a 3D gyrometer (angular speedometer), respectively. With only the biological accelerometers in the head, a distinction between head tilt and head linear acceleration would not be possible. In combination with the gyrometer, however, such a distinction is possible (see Mergner and Glasauer, 1999; Zupan et al., 2002). This 'canal-otolith interaction' represents a sensor fusion which yields from the two input organs three signals: (i) head rotational velocity, (ii) head angle with respect to gravitational vector, and (iii) head linear acceleration. Simulations of this sensor fusion predicted, in accordance with experimental findings, improvements of transmitted information. Improvements are required because signal transfer in the biological gyrometer from angular acceleration, the stimulus, to velocity, the signal in primary canal afferent nerve fibers, is not ideal. Mathematically this transfer corresponds to an incomplete integration.

For the measurement of the a-p (vertical) body rotations in our robot, we assume that the vestibular sensor (combination of one gyrometer and two accelerometers) yields a signal of body rotational velocity and body angle in space.

3.2 Joint angle sense (proprioception)

We know today that the position sense of a limb involves not only spindle receptors in the muscles, but also receptors in the skin and in the capsules of the joints. When tested with psychophysical means, one finds that the brain combines these many signals into a rather accurate position sense over a broad dynamic range (true with the small angles used here). The technical equivalent of this sensor would be a goniometer (angle measuring device; often simply realized in the form of a pivotable potentiometer). From the angular position signal we may then derive in addition the corresponding velocity signal (by differentiation). Physiological and psychophysical evidence clearly indicates such an additional velocity signal. A simple way to demonstrate this is to activate, from the outside, primary spindle afferents by muscle vibration of an immobilized arm, for instance. The arm is then perceived as moving (but, paradoxically, not as displaced, which may indicate a functional segregation of central velocity and position pathways).

In the robot, we use a goniometer system with a position and velocity output in the ankle joint to measure body lean with respect to the feet and thus to the support surface.

3.3 Visual orientation cues

Vision provides information about the relative motion (velocity) between body and the visual surroundings ('scene'; compare literature on optokinetic reflex, OKR). Furthermore, vertical and horizontal items in the scene are used to establish a notion of the ‘ visual vertical’, i.e. of the body angle with respect to the vertical elements in the scene (with an orientation parallel to that of trees and orthogonal to the horizon).

Corresponding technical sensors (cameras and image processing devices) which can extract those signals are nowadays available. We will not consider technical details here and just take for granted that one can equip the robot with two visual devices, one which measures body-to-scene velocity and another one that measures the angle between body and visual vertical.

3.4 Somatosensory plantar pressure receptors

The sum of all reaction forces at the level of the foot soles and the support surface can mathematically be treated as acting at one point, called the Center Of Pressure (COP; in posturography, COP shifts arising during body sway are measured with the help of a 2D force platform). Constituents of the COP are (i) the gravitational vector of the body’s Center Of Mass (COM), (ii) the active torque that is produced in the ankle joint, and (iii) effects of external contact forces (such as a push against, or a pull on the body) to thc extent that they are transmitted to the feet. The receptors appear to be located in deep structures of the foot arch, whereas more superficial mechanoreceptors in the foot sole skin appear to be used to analyze the texture of the support surface, slip, etc.

Technically, one can mimic a COP measuring sensor simply by taking the difference between the signals from pressure sensors under forefoot and heel.

4. BASIC CONCEPTS OF SENSOR FUSION

It is well known that the use of several sensors instead of one sensor is advantageous for us when we try to solve spatiotemporal tasks. Most often considered in this respect is the redundancy effect, a long-known phenomenon (it yields an improvement of temporal and spatial accuracy when one tries to point as fast as possible to a target when this emits both visual and auditory cues rather than only one or the other cue, for instance). But we focus here on other aspects of sensor fusion, which have recently emerged from behavioral and psychophysical studies.

4.1 Complementary dynamics

Gaze stabilization[6] is known to depend on both vestibular and visual mechanisms. These are the VOR (vestibulo-ocular reflex) and the OKR (optokinetic reflex). If we restrict ourselves for a moment to horizontal head rotations and refer to known neural signals in the brainstem, the vestibular contribution will be a high-pass filtered head-in-space velocity signal, while the optokinetic contribution is known to represent mainly a low-pass filtered velocity signal with saturation. The main constraints that appear to determine the dynamics of these reflexes are given by (i) the incomplete acceleration-to-velocity integration of the vestibular signal (see above) and (ii) a considerable delay time in the optokinetic feedback loop (70 ms), which entails the danger of disturbing the system's function by oscillations upon fast and high-frequency visual stimuli (> 1 Hz). In computer simulations we have shown that simple summation of the two central signals can explain the experimental finding that a combination of both reflexes during head rotations yields broad band-pass dynamics of gaze stabilization (with a dominance of the OKR over the VOR in the overlapping range, however; see Schweigart et al., 1995, 1997).[7] This kind of sensor fusion where two sensors complement each other in their dynamic characteristics is often thought to apply also to perceptual or postural aspects of spatial behavior, which appears not to be the case, however (see below).

4.2 Selection of more accurate signal (a sensory re-weighting mechanism)

Why should the sensor fusion mechanism just described for the OKR-VOR not be applicable also for other spatial orientation functions? At least, the constraints for the two signals appear to be the same as above (since the visual information pick-up involves eye movements and thus the visual feedback delay and also the vestibular signal would be the same). A recent study compared perception of visual target motion and target position and came to another conclusion, however (Schweigart et al., 2003). As expected, visual target velocity estimates during combined eye smooth pursuit and optokinetic stimulation were found to be compromised by the non-ideal dynamics of eye velocity control by OKR and pursuit (which bears similar

6 Gaze position reflects the sum of eye and head positions.

7 This fits in with the ecological facts that active head rotations can be fast, while motion of extended visual stimuli, the scene, tends to be slow. The subsequent transformation from internal velocity signals into eye position are not considered here.

dynamics as the OKR). In contrast, the corresponding position estimates showed essentially ideal dynamics. They were related to an almost veridical eye position signal. This could be explained by the fact that the dynamic shortcomings of the eye velocity control are compensated for by saccadic eye movements (or 'fast phases' of the reflexes) and central mechanisms. The ideal eye position signal then is combined with an essentially ideal visual target-on-retina signal. This is derived directly from the visual sensor and shows dynamics up to almost 3 Hz (which we take as 'essentially ideal' in the present context of body excursions).

Note that the visual cues signal only relative motion (unlike the vestibular cues that signal absolute motion). Yet, our intuitive notion is that we use predominantly visual cues for the control of our dynamic behavior in space, and this even with fast movements. This notion received experimental support from a psychophysical study on visual-vestibular interaction (Mergner et al., 2000). In this study, subjects estimated horizontal self-motion during various combinations of optokinetic and vestibular stimuli. The formalisms by which we were able to describe our findings suggested that the perception arises from a sensor fusion that occurs in two processing steps. In the *first step*, a version of the visual signal (head-to-visual scene signal, *hv*) is subtracted from the vestibular head-in-space signal (λhs; λ is taken to represent high-pass characteristics which reflect the aforementioned imperfect acceleration-to-velocity integration). The *hv* signal, essentially ideal at input site, is centrally branched off for this subtraction and fitted to the dynamics of the vestibular signal (i.e. it receives centrally the same high-pass characteristics; λhv). During head rotations in a stationary scene, the two signals always cancel each other, since both have the same magnitude, but opposite signs. The sum is actually an internal estimate of visual scene motion in space (*vs*; $vs = \lambda hs - \lambda hv$). The *second step* is then to yield a self-motion estimate of head-in-space (*hs'*) by superimposing the ideal *hv* signal on the *vs* estimate in the form of $hs' = hv - vs$. Note that whenever the scene is experienced as stationary, the head-in-space self-motion perception is purely visual ($hs' = hv$, if $vs = 0°$). *In other words, first we use vestibular information to check whether the visual scene is stationary and, given this, then relate our self-motion always to the scene.*

A prediction from this concept would be that the dynamic behavior of a patient who shows chronic bilateral vestibular loss and therefore relies on visual input alone is essentially indistinguishable from that of a healthy subject in situations with a stationary visual scene - which we know to be true. Differences between the two subjects should emerge, however, when the visual scene is moving - which is true as well (see Mergner et al., 2000).

The above mechanism described the experimental finding well, but a still better fit between simulated and experimental data was achieved when assuming a still further sensor fusion mechanism on a high processing level (Mergner et al., 2000). At this level, a measure of a 'visual-vestibular conflict' is created whenever the scene is perceived as moving. This measure then performs a sensory re-weighting in that it weakens, by means of a throttle-like mechanism, the visual input and thereby weights more and more the vestibular cue, in a graded way. The mechanism appears to also react to cognitive factors such as the 'known' or 'expected' kinematic state of the scene (which differs, conceivably, when we move from amidst swaying bushes to enter a concrete building). This would represent a so-called 'top-down' mechanism.

We refrain here from further details and simply point out that there exist sensor fusion mechanisms that can replace one sensory signal by another. This is here a substitution of the vestibular signal by the visual signal following rules that comply well to ecological facts such as that the visual scene normally is stationary.

4.3 Coordinate transformations

Since the body segments are mechanically linked with each other and the inter-segmental angles are registered by proprioception, a sensor signal arising in any of the segments can in principle also be used by all the other segments (see Mergner et al., 1997). We give two examples: When we watch the flight of a bird across a homogenous sky, we may wonder in which direction it flies and refer to the earth surface as a space reference (Fig. 16-4 a). While standing on the earth surface, we may register the retinal bird image with respect to the eye (angle BE; retinal signal) and furthermore take into account the eye-in-head angle (EH; efference copy signal), head-on-trunk (HT; proprioceptive), trunk-on-legs (TL; proprioceptive), and legs-on-feet (LF; proprioceptive) angles. This represents a coordinate transformation of the earth reference to the retinal signal via feet, legs, trunk, head, and eye ('up-transformation' or 'up-channeling').

Analogous transformations apply when we want to use vestibular information (signal head-in-space angle, HS) to know the inclination of the body support surface in space (simplified as foot-in-space, FS; Fig. 16-4 b). The mechanical linkage between the body segments and the proprioceptive signals of the corresponding inter-segmental angles allows for this 'down-channeling' of the vestibular space reference to the support, as schematically depicted in Figure 16-4 b. In this way, one knows at any moment the angle of the body segments and of the support *in space* with the help of vestibular information. (Note that the term 'down-channeling' is used here literally, unlike the term 'top-down' which we used in the figurative sense; see 4.2).

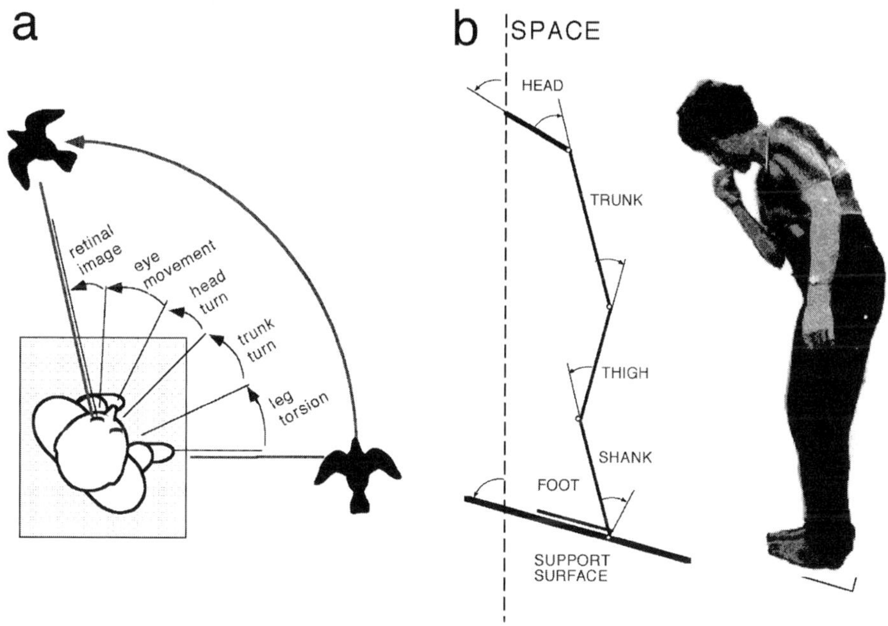

Figure 16-4. Examples of sensory coordinate transformations. **a** Watching a bird fly. This is here performed by turning eyes, head, trunk and legs with respect to the body support surface, taken to be the reference. **b** Knowing the orientation of the head in space (e.g. through vestibular signals) and the angles between the body segments means that we also continuously know the orientation of the other segments in space and here, during standing with firm foot-ground contact, also of the body support surface.

4.4 Internal stimulus reconstruction

The last described coordinate transformation (Fig. 16-4 b) contains an interesting aspect. Let a vestibular signal arise, for instance, in response to either a head-on-trunk rotation, body-on-foot rotation, or support surface rotation. The brain would know about those stimuli because the vestibular signal then would be associated with a corresponding neck proprioceptive signal, a leg proprioceptive signal, or no proprioceptive signal, respectively (and it would be able to decompose the relative contributions of the stimuli in case they are combined). Note that an isolated vestibular signal here indicates a whole-body motion due to support surface motion (whereas for vestibulo-oculomotor physiologists it always indicates a head motion, independently of how the other body segments are moving). Indeed, in corresponding psychophysical experiments (Mergner et al., 1991, 1993) we learned that human self-motion perception does not inform about the physiological stimuli (e.g. vestibular), but rather about the underlying physical stimulus, deriving this information from intersensory interactions.

To make this point more intuitive, we briefly describe results of one of these studies (Mergner et al., 1991). We presented our subjects with horizontal body rotations on a turning chair and independently from this rotated their heads by means of a rotation device mounted on the chair. Subjects' perception of head-in-space rotation was vestibularly determined during *en-bloc* body rotation (identified on the basis of typical high-pass frequency characteristics of the response data and high detection thresholds). In contrast, when the head was rotated on stationary trunk (torso), the head-in-space estimates were proprioceptive (the data showed broad band-pass characteristics and low detection thresholds, typical for proprioception). Other stimulus combinations yielded intermediate frequency and threshold response curves.

All these responses could be mimicked in model simulations, when we assumed that the brain reconstructed the physical events in the way we had created them during the experiment, i.e. by applying trunk(chair)-in-space rotations and superimposing on it head-on-trunk rotations (without subjects being consciously aware of the setup). In the model, this corresponded to (1) a vestibular-proprioceptive interaction that estimates the chair-in-space rotation, and (2) a superposition of a proprioceptive head-on-trunk signal on the chair-in-space estimate. This interpretation of the findings turned out to be superior to alternative explanations as shown in a later study, in which we included into our analyses response variability as a measure of internal noise (Mergner et al., 2001). An additional finding in the latter study was, interestingly, that the noise of the vestibular signal is much higher than that of the proprioceptive signal, especially at low stimulus frequencies.

Note that this vestibular-proprioceptive fusion shows an analogy to the above described visual-vestibular fusion. There, the brain first checks whether the visual scene is stationary and, given this, then uses the scene as a space reference for the self-motion perception, so that the vestibular cue no longer directly contributes to the perception. Here, *the brain checks whether the support surface is stationary and, given this, then uses the support as a space reference for the self-motion perception.* We assume that it is the large noise of the vestibular signal which leads the brain to substitute this signal by other sensory signals.

Thus, the above three sensor-fusion principles - stimulus selection (4.2.), coordinate transformation (4.3.), and physical stimulus reconstruction (4.4.) - are here interrelated. These principles were used to create a concept of how the various sensors in humans form a nested structure of coordinate systems for continuous sensorimotor control of a multi-segment human body (Mergner, 2002).

What leads us to assume that the sensor fusion principles observed for perception also apply to sensorimotor control of action? One argument could be that the estimates of the physical stimuli are also used for sensorimotor control simply because they are available. Furthermore, congruency between the sensory aspects of perception and action is desired (for known discrepancies, see Mergner, 2002). And estimates of physical stimuli are required for optimal motor responses. Take the aforementioned head-on-trunk, body-on-foot, and support-in-space rotations and their distinction with the help of proprioceptive signals. Each of them requires a specific postural response, whereas fixed, stereotyped responses, e.g. in the form of the vestibulo-spinal reflex, would not be useful.

We now return to our 'waiter on the ship' scenario and try to apply, in a parsimonious way, the sensor fusion principles to it. We refrain here from considering further the visual sensor, by simply stating that in certain situations it replaces the vestibular one. But we proceed under the constraint that the brain is to use the less accurate vestibular signal because in our scenario "some large fast-moving objects are currently crossing the waiter's visual field". Thus, in the following we restrict ourselves to three sensors (vestibular, proprioceptive, plantar somatosensory) and try to control with their help the 'inverted pendulum' in Figure 16-3. Before doing so, however, we briefly mention that certain aspects of the information flow in our brains are not intuitive.

4.5 What the brain is hiding from us

Usually we do not perceive consciously the impact that our bodies receive from the gravitational force field, the centrifugal forces, and the Coriolis forces (possibly, this fact explains why those forces were not well understood until Newton). One aspect is that they are not intuitive because they do not occur through contact with external bodies. Another aspect is that they usually are predictable, at least during pro-active movements, because they are either constantly present, such as gravity, or occur in association with certain kinematic events, such as a motion that contains circular trajectories and thus generates centrifugal forces. This allows that these forces are learned. Their effects then tend to be neutralized through compensatory active forces. For example, we do not consciously experience the relatively large forces required to compensate, during a horizontal arm extension, for the gravitational impact on the arm or the centrifugal forces affecting the arm, when we extend it during body turns. Novel forces (such as Coriolis forces occurring during pointing hand movements in rotating room experiments, for instance) are initially consciously perceived and distort the movements (Lackner and DiZio, 1994). But upon repetition, they are no longer consciously perceived and the necessary counter-forces are rapidly learned and then produced automatically. We have suggested earlier that the neural processing of our bodies' kinematics and their planning take place in the cerebral cortex, whereas the calculations for the accompanying compensatory active forces are delegated to other brain sites such as the cerebellum (Mergner, 2002). In the present context of our 'waiter on the ship' scenario, we refrain from considering predictive mechanisms, but proceed with simply stating that the force field gravity is sensed by the vestibular system.

Also not intuitive for us is the presence of signal noise in the nervous system. To give an example: Noise in position and motion sensing, central processing, and muscle activity during a hand pointing task with eyes closed would be the reason for the variability of the achieved pointing performance. The vestibular signals appear to be especially noisy. This is well attested by the large variability of the VOR in the form of slow fluctuations, for instance. Furthermore, vestibular-contingent target localization is far more variable than its proprioceptive counterpart, especially at low stimulus frequencies, as mentioned above (Mergner et al., 2001). The large noise in the vestibular signal appears to have its origin mainly in the canal system and a poor signal-to-noise ratio in the transduction process, which entails the aforementioned imperfect acceleration-to-velocity integration (and thus a high-pass filtered velocity signal in the canal primary afferent nerve fibers). As is well known, such an integration accentuates low-frequency parts of the noise (which could explain why we use only the high-pass part of the canal signal). The accentuation of the low-frequency parts of the noise becomes even more extreme due to the subsequent velocity-to-position integration in the brain and on its way to the actuators such as the eye muscles, with the effect that the VOR shows the pronounced slow fluctuations. The same is true for other vestibular-contingent responses. Yet, the noise is not intuitive in perception. When we sit motionless in the dark, we do not experience fluctuating head or body motions (auto-kinesia may occasionally be experienced, but is an exception). This perceptual stability at rest can be related to a relatively high detection threshold of the canal velocity signal used for the perception (Mergner et al., 1991). But during vestibular motion stimuli, the slow fluctuations are brought above the threshold and thus show up in the form of a corresponding low-frequency response variability (e.g., in the form that estimates of repeated body motions vary somewhat in their magnitude).

Furthermore, only part of the above described coordinate transformations are intuitive, while others are not. When we stand on a moving support surface, we experience our self-motion in space as a consequence of the support motion in space (note that a proprioceptive inter-segmental stabilization compensates for the inertia of our body segments). Thus, we have no problem identifying the body motion as resulting from the support motion. However, the vestibular-proprioceptive interaction by which the vestibular space reference is 'down-channeled' to the internal representation of feet and foot support, by which the support-in-space signal is generated (see 4.3), is not consciously perceived. Experimentally, we had to infer this from the response across various stimulus combinations.

How can we proceed with our partly non-intuitive concept although much more complexity is waiting for us when we will consider, in the future, sensorimotor control of a multi-segment body? We hold that the answer is simple: One just can take physics as a guideline and internally reconstruct, with the help of the sensor signals, what physics is doing externally. Other aspects can then be constructed around this framework (such as the aspect that the vestibular sensor is located in the head, a fact which requires that, during stance, we 'down-channel' the vestibular space reference from the head to the feet and their support). We hold furthermore that the reconstruction of physics is continuously required during movements of the body or its segments, and this not only to cope with gravity and other force fields and with the segments' inertia. Consider, for instance, an arm movement. The forces required to cope with the arm's inertia during acceleration and deceleration are transmitted to the support surface and there require a buttress (e.g. in the form of friction) – a fact which we also usually do not perceive (unless we fall because of very slippery ground).

5. SENSORY FEEDBACK AND SENSORIMOTOR CONTROL IN OUR SCENARIO

The internal sensory reconstruction of the external stimuli in our scenario is shown in Figure 16-5, which can be taken to represent the direct continuation of the model in Figure 16-3. As a reminder, the physical stimuli in Figure 16-3 were gravity (T_{grav}, which adds a perturbing torque to the ankle torque, T_{ank}), a contact force in the form of a pull on the body ('external torque', T_{ext}), and a support surface tilt (producing a foot-in-space angular excursion, FS). The previously described sensors that are used to register and internally reconstruct these stimuli are shown in Figure 16-5 in the following form:

1. ***Proprioceptive system (PROP).*** We assume that this system shows ideal transfer characteristics (symbol 1 in PROP box) and yields from its input the a-p body-to-foot angle, BF, internal estimates of position and velocity, bf and b•f (in ° and °/s, respectively).
2. ***Vestibular system (VEST).*** Its frequency characteristics are taken to be essentially ideal in the vertical rotational plane considered here, due to

the canal-otolith interaction (1 in VEST box). It yields from the input the a-p body-in-space angle, BS, internal estimates of position and velocity, bs and b•s (in ° and °/s, respectively).

3. ***The somatosensory system (SOMAT).*** This system measures the a-p shifts of the COP (in m). As we have described elsewhere (Maurer et al. 2006), this measure is here transformed into an estimate of the ankle torque and, after central processing into an equivalent of body lean angle (together with the estimate of T_{grav}; see below).

The sensory feedback loop is based on a default system ('local loop') in the form of a proprioceptive feedback (via the bf signal), which stabilizes the ankle joint (in terms of a body-to-foot stabilization). Voluntary command signals and other set point signals are then fed into this loop, making it perform a given task. A subsequent neural controller (PID, amplifying the feedback signal with a proportional, integrative, and derivative factor) provides the basis to cope with body inertia. The PID values recently have been inferred from experimental data (Peterka, 2002; Maurer et al., 2006), as was the value for the overall delay time of the stance control system (Δt, 150 ms).

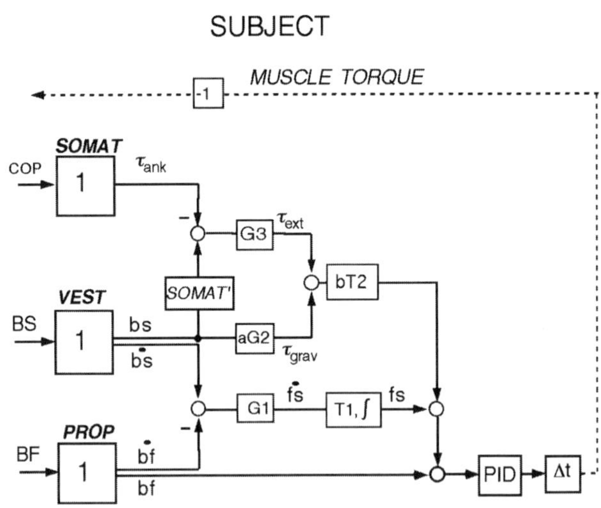

Figure 16-5. Internal sensory reconstruction of external stimuli used for feedback control of inverted pendulum in Figure 16-3. The estimate of support tilt (foot-in-space rotation, FS) is internally reconstructed in the form of fs = bs – bf (velocity estimates of BS and BF, obtained by vestibular, VEST, and proprioceptive, PROP, sensor inputs, respectively). The estimate of the gravitational effect, τ_{grav}, is derived from the vestibular bs signal. The estimate of the external pull stimulus, τ_{ext}, is obtained by comparing an estimate of the ankle torque, τ_{ank}, with a vestibular prediction of it (in box SOMAT'; internal model of plantar somatosensory sensor, SOMAT). Further details in text.

Internal estimates of the external stimuli, which feed into the local loop as set point signals, are created in the following way:

1. ***Estimate of the COM's gravitational torque, τ_{grav}.*** The estimate of this torque is proportional to, and derived from the vestibular body-in-space signal bs (box aG2 contains transforming and gain factor).[8]
2. ***Estimate of support surface tilt.*** The internal estimate of the external FS signal, fs, is derived from vestibular-proprioceptive interaction in the form of $f^{\bullet}s = b^{\bullet}s - b^{\bullet}f$. It represents a coordinate transformation of the vestibular space reference to the support surface (see above, 4.3.). Explained in words: Upon FS tilt, the fs signal makes the local loop to produce a counter BF tilt, so that the body orientation in space BS remains upright (BS = 0°, if BF is made to equal -FS). Before fs is used as a set point signal and thus for feedback in the model, it is fed through a velocity detection threshold and a velocity-to-position integration (box T1, $\int$).
3. ***Estimate of the external contact force, the pull stimulus, τ_{ext}.*** This is extracted from the internal estimate of ankle torque, τ_{ank} (derived from COP). Here we give a simplified explanation of this extraction: A prediction of τ_{ank} can internally be made on the basis of the vestibular bs signal (box SOMAT'), which contains all COP constituents apart from that produced by τ_{ext} (e.g. a pull stimulus). Subtractive summation of τ_{ank} and the prediction therefore yields the estimate τ_{ext}. (Box G3 contains gain factor; box bT2 contains transforming factor and position threshold).

Experimental results obtained in humans whom we presented with pull and support tilt stimuli were well described by simulations using the model of Figures 16-3 and 16-5 (Mergner et al., 2003; Maurer et al., 2006). In these studies, noticeably, also the responses to superposition of the stimuli could be predicted. Furthermore, when adding a visual input, the postural effects of visual motion stimuli were predicted (Mergner et al., 2005). Noticeably, also a superposition of the external stimuli with voluntary lean movements are covered by the model ('superposition criterion', Mergner, 2004). *It is this ability to cope with changing external situations which distinguishes our model from similar approaches in the literature.*

Back to our scenario: Our knowledge base about the waiter's sensorimotor control functions suffices to build a robot that is able to

[8] Note that, if we would assume head motion independent from trunk motion, a coordinate transformation of the vestibular bs signal from the head to the COM in the trunk would be required (a notion which is an 'upgrade' of the idea of von Holst and Mittelstaedt, 1950, of an interaction between vestibulo-spinal and cervico-spinal reflexes during head movements).

maintain an upright body posture despite gravity, 'pushing crowd', and 'swaying deck', and this even during a voluntary body lean. Starting from this point of departure, one could extend the scenario further. The waiter could make coordinate transformations of the vestibular space reference via the upright arm and hand to the haptically perceived beer glass. The scenario still can be further extended by including hip and knee joints. Conceivably, this also requires more extended internal linkages between the sensory representations of the body segments (see Mergner, 2002). Then, for instance, the body lean may lead to walking (a 'controlled way of falling forward', as it is sometimes called) and the physics would become more complicated, accordingly. *Yet, because internally the physics are estimated and these estimates (rather than the 'raw' sensory signals) are used for control, the kinematic and kinetic aspects always are temporally and spatially well coordinated.*

6. FIRST CLINICAL APPLICATIONS

We were encouraged in our modeling approach by recent observations we made in patients with Parkinson's disease. The cardinal symptoms of these patients are hypokinesia and rigidity, but also stance control is impaired. The impairment includes abnormally large and abnormally fast spontaneous body sways. This appears to be paradoxical, because the rigidity entails an increased axial stiffness, which should lead to smaller rather than larger sways. A second apparent paradox is found after treatment (here with dopamine medication and/or subthalamic nucleus stimulation). This leads subjectively, for the patients, as well as on clinical rating scales to an improvement, yet sway amplitude remains abnormally large or even becomes larger (see Maurer et al., 2003).

In a more detailed analysis of these findings (Maurer et al., 2004), the patient's sway revealed an abnormally strong tendency for body oscillations at about 1 Hz. Simulations of our model allowed us to mimic this abnormal sway behavior. This applied when we enhanced the loop gain and assumed an abnormally large internal noise in the system. Thus, a form of resonance in the sensory feedback systems was able to explain the abnormally large and fast sway of the patients (noise components in the order of the 'eigen frequency' of the system produce the resonance tendency). Furthermore, the paradoxical therapy effect could be mimicked. This applied when the loop gain in the model was reduced to normal levels, while the abnormally large noise remained. Even idiosyncrasies of the individual patients' therapy effects could be mimicked.

Meanwhile we have established a method of parameter identification for non-linear models (non-linear because of the aforementioned thresholds in Fig. 16-5) and have successfully applied it to experimental data (Maurer et al., 2006). We hope to use this method now to identify the source of the abnormally large noise in the patients' stance control. From preliminary data it appears that a main source is the vestibular signal and that the ability to replace it by other signals through sensory re-weighting is impaired in the Parkinson's disease patients.

7. PRELIMINARY OBSERVATIONS DURING THE ROBOT CONSTRUCTION

We used the model in Figure 16-5 meanwhile as blueprint for the construction of a robot that is able to cope with the above scenario (we called it "PostuRob"). Details of its construction and performance will be presented elsewhere (Mergner et al., 2006). Here it may suffice to mention that we used 'pneumatic muscles' as actuators and gave them an essentially ideal performance (using air pressure and force controls), but added to them in-series springs to mimic the tendons (spring constant, <1 mgh; where the robot's body mass $m = 95$ kg and COM height above ankle joint $h = 1.0$ m). The control was realized in software as a computer model; this allows us to adjust the control parameters during ongoing task performance. Our focus of interest was on the sensors and the processing of their signals in the model.

The vestibular sensor was built from accelerometers and a gyrometer, in order to mimic the otolith and canal systems in humans. The corresponding signals were fused in form of the canal-otolith interaction mentioned above. Furthermore, goniometers in the ankle joints and pressure receptors in the foot soles were used to mimic proprioception and COP cues, respectively. Internal reconstruction of external stimuli was obtained with the help of these sensors. We found it to be complicated, as expected, by large low-frequency noise in the 'vestibular' signal. The noise stemmed mainly from the gyrometer.

It showed a further low-frequency accentuation in the course of subsequent signal processing (velocity-to-position integration). When we omitted the low-frequency part of the gyrometer signal (by way of a high-pass filtering) and added, instead, a corresponding low-frequency signal derived from the accelerometers (by way of the canal-otolith interaction), the low-frequency noise was largely reduced. A further improvement was achieved when we added a velocity threshold. *Thus, we found a correspondence between biological and technical vestibular systems with respect to the noise and the possibility of its improvement.*

Furthermore, the use of a throttle-like sensory re-weighting mechanism (described above for visual-vestibular interaction, 4.2; not contained in Fig. 16-5) helped us to replace the 'vestibular' signals by 'proprioception' whenever possible (i.e. always when the support base is stationary). Thus, these problems met our predictions from our modeling. Finally, problems arose from the use of the force sensors under the foot sole, in that active torque components in the COP signal entailed the risk of positive feedback. This risk could also be coped with by sensor fusion (see above, 3.4. and 5.).

The robot is able to make voluntary lean movements during concurrent platform tilts and push stimuli. Thus, it fulfills the aforementioned superposition criterion (5.). What would now be different from an overall software simulation? First, we point out that the system still performs successfully after replacing the simplified physics in the model (Fig. 16-3) by hardware and thus restricting the software model simulations to the sensor fusion and control mechanisms (the 'demonstrator' aspect). Second, the sensor signals in the robot generally are not as reproducible and stable as in the simulation, which may affect the processing of the control signals considerably. We learned from this to search for stabilizing mechanisms (see previous paragraph, throttle-like mechanism), but also came to appreciate the robustness of our control algorithms. Third, when we look into the future, we have to reckon with an enormous increase of complexity when trying to simulate multi-segment individuals in rich and variable environments, so that it becomes more and more difficult to find the right simplifications and formalization to capture the relevant aspects of the outside world. Our 'hardware in the loop' simulation approach helps to overcome this problem in that we can restrict the software simulation to the sensorimotor control functions (noticeably, however, there remains the problem of simulating human biomechanics by means of hardware). Furthermore, we do not have to fear that data acquisition involves a very complex experimental setup (with complicated stimulus equipment, measuring devices, etc.), because this applies equally to the experimental and simulated data. *Therefore, we hold that the 'hardware in the loop' simulation approach is particularly suited for our aim to better understand sensorimotor functions and deficits of complex systems by back and forth iterations of simulation and experiment.*

How does our approach relate to the field of bionics where engineers try to use biological principles for technical purposes? The principles we have shown here may become relevant in the future also for the design of hominoid robots other than the ones we envisaged here. Currently, engineers tend to implement mostly gyroscopes into robots for postural control (very accurate, but also very expensive inertial guiding systems, which over longer time periods require position updates, however). It will be important to compare a number of criteria to decide between different control principles, such as manufacturing costs, energy consumption, robustness, flexibility, modularity, time required for implementation and training, etc. Coming back to the point of departure of our project, the medical aim, we would like to mention an important additional aspect. We see our project also as an important step towards the construction of neural protheses and orthoses (exoskeletons in the form of shells around body segments, equipped with motors that guide, perform, or assist segment movements). For instance, we consider it realistic to build, on the basis of our robot, an exoskeleton for equilibrium control during behavior, for substitution of the whole stance system (in paraplegic patients) or parts of it (in case of sensor loss, etc.). *We hold that a control system that is constructed to resemble the human one has a much better chance to fit into human behavior and to gain compliance from patients than an arbitrary technical system.*

Note that the sensorimotor control of our robot so far is only a re-active mechanism, in that it is not yet endowed with predictive feedforward mechanisms during active lean. This does not mean that we downplay such mechanisms. Rather, the reason is that we saw so far no indication for them in our experimental data, which would demand that we implement them (i.e., we would have currently no arguments to use more complex model approaches than needed, such as optimal estimators). Evidence for feedforward mechanisms in re-active stance control have been reported in the literature, but so far is not convincing. Advocates of such mechanisms would have to explain how these may function, e.g. by presenting corresponding dynamic models that would allow us to describe or predict experimental findings.

8. GENERAL CONCLUSIONS FOR SENSORIMOTOR COORDINATION

Sensorimotor control normally requires appropriate temporal and spatial cooperation of several muscles in order to cope with a given behavioral task – an aspect that is called sensorimotor coordination. The principles underlying the coordination are still not well understood. There is considerable work on learning motor patterns related to certain task optimizations, such as minimizing energy consumption or improving movement accuracy and speed, etc. But so far these mechanisms shed little light on the organizing principles of coordination. We hold that the here described sensorimotor control principle, i.e. a control by internal sensory reconstruction of the outside physics, is an instrumental constituent of the coordination. To give an example: A transient platform tilt in our scenario will primarily be responded to by the internal estimate of support surface tilt (fs signal in the model of Figure 16-5, by which the system tends to produce a compensatory torque in the ankle joint such as to make $BF = -FS$ so that BS remains stationary, $BS = BF + FS = 0°$). The tilt compensation is, however, usually not perfect and there will be some BS excursion. This will be accounted for by the gravity compensation at the time when, and to the extent that the internal signal $bs \neq 0°$. Furthermore, the proprioceptive bf signal will be involved, but only at the time when, and to the extent that, body inertia is to be overcome during acceleration.

In our simple scenario we have only one degree of freedom and, yet, there is this need for coordinating the different compensatory actions. With more degrees of freedom, which result when we use more joints, the physics of the situation would be much more complex and the coordination may be distributed over two or more segments and actuators. However, the point we want to make here remains the same: the temporal and spatial order of the movements and postural adjustments are determined by the physics, so that the fact that sensorimotor control is governed by internal estimates of the physics and that these estimates are implemented in form of set point signals (by which also voluntary movements are embedded into the mechanism) solves this aspect of the coordination problem.

ACKNOWLEDGMENTS

Supported by DFG Me 715/5-3, 715/6-1.

REFERENCES

Lackner, J.R., Dizio, P., 1994, Rapid adaptation to Coriolis force perturbations of arm trajectory, *J Neurophysiol* **72**:299-313.

Maurer, C., Mergner, T., Bolha, B., Hlavacka, F., 2000, Vestibular, visual, and somatosensory contributions to human control of upright stance, *Neursci Let* **281**:99-102.

Maurer, C., Mergner, T., Bolha, B., Hlavacka, F., 2001, Human balance control during cutaneous stimulation of the plantar soles, *Neurosci Let* **302**:45-48.

Maurer, C., Mergner, T., Peterka, R. J., 2004, Abnormal resonance behavior of the postural control loop in Parkinson's disease, *Exp Brain Res* **157**:369-376.

Maurer, C., Mergner, T., Peterka, R. J., 2006, Multisensory control of human upright stance, *Exp Brain Res* **171**:231-250.

Maurer, C., Mergner, T., Xie, J., Faist, M., Pollak, P., Lücking, C. H., 2003, Effect of chronic bilateral subthalamic nucleus (STN) stimulation on postural control in Parkinson's disease, *Brain* **126**:1146-1163.

Mergner, T., 2002, The Matryoshka Dolls principle in human dynamic behavior in space - A theory of linked references for multisensory perception and control of action, *Curr Psychol Cogn* **21**:129-212.

Mergner, T., 2004, Meta level concept versus classic reflex concept for the control of posture and movement, *Arch Ital Biol* **142**:175-198.

Mergner, T., Becker, W., 2003, A Modeling Approach to the Human Spatial Orientation System, in: The oculomotor and vestibular systems: their function and disorders, T. Brandt, B. Cohen and C. Siebold, eds., *Ann NY Acad Sci* **1004**:303-315.

Mergner, T., Glasauer, S., 1999, A simple model of vestibular canal-otolith signal fusion, *Ann New York Acad Sci* **871**:430-434.

Mergner, T., Hlavacka, F., Schweigart, G., 1993, Interaction of vestibular and proprioceptive inputs, *J Vest Res* **3**:41-57.

Mergner, T., Huber, W., Becker, W., 1997, Vestibular-neck interaction and transformations of sensory coordinates, *J Vestib Res* **7**:119-135.

Mergner, T., Huethe, F., Maurer, C., Ament, C., 2006, Human equilibrium control principles implemented into a biped humanoid robot, in: *Romansy 16, Robot design, dynamics, and control* (series: CISM International Centre for Mechanical Sciences, no. 487), T. Zielinska and C. Zielinski, eds., Springer Science+Business Media, pp. 271-278.

Mergner, T., Maurer, C., Peterka, R. J., 2003, A multisensory posture control model of human upright stance, in: Neural control of space coding and action production, C. Prablanc, D. Pélisson and Y. Rossetti, eds., *Progr Brain Res* **142**:189-201.

Mergner, T., Nasios, G., Maurer, C., Becker, W.,2001, Visual object localization in space. Interaction of retinal, eye position, vestibular and neck proprioceptive information, *Exp Brain Res* **141**:33-51.

Mergner, T., Schweigart, G., Maurer, C., Blümle, A., 2005, Human postural responses to motion of real and virtual visual environments under different support base conditions, *Exp Brain Res* **167**:535-556.

Mergner, T., Schweigart, G., Müller, M., Hlavacka, F., Becker, W., 2000, Visual contributions to human self-motion perception during horizontal body rotation, *Arch Ital Biol* **138**:139-167.

Mergner, T., Siebold, C., Schweigart, G., Becker, W., 1991, Human perception of horizontal head and trunk rotation in space during vestibular and neck stimulation, *Exp Brain Res* **85**:389-404.

Peterka, R. J., 2002, Sensorimotor integration in human postural control, *J Neurophysiol* **88**:1097-1118.

Schweigart, G., Mergner, T., Barnes, G. R., 2003, Object motion perception is shaped by the motor control mechanism of ocular pursuit, *Exp Brain Res* **148**:350-365.

Schweigart, G., Mergner, T., Becker, W., 1995, Eye stabilization by vestibulo-ocular reflex (VOR) and optokinetic reflex (OKR) in macaque monkey: Which helps which?, *Acta Otolaryngol* **115**:19-25.

Schweigart, G., Mergner, T., Evdokimidis, I., Morand, S., Becker, W., 1997, Gaze stabilization by optokinetic reflex (OKR) and vestibulo-ocular reflex (VOR) during active head rotation in man, *Vision Research* **37**:1643-1652.

von Holst, E., Mittelstaedt, H., 1950, Das Reafferenzprinzip (Wechselwirkungen zwischen Zentralnervensystem und Peripherie), *Naturwissensch* **37**:464-476.

Zupan, L., Merfeld, D. M., Darlot, C., 2002, Using sensory weighting to model the influence of canal, otolith and visual cues on spatial orientation and eye movements, *Biol Cybern* **86**:209-230.

Chapter 17

THE VENTRO-DORSAL STREAM: PARIETO-PREMOTOR NEURAL CIRCUITS AND THEIR ROLE IN PRIMATE COGNITION

Vittorio Gallese
Department of Neuroscience, University of Parma, Italy

Abstract: The aim of the present chapter is twofold. First it aims to show that perception requires action. This is most evident for space and action perception. Second, it aims to show that the distinction of the cortical visual processing into two streams is insufficient and leads to possible misunderstandings on the true nature of perceptual processes. I review empirical findings suggesting that visual processing is carried out along three distinct visual pathways qualified as dorso-dorsal, ventro-dorsal, and ventral streams. The relevant anatomical and functional features of the ventro-dorsal stream are presented and discussed.

Key words: action; intentions; mirror neurons; monkey; neglect; perception; simulation; space.

1. INTRODUCTION

Perception and action have been traditionally considered separate domains, each of them being implemented in separate anatomical and functional brain sectors. These sectors are serially organized: individuals first perceive, then act. In primates, the paradigmatic sensory modality for the study of the mechanisms underlying perception is vision. Visual information is processed both serially and along parallel pathways. Serial organization fits well the classical concept that information travels from the periphery to progressively more complex "association" areas where perception occurs and then proceeds to output channels for action. In contrast, parallel organization of visual processing needs explanation.

A particularly influential account of why there is parallel processing of visual information in the primate visual cortex is that proposed by Ungerleider and Mishkin (1982). According to these authors, the visual cortical areas are organized in two separate streams of visual information. A dorsal stream, which includes visual areas MT, MST, FST, V3A, and V6, and culminates in the inferior parietal lobule, and a ventral stream, which includes visual areas V3 and V4, and culminates in the inferior temporal cortex. The dorsal stream is responsible for perception of space, while the ventral stream for object perception.

An equally influential and radically different view was advanced by Milner and Goodale (Goodale and Milner, 1992; Milner and Goodale, 1995). In accord with Mishkin and Ungerleider (1982) they maintain that there is a fundamental functional difference between the dorsal and ventral stream. They deny, however, that the difference is in the resulting percept (space vs. object). According to Goodale and Milner the difference is in the output characteristics of the two cortical visual streams. The ventral stream is fundamental for perception. The dorsal stream, in contrast, processes visual stimuli to provide high order visual information for the control of action, but it is not involved in perception. A similar view was independently proposed by Jeannerod (1994, 1997). According to Jeannerod, the ventral stream is responsible for the "semantic mode" of object representation, while the dorsal stream is responsible for the "pragmatic mode" of stimulus processing. The semantic mode of object representation refers to object analysis described in object-centered coordinates. The pragmatic mode indicates the type of processing that stimuli have to undergo for action organization. Although the distinction between a semantic and a pragmatic system proposed by Jeannerod appears to be more cautious than that of Milner and Goodale, the essence of the two proposals is very similar.

The aim of the present chapter is two-folded. First it aims to show that perception requires action. This is most evident for some types of visual percept (e.g. space perception and action perception). Second, it aims to show that the distinction of the cortical visual processing into two streams is insufficient and leads to possible misunderstandings on the true nature of perceptual processes.

I will briefly review empirical findings suggesting that visual processing is carried out along three distinct streams (see Figure 17-1). Two of them include the parietal lobe, one includes the inferotemporal lobe. These three streams are qualified as dorso-dorsal, ventro-dorsal and ventral streams (for a detailed analysis of the dorsal streams, see Rizzolatti and Matelli, 2003; Rozzi et al., 2006).

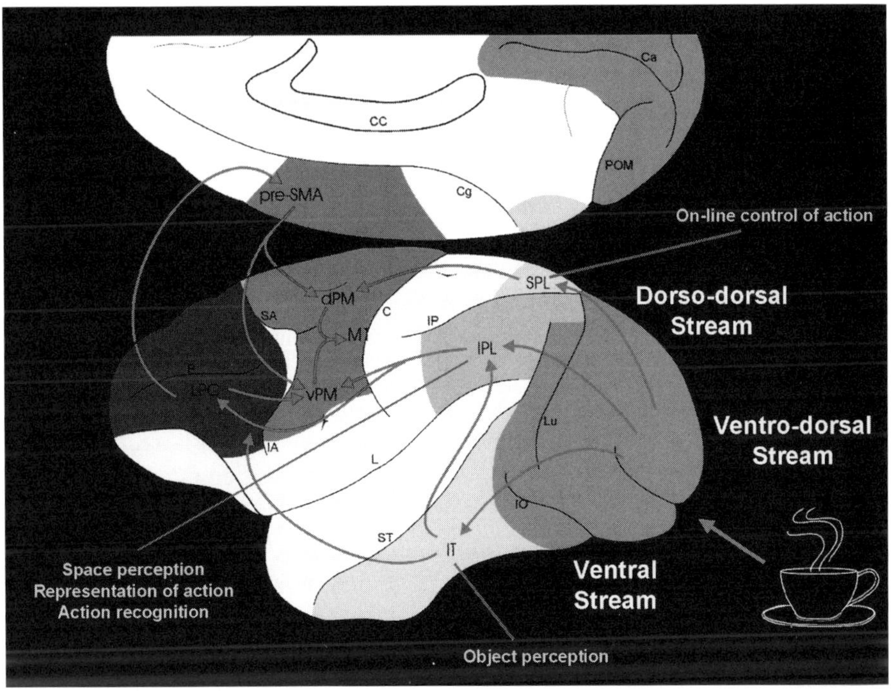

Figure 17-1. The three Visual Streams. Lateral and mesial view of the macaque monkey brain showing the main connectivity along the dorso-dorsal, ventro-dorsal and ventral streams of visual processing and their hypothesized functions. Abbreviations: C=central sulcus; Ca=calcarine fissure; CC=corpus callosus; cg=cingulated gyrus; dPM=dorsal premotor cortex; IA=inferior arcuate sulcus; IO=inferior orbital sulcus; IPL=inferior parietal lobule; IT=inferior temporal cortex; L=lateral fissure; LPC=lateral prefrontal cortex; Lu=lunate sulcus; MI=primary motor cortex; P=principal sulcus; pre-SMA=pre-supplementary motor area; SPL=superior parietal lobule; ST=superior temporal sulcus; VPM=ventral premotor cortex.

The dorso-dorsal stream has the characteristics suggested by Milner and Goodale and Jeannerod when they describe the dorsal stream as a whole. It appears to be possibly the only stream that is not directly related to perception. The ventro-dorsal stream, which will be the main focus of this chapter, is responsible for the organization of actions directed towards objects, but also for space perception and action perception. Finally, the ventral stream is responsible for the organization of actions following object categorization, and for object semantics. In the ventral stream are located the semantic trees linking the analyzed objects to individual's semantic knowledge. All three visual streams terminate into frontal cortical areas endowed with different degrees of complexity.

2. THE DORSAL STREAMS

The intraparietal sulcus is, evolutionary speaking, a very ancient sulcus. This sulcus, that is already present in prosimians, represents a fundamental parietal landmark. It subdivides the posterior parietal lobe into two main sectors: the superior parietal lobule (SPL) and the inferior parietal lobule (IPL). These two sectors receive different cortical inputs and have different connections with the motor cortex and frontal lobe. Early experiments in monkey showed that the superior parietal lobule (SPL) is part of the somatosensory system. It receives information from the primary sensory cortex, and in particular from those areas that code proprioception, and sends inputs to the primary motor cortex (F1) and to the dorsal premotor cortex (area F2). Recent neurophysiological data showed that SPL receives also visual inputs. Neurons activated by visual stimuli have been described in its caudal part (Galletti et al., 1996; Caminiti et al., 1996) and, in particular, in areas V6A and MIP. Both these areas are connected with frontal motor areas. Their main target is the dorsal premotor cortex (area F2) (Caminiti et al., 1996; Matelli et al., 1998).

Like SPL, also IPL (especially its rostral sector) receives somatosensory afferents. In addition, IPL is the main site of convergence of the pathways from the extrastriate visual areas of the dorsal stream. It projects to areas of the ventral premotor cortex (areas F4 and F5) and to the prefrontal lobe. The functional properties of IPL are in accord with the pattern of anatomical connections. IPL neurons are often bimodal responding both to visual and somatosensory stimuli (see also below). Taken together, these findings indicate that the parietal cortex performs two separate analysis of incoming sensory information. The analysis carried out in SPL (dorso-dorsal stream) concerns mostly proprioceptive input, but with an important visual contribution. The analysis performed in IPL (ventro-dorsal stream) consists in the integration of visual, auditory and somatosensory stimuli for action on the external world.

Before reviewing the organization of the ventro-dorsal stream in detail, an important point about the homology between human and other primates parietal lobe organization should be clarified. Human posterior parietal lobe as that of the macaque monkey is formed by two lobules – SPL and IPL – separated by the intraparietal sulcus. As shown by Forster (1936), contrary to the original parcellation of Brodmann, SPL is basically co-extensive with area 5 (5a + 5b), while IPL with area 7 (7a + 7b).

This means that the cytoarchitectonic organization of the posterior parietal lobe is similar in monkeys and humans. This view was confirmed by von Bonin and Bailey (1947) who, adopting the terminology of von

Economo (1929), found that in monkeys as in humans SPL is formed in large part by area PE (area 5) and the IPL by areas PF (7b) and PG (7a).

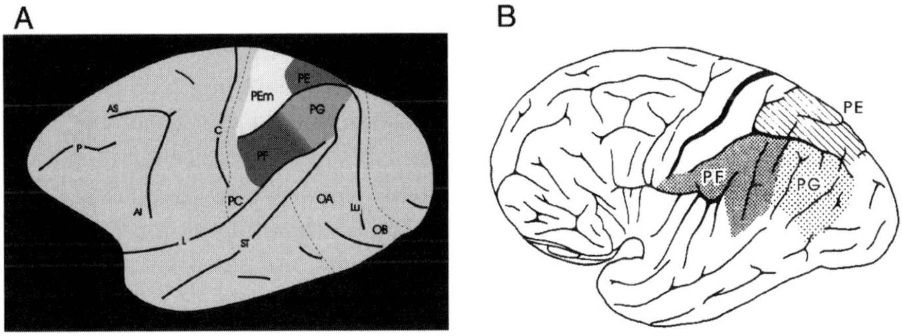

Figure 17-2. A. The inferior parietal lobule: monkey-human homology. Lateral view of the macaque monkey brain showing the cytoarchitectonic parcellation of the superior and inferior parietal lobules according to Von Bonin and Bailey (1947). B. Lateral view of the human brain showing the cytoarchitectonic parcellation of the superior and inferior parietal lobules according to Von Economo (1929).

Thus, when monkey data on area 7 are used to discuss functional properties of human parietal lobe, they should be used in reference to human IPL and not SPL as one can be tempted to do on the basis of the (wrong) Brodmann map. Similarly, the data on monkey area 5 should be used in reference to human SPL.

In the following sections I will describe in detail two parallel parietal-premotor neural circuits, both components of the ventro-dorsal stream: the VIP-F4 network and the PF/PFG-F5 network. It will be shown that these networks are involved in the organization of action in space and space perception and in action understanding, respectively.

3. THE VENTRO-DORSAL STREAM: ACTION IN SPACE AND SPACE PERCEPTION

The cortical circuit formed by area F4, which occupies the posterior sector of the ventral premotor cortex of the macaque monkey, and area VIP (Colby et al., 1993), which occupies the fundus of the intraparietal sulcus, is involved in the *organization of head and arm actions in space.* Single neuron studies showed that in area VIP there are two main classes of neurons responding to sensory stimuli: purely visual neurons and bimodal, visual and tactile neurons (Colby et al., 1993). Bimodal VIP neurons respond independently to both visual and tactile stimuli. Tactile receptive fields are located predominantly on the face. Tactile and visual receptive fields are usually in "register," that is, the visual receptive field encompasses a three-dimensional spatial region (peri-personal space) around the tactile receptive field. Some bimodal neurons are activated preferentially or even exclusively when 3D objects are moved towards or away from the tactile receptive field. About thirty percent of VIP neurons code space in reference to the monkey's body. There are also neurons that have hybrid receptive fields. These receptive fields change position when the eyes move along a certain axis, but remain fixed when the eyes move along another axis (Duhamel et al., 1997).

Consistent with the single neuron data, are the results of lesion studies (Duhamel, personal communication). Selective electrolytic lesion of area VIP in monkeys determines mild but consistent contralesional neglect for peri-personal space. No changes were observed in ocular saccades, pursuit and optokinetic nystagmus. Tactile stimuli applied to the contralesional side of the face also failed to elicit orienting responses.

Single neurons studies showed that most F4 neurons discharge in association with monkey's active movements (Gentilucci et al., 1988). The movements more represented are head and arm movements, such as head turns and reaching. Most F4 neurons respond to sensory stimuli. As neurons in VIP, F4 sensory-driven neurons can be subdivided into two classes: unimodal, purely sensory neurons, and bimodal, somatosensory and visual neurons (Gentilucci et al., 1988; Fogassi et al., 1992, 1996). Tactile receptive fields, typically large, are located on the face, chest, arm and hand. Visual receptive fields are also large. They are located in register with the tactile ones, and similarly to VIP, confined to the peri-personal space (Gentilucci et al., 1983, 1988; Fogassi et al., 1992, 1996; Graziano et al., 1994). Recently, trimodal neurons responding also to auditory stimuli were described in F4 (Graziano et al., 1999).

Several electrophysiological studies have shown that in most F4 neurons visual receptive fields do not change position with respect to the observer's body when the eyes move (Gentilucci et al., 1983; Fogassi et al., 1992, 1996; Graziano et al., 1994). The visual responses of F4 neurons do not signal positions on the retina, but positions in space relative to the observer. The spatial coordinates of the visual receptive fields are anchored to different body parts, and not to a single reference point, and they are coded in egocentric coordinates (Fogassi et al., 1996a, 1996b). Furthermore, visual receptive fields located around a certain body part (e.g., the arm) move when that body part is moved (Graziano et al., 1997).

Empirical evidence in favor of the simulation-based motor nature of space coding derives from the properties of F4 neurons. In principle there are two main possibilities on what these neurons code. The first is that they code space "visually". If this is so, given a reference point the neurons should signal the location of objects by using a Cartesian or some other geometrical system. The alternative possibility is that the discharge of neurons reflects a potential, simulated motor action directed towards a particular spatial location. This simulated potential action would create a motor space. When a visual stimulus is presented, it evokes directly the simulation of the congruent motor schema which, regardless of whether the action is executed or not, maps the stimulus position in motor terms.

Arguments in favor of the visual hypothesis are the tight temporal link between stimulus presentation and the onset of neural discharge, the response constancy, and the presence of what appears to be a visual receptive field. If, however, there is a strict association between motor actions and stimuli that elicit them, it is not surprising that stimulus presentation determines the effects just described. More direct evidence in favor of a motor space came from the study of properties of F4 neurons in response to moving stimuli. According to the visual hypothesis, each set of neurons, when activated should specify the object location in space, regardless of the stimulation's temporal dimension. A locus 15 cm from the tactile origin of the visual receptive field should remain 15 cm from it regardless of *how* the object reaches this position. The spatial map, as expressed by receptive field organization, should be basically static. In contrast, in the case of motor space, because time is inherent to movement, the spatial map may have dynamic properties and may vary according to the change in time of the object's spatial location. The experiments of Fogassi et al. (1996) showed that this is indeed the case. The visual receptive field extension of F4 neurons increases in depth when the speed of an approaching stimulus increases.

The notion that spatial awareness is linked to movement is pretty old. Von Helmoltz (1896) proposed the notion that the "a-priori" nature of our representation of space depends on the fact that it is generated by active exploratory behavior. Indeed, as it has been argued elsewhere (see Rizzolatti et al., 1997), a strong support to the notion that spatial awareness derives from motor activity is the demonstration of the existence of peri-personal space. From a purely sensory point of view, there is no principled reason why our eyes should select light stimuli coming exclusively from a space sector located around our body. Light stimuli arriving from far or from near should be equally effective. However, if we consider that peri-personal stimuli occupy the space where the targets of the actions performed by hands and mouth are mostly located, it becomes clear why space is mapped in motor terms.

It is interesting to note the closeness of the view emerging from single-neuron recordings, and the philosophical perspective offered by phenomenological philosophers on space perception (see also Zahavi, 2002). As Merleau-Ponty (1962, p. 243) wrote, space is "…not a sort of ether in which all things float.... The points in space mark, in our vicinity, the varying range of our aims and our gestures." Furthermore, it is interesting to note that Husserl wrote that every thing we see, we simultaneously also see it as a tactile object, as something which is directly related to the lived body, but not by virtue of its visibility (Husserl, 1989). The body entertains a dual reality of spatial externality and internal subjectivity. The perspectival spatial location of our body provides the essential foundation to our determination of reality. But in contrast to what Husserl considered the *physiological* definition of the body – by considering it a material object – contemporary neurophysiological research suggests that a part of the body, the sensory-motor system, is also responsible for the phenomenal awareness of the body's relations with the world.

Why is action important in spatial awareness? Because what integrates multiple sensory modalities within the F4-VIP neural circuit is action embodied simulation (Gallese, 2005a, 2005b, 2006). Vision, sound and action are parts of an integrated system; the sight of an object at a given location, or the sound it produces, automatically triggers a "plan" for a specific action directed toward that location. What is a "plan" to act? It is a simulated potential action.

The characterization so far provided of this cortical network would seem at first sight to be fully consistent with the control of body actions within peri-personal space. If, however, we consider the results of lesion of this network, a different picture emerges. Unilateral lesion of the ventral premotor cortex of the monkey, including area F4, produces two series of deficits: motor deficits and perceptual deficits (Rizzolatti et al., 1983; see also Rizzolatti et al., 2001). Motor deficits consist in a reluctance to use the contralesional arm, spontaneously or in response to tactile and visual stimuli, and in a failure to grasp with the mouth food presented contralateral to the side of the lesion. Perceptual deficits concern neglect of the contralesional peripersonal space, and of the personal (tactile) space. A piece of food moved in the contralesional space around the monkey's mouth does not elicit any behavioral reaction. Similarly, when the monkey is fixating a central stimulus, the introduction of food contralateral to the lesion is ignored. In contrast, stimuli presented outside the animal's reach (far space) are immediately detected.

Neglect in humans occurs after lesion of the IPL and, less frequently, following damage of the frontal lobe, and in particular following lesions of area 6, 8, and 45 (see Bisiach and Vallar, 2000). The most severe neglect in humans occurs after lesion of the right IPL. In the full-fledged unilateral neglect, patients may show a more or less complete deviation of the head and eyes towards the ipsilesional side. Routine neurological examination shows that patients with unilateral neglect typically fail to respond to visual stimuli presented in the contralesional half field and to tactile stimuli delivered to the contralesional limbs. As in monkeys, also in humans neglect may selectively affect the extrapersonal and the peripersonal space. In humans, this dissociation was first described by Halligan and Marshall (1991). They examined a patient with severe neglect using a line bisection task. In this task the subject is usually required to mark the midpoint of a series of lines scattered all over a sheet of paper. The task was executed in the near space and in the space beyond hand reaching distance using a laser pen that the patient held in his right hand. The results showed that when the line was bisected in the near space the midpoint mark was displaced to the right, as typically occurs in neglect patients. However, the neglect dramatically improved or even disappeared when the testing was carried out in the far space. A similar dissociation was reported by Berti and Frassinetti (2000). Other authors described the opposite dissociation: severe deficits in tasks carried out in the extrapersonal space, slight or no deficit for tasks performed in the peripersonal space (see Shelton et al., 1990; Cowey et al., 1994, 1999). The lesions causing neglect in humans are usually very large, thus while the findings of separate systems for peripersonal and extrapersonal space are robust and convincing, any precise localization of the two systems in humans is at the moment impossible.

In conclusion, lesions of IPL and its frontal targets both in monkeys and humans determine body awareness deficits. Furthermore, it must be stressed that not only does IPL appear to play a fundamental role in body and spatial awareness, but it is also necessary for the *awareness of the quality of objects* presented within peripersonal space. Evidence in favor of this point of view comes from a series of clinical and neuropsychological studies. Marshall and Halligan (1988) reported the case of a lady who, due to a severe visual neglect, explicitly denied any difference between the drawing of an intact house and that of the same house when burning, if the relevant features for the discrimination were on the neglected side. However, when forced to choose the house where she would prefer to live, she consistently choose the intact one, showing in this way an implicit knowledge of the content she was unable to report. Berti and Rizzolatti (1992) confirmed these findings in a systematic way. In their experiments patients with severe unilateral neglect were asked to respond as fast as possible to target stimuli presented within the intact visual field by pressing one of two keys according to the category of the target (fruits and animals). Before showing these stimuli, pictures of animals and fruits were presented to the neglected field as priming stimuli.

The patients denied of seeing these priming stimuli. Yet, their responses to the stimuli shown in the intact field were facilitated by the primes. This occurred not only in "highly congruent conditions", that is when the prime stimulus and the target were physically identical (e.g. a dog), but also when prime and stimulus constituted two elements of the same semantic category, though physically dissimilar (e.g. a dog and an elephant).

These findings demonstrate that neglect patients are able to process stimuli presented within the neglected field up to a categorical semantic level of representation. However, they are not aware of them in the absence of IPL processing. This implies that the parieto-premotor circuits of the ventro-dorsal stream must be intact for achieving awareness even of those stimuli, such as fruits or animals that are mostly analyzed in the ventral stream.

Lesions of sensory-motor circuits, whose primary function is that of controlling movements of the body or of body parts towards or away from objects, produce deficits that do not exclusively concern the capacity to orient towards objects or to act upon them. These lesions produce also deficits in body, space, and object awareness.

4. THE VENTRO-DORSAL STREAM: ACTION UNDERSTANDING

Our social world is inhabited by a multiplicity of acting individuals. Much of our social competence depends on our capacity for understanding the meaning of the actions we witness and of the intentions producing those actions. What makes our perception of actions different from our perception of the inanimate world is the fact that there is something shared between the first and third person perspective of actions; the observer and the observed are both human beings endowed with a similar brain-body system making them act alike (Gallese, 2001).

Even more importantly, the observer and the observed share the same action-related neural circuits. The discovery of mirror neurons in the ventral premotor cortex of the monkey has triggered new perspectives on the neural mechanisms at the basis of action understanding. This will be the focus of the next sections.

5. THE MIRROR NEURON SYSTEM FOR ACTIONS IN MONKEYS AND HUMANS: EMPIRICAL EVIDENCE

More than a decade ago a new class of motor neurons was discovered in the ventral premotor cortex of the macaque monkey: mirror neurons. These neurons discharge not only when the monkey executes goal-related hand and/or mouth actions like grasping objects, but also when observing other individuals (monkeys or humans) executing similar actions (di Pellegrino et al., 1992; Gallese et al., 1996; Rizzolatti et al., 1996; Ferrari et al., 2003). Neurons with similar mirroring properties, matching action observation and execution have also been discovered in a sector of the posterior parietal cortex reciprocally connected with area F5 (see Rizzolatti et al., 2001; Gallese et al., 2002; Fogassi et al., 2005). It has been proposed that this "motor resonance" may underpin a direct form of action understanding (Gallese et al., 1996; Rizzolatti et al., 1996; Rizzolatti et al., 2001; Gallese et al., 2004; Rizzolatti and Craighero, 2004), by exploiting embodied simulation, a specific mechanism by means of which the brain/body system models its interactions with the world (Gallese 2001, 2003a, 2003b, 2005a, 2005b, 2006).

A series of experiments by Umiltà et al. (2001) showed that F5 mirror neurons become active also during the observation of partially hidden actions, when the monkey can predict the action outcome, even in the absence of the complete visual information about it (Umiltà et al., 2001). Macaque monkey's mirror neurons therefore represent actions made by others not exclusively on the basis of their visual description, but also on the basis of the anticipation of the final goal of the action, by means of the activation of its motor representation in the observer's premotor cortex.

In another series of experiments it has been shown that a particular class of F5 mirror neurons ("audio-visual mirror neurons") respond not only when the monkey executes and observes a given hand action, but also when it just hears the sound typically produced by the action (Kohler, 2002). These neurons respond to the sound of actions and discriminate between the sounds of different actions, but do not respond to other similarly interesting sounds such as arousing noises, or monkeys' and other animals' vocalizations. In sum, the different modes of presentation of events as intrinsically different as sounds, images, or willed acts of the body, are nevertheless bound together within a simpler level of semantic reference, underpinned by the same network of audio-visual mirror neurons. The presence of such neural mechanism within a non-linguistic species can be interpreted as the neural correlate of the dawning of a conceptualization mechanism (see Gallese, 2003c; Gallese and Lakoff, 2005).

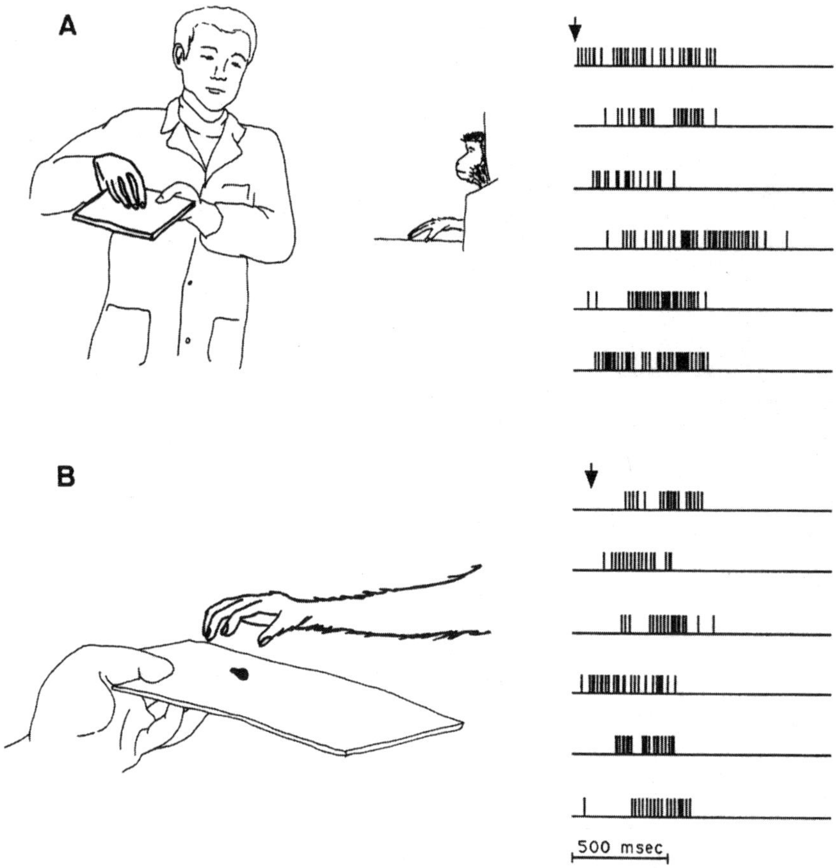

Figure 17-3. Example of a typical F5 mirror neuron. A. Experimental situation (Action observation) and six consecutive rasters showing the response of a mirror neuron during grasping observation. B. Experimental situation (Action execution) and six consecutive rasters showing the response of the same mirror neuron during active grasping execution. Arrows indicate the onset of observed and executed grasping. (Modified from di Pellegrino et al., 1992).

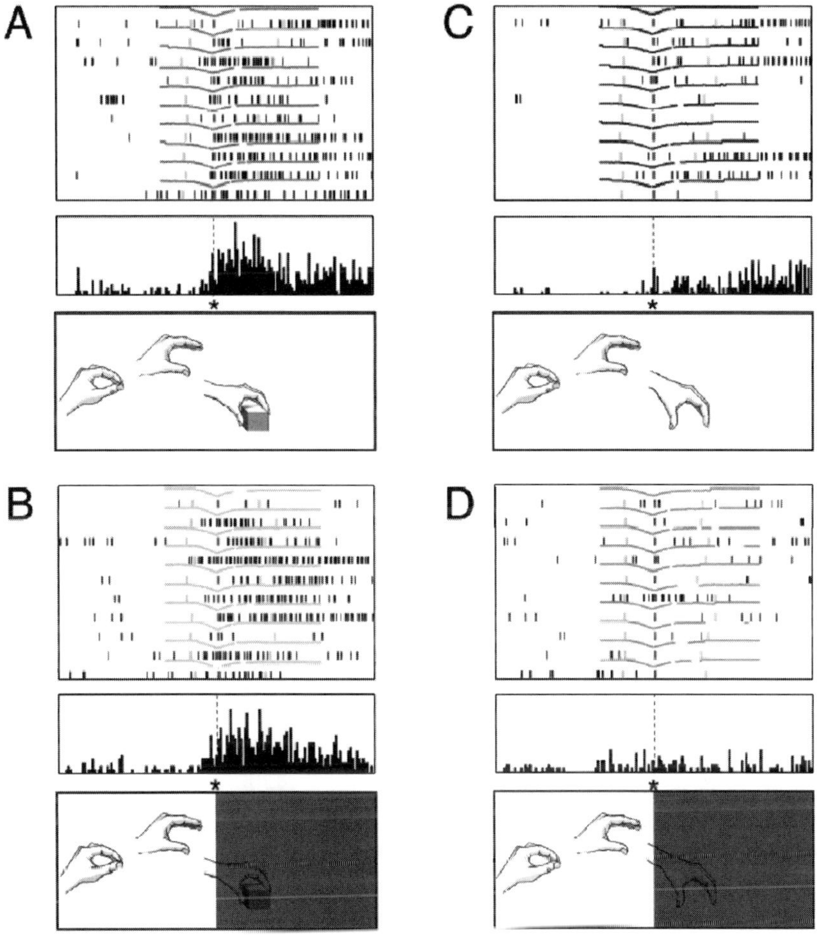

Figure 17-4. Example of an F5 mirror neuron responding to action observation in Full vision and in Hidden condition. The lower part of each panel illustrates schematically the experimenter's action as observed from the monkey's vantage point: the experimenter's hand starting from a fixed position, moving toward an object and grasping it (panels A and B), or mimicking grasping (panels C and D). The behavioral paradigm consisted of two basic conditions: Full vision condition (A) and Hidden condition (B). Two control conditions were also performed: Mimicking in full vision (C), and Mimicking hidden (D). In these last two conditions the monkey observed the same movements as in A and B, but without the target object. The black frame depicts the metallic frame interposed between the experimenter and the monkey in all conditions. In panels B and D the gray square inside the black frame represents the opaque sliding screen that prevented the monkey from seeing the experimenter's action performed behind it. The asterisk indicates the location of a marker on the frame. In hidden conditions the experimenter's hand started to disappear from the

monkey's vision when crossing the marker position. The upper part of each panel shows rasters displays and histograms of ten consecutive trials recorded during the corresponding experimenter's hand movement illustrated in the lower part. Above each raster kinematics recordings (black traces) of the experimenter's hand are shown. The black trace indicates the experimenter's hand movements recorded using a motion analysis system. The illustrated neuron responded to the observation of grasping and holding in Full vision (A) and in the Hidden condition (B), in which the interaction between the experimenter's hand and the object occurred behind the opaque screen. The neuron response was virtually absent in the two conditions in which the observed action was mimed (C and D). Histograms bin width = 20 ms. Ordinates: spikes/s; abscissae: time (Modified from Umiltà et al., 2001).

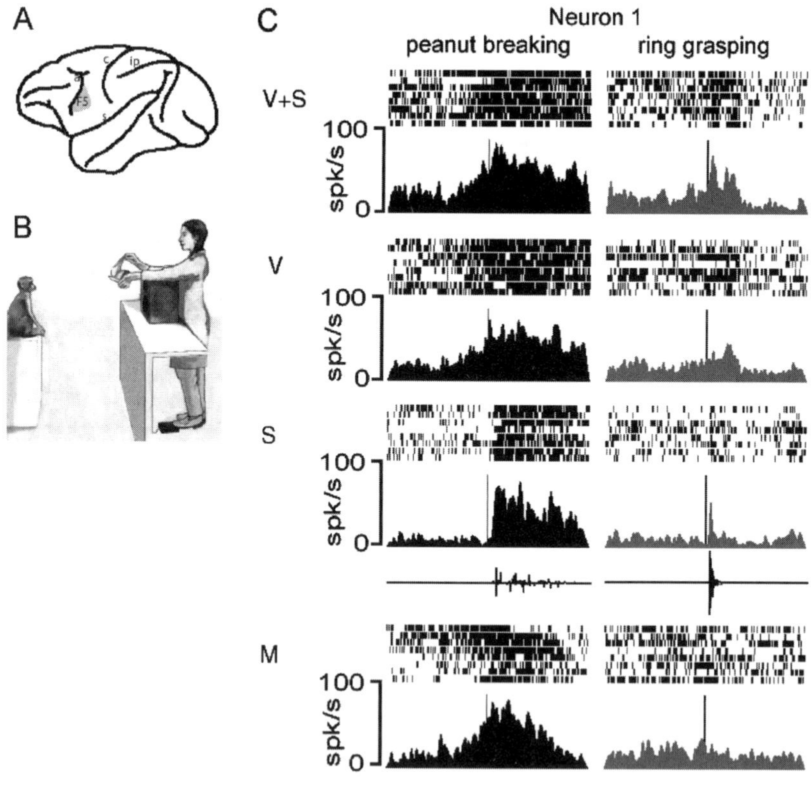

Figure 17-5. Example of one F5 audio-visual mirror neuron. A. Lateral view of the macaque brain with the location of area F5 shaded in gray. Major sulci: a = arcuate, c = central, ip = intraparietal, s = sylvian sulcus. B. Schematic view of the experimental situation. C. Response of a neuron (Neuron 1) discriminating between two actions in all modalities. Rasters

are shown together with spike density functions for the best (black) and one of the less effective actions (grey). V+S, V, S and M stand for Vision and Sound, Vision only, Sound only and Motor conditions, respectively. The vertical lines indicate the time at which the sound occurred or would have occurred (V). The traces under the spike-density functions in the sound-only conditions are oscillograms of the sounds played back to test the neurons. This neuron discharged when the monkey broke a peanut (row 'M') and when the monkey observed the experimenter making the same action (rows V and V+S). The same neuron also responded when the monkey only heard the sound of a peanut being broken without seeing the action (row 'S'). When the monkey grasped a ring ('M'), Neuron 1 responded much less, demonstrating the motor specificity of the neuron. Also both the vision and the sound of an experimenter grasping the ring determined much smaller responses. A statistical criterion yielded both auditory and visual selectivity for this neuron. (Adapted from Kohler et al., 2002).

Several studies using different experimental methodologies and techniques have demonstrated also in the human brain the existence of a mirror neuron system matching action perception and execution (for review, see Rizzolatti et al., 2001; Gallese, 2003a, 2003b, 2006; Rizzolatti and Craighero, 2004; Gallese et al., 2004). During action observation there is a strong activation of premotor and parietal areas, the likely human homologue of the monkey areas in which mirror neurons were originally described. The mirror neuron matching system for actions in humans is somatotopically organized, with distinct cortical regions within the premotor and posterior parietal cortices being activated by the observation/execution of mouth, hand, and foot related actions (Buccino et al., 2001).

More recently, it has been shown that the mirror neuron system for actions in humans is directly involved in imitation (Iacoboni et al., 1999; Buccino et al., 2004a) and in the perception of communicative actions (Buccino et al., 2004b). Furthermore, the premotor cortex containing the mirror system for action is involved in processing action-related sentences (Hauk et al., 2004; Tettamanti et al., 2005; Buccino et al., 2005; see also Pulvermuller, 2002).

6. THE MIRROR NEURON SYSTEM FOR ACTIONS AND THE UNDERSTANDING OF INTENTIONS

What does the presence of mirror neurons in different species of primates such as macaques and humans tell us about the evolution of social cognition? Monkeys may exploit the mirror neuron system to optimize their social interactions. At least, the evidence collected so far seems to suggest that the mirror neuron system for actions is enough sophisticated to enable its exploitation for social purposes. This matching mechanism can support social facilitation in monkeys. It has been recently shown that the observation and hearing of noisy eating actions facilitates eating behavior in pigtailed macaque monkeys (Ferrari et al., 2005).

Another recently published study shows that pigtailed macaque monkeys recognize when they are imitated by a human experimenter (Paukner et al., 2005). Pigtailed macaques preferentially look at an experimenter imitating the monkeys' object-directed actions compared with an experimenter manipulating an identical object but not imitating their actions. Since both experimenters acted in synchrony with the monkeys, the monkeys based this preference not on temporal contingency, but took into account the structural components of the experimenters' actions.

It may well be the case, as repeatedly argued, that macaque monkeys are not capable of motor imitation – though recent evidence by Subiaul et al. (2004) shows that they are capable of cognitive imitation. The study by Paukner et al. (2005) nevertheless shows that macaque monkeys do entertain the capacity to discriminate between very similar goal-related actions on the basis of their degree of similarity with the goal-related actions the monkeys themselves have just executed. This capacity seem to be rather sophisticated, in that it implies a certain degree of meta-cognition in the domain of purposeful actions.

But monkeys do not entertain full-blown mentalization. Thus, what makes humans different? At present we can only make hypotheses about the relevant neural mechanisms underpinning the still poorly understood mentalizing abilities of humans. In particular, we do not have a clear neuroscientific model of how humans understand the intentions promoting the actions of others they observe.

When an individual starts a movement aimed to attain a goal, such as picking up a pen, he/she has clear in mind what he/she is going to do, for example writing a note on a piece of paper. In this simple sequence of motor acts the final goal of the whole action is present in the agents' mind and is somehow reflected in each motor act of the sequence. The action intention, therefore, is set before the beginning of the movements. This also means that when we are going to execute a given action we can also predict its consequences.

However, in social contexts a given action can be originated by very different intentions. Suppose one sees someone else grasping a cup. Mirror neurons for grasping will most likely be activated in the observer's brain. A simple motor equivalence between the observed action and its motor representation in the observer's brain, though, can only tell us what the action is (it's a grasp) and not why the action occurred. This has led to argue against the relevance of mirror neurons for social cognition, and in particular, for determining the intentions of others (see Jacob and Jeannerod, 2005).

We should ask ourselves: What does it mean to determine the intention of the action of someone else? I propose a deflationary answer. Determining why action A (e.g., grasping a cup) was executed can be equivalent to detecting the goal of the still not executed and impending subsequent action (e.g., drink from the cup).

In a recent fMRI study (Iacoboni et al., 2005) these issues were experimentally addressed. Subjects watched three kinds of stimuli: grasping hand actions without a context, context only (a scene containing objects), and grasping hand actions embedded in contexts. In the latter condition the context suggested the intention associated with the grasping action (either drinking or cleaning up). Actions embedded in contexts, compared with the other two conditions, yielded a significant signal increase in the posterior part of the inferior frontal gyrus and the adjacent sector of the ventral premotor cortex where hand actions are represented. Thus, premotor mirror areas – areas active during the execution and the observation of an action – previously thought to be involved only in action recognition – are actually also involved in understanding the "why" of action, that is, the intention promoting it. For simple actions as those employed in this study, the ascription of intentions occurs by default and it is underpinned by the mandatory activation of an embodied simulation mechanism (Gallese, 2005a, 2005b, 2006).

The neurophysiological mechanism at the basis of the relationship between intention detection and action prediction was recently discovered. Fogassi et al. (2005) described a class of parietal mirror neurons whose discharge during the observation of an act (e.g., grasping an object), is conditioned by the type of not yet observed subsequent act (e.g., bringing the object to the mouth) specifying the overall action intention. This study shows that the inferior parietal lobe of the monkey contains mirror neurons discharging in association with the execution/observation of motor acts (grasping) only when they are embedded in a specific action aimed at different goals. It must be emphasized that the neurons discharge before the monkey observes the experimenter starting the second motor act (bringing the object to the mouth or placing it into the cup). Single motor acts are dependent to each other as they participate to the overarching distal goal of an action, thus forming pre-wired intentional chains, in which each next motor act is facilitated by the previously executed one.

This suggests that in addition to recognizing the goal of the observed motor act, mirror neurons allow the observing monkey to predict the agent's next action, henceforth its overall intention. It is possible to interpret this mechanism as the neural correlate of the dawning of more sophisticated intention understanding abilities, as those characterizing our species.

The mechanism of intention understanding just described appears to be rather simple: depending on which motor chain is activated, the observer is going to activate the motor schema of what, most likely, the agent is going to do. How can such a mechanism be formed? We hypothesize that the statistical frequency of action sequences, as they are habitually performed or observed in the social environment, can constrain preferential paths of inferences/predictions. This can be accomplished by chaining together different motor schemata. At the neural level this would be equivalent to the chaining of different populations of mirror neurons coding not only the observed motor act, but also those that in a given context would normally follow. Ascribing intentions would therefore consist in predicting a forthcoming new goal.

7. A NEW PERSPECTIVE ON ACTION AND PERCEPTION

Perception can be accounted for only if one takes into consideration the relationship between the agent and his/her environment. Actions are fundamental for building a meaningful description of the visual world. Only through action visual experience can be "validated" and acquire a meaning for the individual. This is true, in particular, for space perception and action understanding.

The influence of action on perception is – as we have seen – instantiated by space perception. The conventional view about space is that space is unitary, and that the brain has a center specifically devoted to space representation. According to this view, this center is used for all purposes, for walking, for reaching objects, or for describing a scene verbally. Classically, this putative multi-purpose center was considered to be located in the parietal lobe (Critchley, 1953; Hyvärinen, 1982; Ungerleider and Mishkin, 1982). As already mentioned, neuropsychological evidence indicates that lesions of parietal-premotor sensory-motor circuits, whose primary function is that of controlling orienting movements, as well as movements of the body or of body parts towards objects, produces deficits which do not exclusively concern the capacity to orient towards objects or to act upon them. These lesions produce also deficits in space perception.

It is likely that the perception of space derives from the simultaneous activity of several sensory-motor circuits. The lesion of only one of them is sufficient to decrease the salience of the stimuli in the contralesional field in such a way that the subject denies their presence. The recovery from neglect frequently observed even with relatively large lesions, is most likely due to compensatory effects of the residual circuits.

In the case of action understanding the issue is how to build a meaningful account of the action made by others when observed. It is obvious that there should be a system that analyzes and describes in pictorial ways the actions of others. However, the view that such "pictorial" analysis is per se *sufficient* to provide an understanding of the observed action must be questioned. Without a reference on personal knowledge, this description is devoid of any meaning for the observing individual. This impasse can be solved if there is a system matching the observed action on an internal motor representation of the same action. The mirror system for actions has precisely the properties that such a system must have. It internally represents the action and it responds when another agent performs a similar action. According to this view, action understanding is viewed as an embodied function. It relies on a neural circuit involved in the control of action.

It is important to stress that this motor-centered view of perception allows one to explain a series of psychological data on the effect of motor function on stimulus perception, which would be very difficult to account for according to the conventional view on perception. There is a large body of literature demonstrating motor effects on perception (for a review, see Viviani, 2002). For reasons of space I will limit myself here to discuss only some psychological experiments more closely related to the points addressed above.

Craighero et al. (1999) demonstrated that preparation of a grasping movement affects detection and discrimination of a visual stimulus. In this experiment normal subjects were required to grasp a bar after the presentation of a visual stimulus whose orientation was either congruent or incongruent with that of the bar. The results 24 showed that grasping preparation enhances the detection of a visual stimulus whose intrinsic properties are congruent with those of the object to be grasped. The role played by handedness in performing perceptual tasks is another example of the involvement of motor processes in perceptual functions. De Sperati and Stucchi (1997) showed that right- and left-handed normal subjects used an internal simulation of the movement of their dominant hand in order to discriminate between observed screwing and unscrewing screwdrivers. In another series of experiments Gentilucci et al. (1998a, 1998b), asked normal subjects to judge the handedness of pictures of hands and fingers assuming different postures. The results showed that the presentation of postures that hand and fingers commonly assume at rest, or when interacting with objects, facilitated the responses with respect to the presentation of less usual hand-finger postures, even when the latter were richer in visual cues useful for handedness recognition (see also Parsons 1987, 1994; Parsons et al., 1995 for similar results). Once again motor knowledge was employed to solve a perceptual task.

The data reviewed here show that different and parallel parieto-premotor networks create internal representations of actions. These representations may be used for various purposes. One is action generation, but others are space perception, action and intention understanding. These functions are unlikely to be the only cognitive functions that the parieto-premotor systems perform. Their capacity to "validate" experience renders them unique for acquiring knowledge – even abstract knowledge (see Lakoff and Johnson, 1980, 1999; Gallese and Lakoff, 2005) about the external world. A challenge for future research will be to investigate the role played by parieto-premotor sensorimotor circuits in the most sophisticated cognitive endowments of our species, language and thought.

ACKNOWLEDGMENTS

This work has been supported by MIUR and, as part of the Program EUROCORES OMLL, has been supported by grants to V.G. from the Italian C.N.R.

REFERENCES

Berti, A., and Frassinetti, F., 2000, When far becomes near: re-mapping of space by tool use, *J Cognitive Neurosci* **12**:415-420.

Berti, A., and Rizzolatti, G., 1992, Visual processing without awareness: Evidence from unilateral neglect, *J Cognitive Neurosci* **4**:345-351.

Bisiach, E., and Vallar, G., 2000, Unilateral neglect in humans, in: *Handbook of Neuropsychology*, 2nd ed., Vol. I, F. Boller, J. Grafman, and G. Rizzolatti, eds., Elsevier Science B.V., Amsterdam, pp. 459-502.

Buccino, G., Binkofski, F., Fink, G. R., Fadiga, L., Fogassi, L., Gallese, V., Seitz, R. J., Zilles, K., Rizzolatti, G., and Freund, H-. J., 2001, Action observation activates premotor 29 and parietal areas in a somatotopic manner: an fMRI study, *Eur J Neurosci* **13**: 400-404.

Buccino, G., Lui, F., Canessa, N., Patteri, I., Lagravinese, G., Benuzzi, F., Porro, C. A., and Rizzolatti, G., 2004a, Neural circuits involved in the recognition of actions performed by nonconspecifics: An fMRI study, *J Cognitive Neurosci* **16**:114-126.

Buccino, G., Vogt, S., Ritzl, A., Fink, G. R., Zilles, K., Freund, H-. J., and Rizzolatti, G., 2004b, Neural circuits underlying imitation learning of hand actions: an event-related fMRI study, *Neuron* **42**:323-334.

Caminiti, R., Ferraina, S., and Johnson, P. B., 1996, The sources of visual information to the primate frontal lobe: a novel role for the superior parietal lobule, *Cereb Cortex* **6**:319-328.

Colby, C. L., Duhamel, J. -R., and Goldberg, M. E., 1993, Ventral intraparietal area of the macaque: anatomic location and visual response properties, *J Neurophysiol* **69**:902-914.

Cowey, A., Small, and M., Ellis, S., 1994, Left visuo-spatial neglect can be worse in far than near space, *Neuropsychologia* **32**:1059-1066.

Cowey, A., Small, M., and Ellis, S., 1999, No abrupt change in visual hemineglect from near to far space, *Neuropsychologia* **37**:1-6.

Craighero, L., Fadiga, L., Rizzolatti, G., and Umiltà, C. A., 1999, Movement for perception: a motor-visual attentional effect, *J Exp Psychol* **25**:1673-1692.

Critchley, M., 1953, *The parietal lobes*, Hafner Press, New York.

De' Sperati, C., and Stucchi, D., 1997, Recognizing the motion of a graspable object is guided by handedness, *Neuroreport* 8:2761-2765.

Di Pellegrino, G., Fadiga, L., Fogassi, L., Gallese, V., and Rizzolatti, G., 1992, Understanding motor events: a neurophysiological study, *Exp Brain Res* **91**:176-180.

Duhamel J. –R., Bremmer, F., Ben Hamed, S., and Graf, W., 1997, Spatial invariance of visual receptive fields in parietal cortex neurons, *Nature* **389**:845-848.

Ferrari, P. F., Gallese, V., Rizzolatti, G., and Fogassi, L., 2003, Mirror neurons responding to the observation of ingestive and communicative mouth actions in the monkey ventral premotor cortex, *Eur J Neurosci* **17**:1703-1714.

Ferrari, P. F., Maiolini, C., Addessi, E., Fogassi, L., and Visalberghi, E., 2005, The observation and hearing of eating actions activates motor programs related to eating in macaque monkeys, *Behav Brain Res* **161**:95-101.

Foerster, O., 1936, Motorische Felder und Bahnen. Sensible corticale Felder, in: *Handbuch der Neurologie*, O. Bumke and O. Foerster, eds., Vol. 6.6., Springer, Berlin, pp. 1-448.

Fogassi, L., Ferrari, P. F., Gesierich, B., Rozzi, S., Chersi, F., and Rizzolatti, G., 2005, Parietal lobe: From action organization to intention understanding, *Science* **302**:662-667.

Fogassi L., Gallese, V., Di Pellegrino, G., Fadiga, L., Gentilucci, M., Luppino, G., Matelli, M., Pedotti, A., and Rizzolatti, G., 1992, Space coding by premotor cortex, *Exp Brain Res* **89**:686-690.

Fogassi, L., Gallese, V., Fadiga, L., Luppino, G., Matelli, M., and Rizzolatti, G., 1996, Coding of peripersonal space in inferior premotor cortex (area F4), *Journal of Neurophysiology* **76**:141-157.

Gallese, V., 2001, The "Shared Manifold" Hypothesis: from mirror neurons to empathy, *J Consciousness Stud* **8**(5-7):33-50.

Gallese, V., 2003a, The manifold nature of interpersonal relations: The quest for a common mechanism, *Phil Trans Royal Soc London* **358**:517-528.

Gallese, V., 2003b, The roots of empathy: The shared manifold hypothesis and the neural basis of intersubjectivity, *Psychopatology* **36**(4):171-180.

Gallese, V., 2003c, A neuroscientific grasp of concepts: From control to representation, *Phil Trans Royal Soc London* **358**:1231-1240.

Gallese, V., 2005a, Embodied simulation: from neurons to phenomenal experience, *Phenomenology and the Cognitive Sciences* **4**:23-48.

Gallese V., 2005b, "Being like me": Self-other identity, mirror neurons and empathy, in: *Perspectives on Imitation: From Cognitive Neuroscience to Social Science*, Vol. 1, S. Hurley and N. Chater, eds., MIT Press, Boston, MA, pp. 101-118.

Gallese, V., 2006, Intentional attunement: A neurophysiological perspective on social cognition and its disruption in autism, *Cognitive Brain Res* **1079**(1):15-24.

Gallese, V., and Goldman, A., 1998, Mirror neurons and the simulation theory of mindreading, *Trends Cogn Sci* **12**:493-501.

Gallese, V., and Lakoff, G., 2005, The Brain's Concepts: The Role of the Sensory-Motor System in Reason and Language, *Cogn Neuropsychol* **22**:455-479.

Gallese, V., Fadiga, L., Fogassi, L., and Rizzolatti, G., 1996, Action recognition in the premotor cortex, *Brain* **119**:593-609.

Gallese, V., Fadiga, L., Fogassi, L., and Rizzolatti, G., 2002, Action representation and the inferior parietal lobule, in: *Common Mechanisms in Perception and Action: Attention and Performance*, Vol. XIX, W. Prinz and B. Hommel, eds., Oxford University Press, Oxford, pp. 334-355.

Galletti, C., Fattori, P., Battaglini, P. P., Shipp, S., and Zeki, S., 1996, Functional demarcation of a border between areas V6 abd V6A in the superior parietal gyrus of the macaque monkey, *Eur J Neurosci* **8**:30-52.

Gentilucci, M., Scandolara, C., Pigarev, I. N., and Rizzolatti ,G., 1983, Visual responses in the postarcuate cortex (area 6) of the monkey that are independent of eye position, *Exp Brain Res* **50**:464–468.

Gentilucci, M., Fogassi, L., Luppino, G., Matelli, M., Camarda, R., and Rizzolatti, G., 1988, Functional organization of inferior area 6 in the macaque monkey: I. Somatotopy and the control of proximal movements, *Expl Brain Res* **71**:475-490.

Gentilucci, M., Daprati, E., and Gangitano, M., 1998a, Implicit visual analysis in handedness recognition, *Conscious Cogn* **7**:478-493.

Gentilucci, M., Daprati, E., and Gangitano, M., 1998b, Right-handers and left-handers have different representations of their own hand, *Cogn Brain Res* **6**:185-192.

Goodale, M. A., and Milner, A. D., 1992, Separate visual pathways for perception and action, *Trends Neurosci* **15**:20-25.

Graziano, M. S. A., Yap, G. S., and Gross, C. G., 1994, Coding of visual space by premotor neurons, *Science* **266**:1054-1057.

Graziano, M. S. A., Hu, X., and Gross, C. G., 1997, Visuo-spatial properties of ventral premotor cortex *J Neurophysiol* **77**:2268-2292.

Graziano, M. S. A., Reiss, L. A. J., and Gross, C. G., 1999, A neuronal representation of the location of nearby sounds, *Nature* **397**:428-430.

Grèzes, J., Costes, N., and Decety, J., 1998, Top-down effect of strategy on the perception of human biological motion: a PET investigation, *Cogn Neuropsychol* **15**:553-582.

Halligan, P. W., Marshall, J. C., 1991, Left neglect for near but not far space in man, *Nature* **350**:498-500.

Helmoltz, H., von, 1896, Die Tatsache der Wahrnehmung, in: *Vorträge und Reden*, Vol. II, Braunschweig.

Husserl, E., 1989, *Ideas Pertaining to a Pure Phenomenology and to a Phenomenological Philosophy, Second Book: Studies in the Phenomenology of Constitution*, Kluwer Academic Publishers, Dordrecht.

Hyvärinen, J., 1982, Posterior parietal lobe of the primate brain, *Physiological Reviews* **62**:1060–1129.

Iacoboni, M., Woods, R. P., Brass, M., Bekkering, H., Mazziotta, J. C., and Rizzolatti, G., 1999,

Cortical mechanisms of human imitation, *Science* **286**:2526-2568.

Iacoboni, M., Molnar-Szakacs, I., Gallese, V., Buccino, G., Mazziotta, J., and Rizzolatti, G., 2005, Grasping the intentions of others with one's owns mirror neuron system, *PLoS Biol* **3**:529-535.

Jacob, P., and Jeannerod, M., 2005, The motor theory of social cognition: a critique, *Trends Cogn Neurosci* **9**:21-25.

Jeannerod, M., 1994, The representing brain: neural correlates of motor intention and imagery, *Behav Brain Sci* **17**:187-245.

Jeannerod, M., 1997, *The Cognitive Neuroscience of Action*, Blackwell, Oxford.

Lakoff, G., and Johnson, M., 1980, *Metaphors We Live By*, University of Chicago Press, Chicago.

Lakoff, G., and Johnson, M., 1999, Philosophy in the Flesh: The Embodied Mind and its Challenge to Western Thought, Basic Books, New York.

Luppino, G., Murata, A., Govoni, P., and Matelli, M., 1999, Largely segregated parietofrontal connections linking rostral intraparietal cortex (areas AIP and VIP) and the ventral premotor cortex (areas F5 and F4), *Exp Brain Res* **128**:181-187.

Marshall, J. F., Halligan, P.W., 1988, Blindsight and insight in visuo-spatial neglect, *Nature* **336**:766-767.

Matelli, M., Govoni, P., Galletti, C., Kutz, D. F., and Luppino, G., 1998, Superior area 6 afferents from the superior parietal lobule in the macaque monkey, *J Comp Neurol* **402**:327-352.

Merleau-Ponty, M., 1962, *Phenomenology of Perception*, (translated from the French by C. Smith), Routledge, London.

Milner, D., and Goodale, M. A., 1995, *The visual Brain in Action*, Oxford University Press, Oxford.

Pandya, D. N., and Seltzer, B., 1982, Intrinsic connections and architectonics of posterior parietal cortex in the rhesus monkey, *J Comp Neurol* **204**:196-210.

Parsons, L. M., 1987, Imaged spatial transformations of one's hands and feet, *Cog Psychol* **19**:178-241.

Parsons, L. M., 1994, Temporal and kinematic properties of motor behavior reflected in mentally simulated action, *J Exp Psychol* **26**:709-730.

Parsons, L. M., Fox, P. T., Downs, J. H., Glass, T., Hirsch, T. B., Martin, C. C., Jerabek, P. A., and Lancaster, J. L., 1995, Use of implicit motor imagery for visual shape discrimination as revealed by PET, *Nature* **375**:54-58.

Paukner, A., Anderson, J. R., Borelli, E., Visalberghi, E., and Ferrari, P. F., 2005, Macaques (Macaca nemestrina) recognize when they are being imitated, *Biol Lett* **1**:219-222.

Rizzolatti, G., Craighero L., 2004, The mirror neuron system, *Ann Rev Neurosci* **27**:169-192.

Rizzolatti, G. and Matelli, M., 2003, Two different streams for the dorsal visual system: anatomy and functions, *Exp Brain Res* **153**:146-157.

Rizzolatti, G., Berti, A., and Gallese, V., 2000, Spatial neglect: neurophysiological bases, cortical circuits and theories, in: *Handbook of Neuropsychology*, 2nd ed., Vol. I, F. Boller, J. Grafman, and G. Rizzolatti, eds., Elsevier Science B.V., Amsterdam, pp. 503-537.

Rizzolatti, G., Fogassi, L., and Gallese, V., 2001, Neurophysiological mechanisms underlying the understanding and imitation of action, *Nat Rev Neurosci* **2**:661-670.

Rizzolatti, G., Matelli, M., and Pavesi, G., 1983, Deficits in attention and movement following the removal of postarcuate (area 6) and prearcuate (area 8) cortex in macaque monkeys, *Brain* **106**:655–673.

Rizzolatti, G., Fadiga, L., Fogassi, L., and Gallese, V., 1997, The space around us, *Science* **277**:190-191.

Rizzolatti, G., Fadiga, L., Gallese, V., and Fogassi, L., 1996a, Premotor cortex and the recognition of motor actions, *Cognitive Brain Res* **3**:131-141.

Rozzi, S., Calzavara, R., Belmalih, A., Borra, E., Gregoriou, G. G., Matelli, M., and Luppino, G., 2006, Cortical Connections of the Inferior Parietal Cortical Convexity of the Macaque Monkey, *Cerebral Cortex,* in press.

Rizzolatti, G., Camarda, R., Fogassi, M., Gentilucci, M., Luppino, G., and Matelli, M., 1988, Functional organization of inferior area 6 in the macaque monkey: II. Area F5 and the control of distal movements, *Exp Brain Res* **71**:491-507.

Shelton, P. A., Bowers, D., and Heilman, K. M., 1990, Peripersonal and vertical neglect, *Brain* **113**:191-205.

Subiaul, F., Cantlon, J. F., Holloway, R. L., and Terrace, H. S, 2004, Cognitive imitation in Rhesus macaque *Science* **305**:407-410.

Ungerleider, L. G., and Mishkin, M., 1982, Two visual systems, in: *Analysis of visual behavior*, D. J. Ingle, M. A. Goodale, and R. J. W. Mansfield, eds., MIT Press, Cambridge, MA, pp. 549-586.

Vallar, G., and Perani, D., 1987, The anatomy of spatial neglect in humans, in: *Neurophysiological and Neuropsychological Aspects of Spatial Neglect*, M. Jeannerod, ed., North-Holland Co./Elsevier Science Publishers, Amsterdam, pp. 235-258.

Viviani, P., 2002, Motor-perceptual interactions: assessing the conceptual implications of some known phenomena, in: *Common Mechanisms in Perception and Action: Attention and Performance*, W. Prinz and B. Hommel, eds., Vol. XIX., Oxford University Press, Oxford, pp. 406-442.

Von Bonin, G., Bailey, P., 1947, *The neocortex of macaca mulatta*, University of Illinois Press, Urbana Il., pp. 136.

Von Economo, C., 1929, *The Cytoarchitectonics of the Human Cerebral Cortex*, Oxford University Press, London, pp. 186.

Zahavi, D., 2002, First-person thoughts and embodied self-awareness: Some reflections on the relation between recent analytical philosophy and phenomenology, *Phenomenology and the Cognitive Sciences* **1**:7–26.

Chapter 18

MIND OVER MATTER? IMAGINED BODY MOVEMENTS AND THEIR NEURONAL CORRELATES

Fred W. Mast[1], Laura Bamert[1] and Nathaniel Newby[2]
[1]Department of Psychology, University of Lausanne, Switzerland; [2]Johnson Space Center, Houston, USA

Abstract: We use mental imagery not only to anticipate future perception but also for our own movements. In this chapter we review the most recent literature in the domain of motor imagery, with particular emphasis on clinical findings. A wealth of evidence suggests that imagined movements of body parts draw - at least partly - on mechanisms associated with actual execution of the same movements. It is thus also possible to improve one's motor performance via mental imagery techniques; motor imagery is widely used for athletes and researchers began to study its beneficial effects during rehabilitation in patients after cerebral lesions. Moreover, motor imagery is not restricted to single parts of the body. The body as whole can be rotated in imagery, for example, when we need to make a spatial judgment from another (or someone else's) perspective. To better understand the mechanisms underlying whole body rotations in imagery we suggest investigating more specifically the yet rather neglected vestibular cortical projections, and discuss their possible role in cognitive tasks. We present an applied example showing how motor imagery training can change perception of movement.

Key words: mental imagery; motor imagery; mental rotation; vestibular system; artificial gravity.

In this chapter we will elaborate on the mechanisms involved in real and imagined movements, including movements of the entire body. In fact, it is striking that motor imagery shares a number of properties associated with actual movements. For example, imagining body part movements activates many of the same areas that are involved in the execution of the same movement (hand: Porro et al., 2000; Lotze et al., 1999; Thobois et al., 2000;

foot: Alkhadi et al., 2005; tongue, toes, and fingers: Ehrsson et al., 2003) even though the overlap may only be partial (Wolbers et al., 2003; Deiber et al., 1998; Jeannerod and Frak, 1999; Gerardin et al., 2000). Furthermore, imagined and executed movements are both controlled by the contralateral hemisphere (Wolbers et al., 2003; Dominey et al., 1995; Sirigu et al., 1996; Naito et al., 2002; Alkadhi et al., 2005). Instead of adding yet other commonalities at this point we shall restrict ourselves in order to focus more specifically on the underlying mechanisms shared by real and imagined movements. Particular emphasis will be given to neuroimaging and clinical research.

In terms of the latter, a wealth of knowledge is available from studies with congenital, traumatic, or pathological impairments of motor functions. The logic for analyzing these data is straightforward. If imagery and motor functions are intimately intertwined, lesions of the latter will affect - at least partly - the ability to imagine a motor action. A recent example by Alkadhi and colleagues (2005) compared brain activation in paraplegic patients and healthy controls. Both groups performed a task involving imagined movements of the foot. The results revealed for patients and healthy subjects activity in several brain areas known to be involved in the execution and control of real movements such as the supplementary motor area (SMA), pre-SMA, cingulate motor area (CMA), dorsal premotor area (PMd), ventral premotor area (PMv), superior and inferior parietal regions, secondary sensory cortex (S2), insular and pre-frontal cortex. The primary sensory foot representation and motor cortices were activated in the contralateral hemisphere only, which again strengthens the claim that neural mechanisms associated with real movements are also drawn upon when the same movements are imagined (see also Dominey et al., 1995; Sirigu et al., 1996; Porro et al., 2000; Naito et al., 2002). Although spinal lesions effectively prevent motor execution, it has to be noted that paraplegic patients were able to move prior to the injury and are still aware of the presence of their limbs.

SENSORIMOTOR REPRESENTATIONS AND IMAGERY

Motor programs stored in memory may have an impact on preserved abilities to use motor imagery even though paraplegic patients are no longer able to execute the same movement. To further investigate the role of memory, Nico and colleagues (2004) compared performance of upper limb amputees, patients with congenital deletion of the forearm, and patients with a non-functioning limb due to unilateral lesions of the brachial plexus. The patients saw line drawings of a right or left human hand in either a natural or unnatural posture, and they had to judge whether it is a right or a left hand.

Interestingly, some amputees reported a re-activation of extinguished phantom limb sensations while performing this task, suggesting that motor commands to the missing limb are elicited by the motor imagery required to make the laterality judgment. Dominant limb amputation resulted in much longer response times and the patient will use either a visual strategy (rotating the visual hand stimulus) or another motor strategy (rotation of the non-dominant hand). Interestingly, patients with congenital deletion were not affected when they viewed unnatural joint positions. They never had any experience with actual movements, and thus no prior memories could possibly influence mental imagery of movements. Prior experience is therefore crucial for motor imagery even though there is also scarce evidence showing that - to some extent - new movements can be trained via mental imagery (Mulder et al., 2004). In this context it is also of interest that we are currently studying the effects of motor imagery in stroke patients; preliminary results yield promising evidence for improved motor performance as a result of repeated motor imagery training (Grabherr et al., 2006). In the sample we studied, the gain in the imagery training group exceeded the gain in the execution training group, and this again suggests that neuroplastic changes are specific to the form of training (Nyberg et al., 2006).

Plenty of evidence suggests that motor areas associated with movement control and movement execution overlap with those used when we imagine the same movements. Interestingly, however, such motor activation is not restricted to mental imagery of movements. In the following section, we present findings showing that motor activations occur during imagery tasks without any obvious motor involvement.

MOTOR PROCESSES AND MENTAL ROTATION

Tomasino and colleagues (Tomasino et al., 2005a) tested a patient with a lesion of the right inferior brachial plexus in a mental rotation task. The patient was instructed either to use a motor strategy (i.e. imagine himself rotating the objects by hand), or to use a visual strategy (i.e. imagine someone else or an external force rotating the objects). To relieve severe neuropathic pain refractory to medical treatment, the patient was implanted with an electrode grid covering the left primary motor and sensory cortices. In the absence of stimulation, the patient's performance did not differ from that of healthy controls. During electrode stimulation, however, the response times increased when the patient used the motor strategy. This drop in performance was found for tools, cubes, and hands and therefore suggests that the primary motor cortex is also involved during mental rotation of objects other than body parts. Several studies (Wraga et al., 2003; Lamm et al., 2001; Richter et al., 2000) found a motor activation when subjects mentally rotated visual objects. It is noteworthy, however, that some studies (Barnes et al., 2000; Harris et al., 2000; Jordan et al., 2001) have failed to confirm a significant activation in the primary motor cortex during mental rotation. It is therefore interesting to further investigate more specifically the conditions under which the primary motor cortex is or is not involved when people use motor imagery to solve a given task. For example, an fMRI study by Hanakawa and colleagues (2003) revealed that the activity of the primary motor cortex during a finger-tapping task in imagery seems to be related only to the motor preparation phase. The involvement of this area during the imagined finger movements is not revealed by the time-course analysis of activity.

Even though there clearly is a need for more specific research on motor imagery two factors appear to be crucial for the involvement of motor areas in mental rotation tasks. First, the nature of the stimulus matters, that is, hands, for example, can automatically trigger a motor activation (Tomasino et al., 2005b), and second, the strategy that subjects apply to solve the task may or may not engage motor areas (e.g., Kosslyn et al., 2001).

SELF-MONITORING OF MOTOR INTENTIONS

A major brain structure involved in motor tasks is the parietal cortex. It serves a crucial role in transforming sensory input into motor output. With respect to motor imagery two competing explanations on the role of the parietal cortex have been proposed (Sirigu et al., 1996). One explanation suggests a strong involvement in movement simulation whereas the other claims that the parietal cortex is responsible for monitoring motor outflow via an efference copy received from downstream motor areas. The latter is supported by a recent study (Sirigu et al., 2004) comparing parietal and cerebellar patients. Specifically, patients were asked to indicate both the moment when a finger act was initiated, and the moment when they had decided to move. Parietal patients were unable to discriminate the onset of movement from the moment in which they decided to initiate the movement (i.e., the two measures collapsed), whereas cerebellar patients were able to make this discrimination. Moreover, in controls and cerebellar patients about 1.5 seconds prior to movement onset, there was a progressive and negative rise in the cortical potentials from contralateral motor areas. The EEG activation was significantly weaker in parietal patients. This result suggests that the parietal cortex is involved in a conscious self-monitoring of motor intentions, possibly via a downstream efference copy of the motor commands. This also relates to clinical studies showing a lack of awareness as a result of parietal lesions (e.g., anosognosia).

Yet another issue concerns the role the parietal cortex may play in the inhibition process postulated by Jeannerod (2001), which, according to the simulation theory, prevents covert action from being executed. It is hard to find decisive empirical evidence for its existence but there are clinical studies, which support the view of an active motor inhibition during motor imagery. When a patient with a bilateral posterior parietal lesion was instructed to imagine a sequence of finger movements he was unable to prevent them from being executed (Schwoebel et al., 2002) even though there was no doubt that he understood the instructions. What this patient imagines is no longer private or hidden to others but rather becomes observable as overt action. The patient acts out what he imagines. Is this due to the lack of a parietal inhibition process, which serves to block the imagined movements from being executed? So far, there seems to be no definitive answer to this question. A distinct neuronal correlate of the efference copy has not yet been identified. In this context, it is also noteworthy that Alkadhi and colleagues (2005) failed to find specialized regions involved in active movement suppression when healthy subjects were engaged in motor imagery.

IMAGERY AND MOTOR PERFORMANCE

Additional behavioral evidence that motor imagery and motor execution share a common set of brain processes comes from an independent line of research. Several studies have shown that motor skills can be improved through mental imagery techniques. In a review paper, Jeannerod and Decety (1995) discuss how mental imagery training can influence motor performance in terms of muscular strength, reduction of variability and increase in temporal consistency. For example, Yue and Cole (1992) have presented a convincing result showing that imaginary fifth finger abductions led to an increased level of muscular strength. Even though it is conceivable that the effects of motor imagery also involve peripheral mechanisms, the activity of muscle groups associated with imagined movements is presumably too weak to be measured with certainty. The increase in muscle strength observed by Yue and Cole (1992) was not due to a gain in muscle mass but is rather based on higher-level changes in cortical maps or a more efficient recruitment of motor units.

Yet other studies on motor imagery showed an enhancement of the movement range (Mulder et al., 2004) or increased accuracy (Yaguez et al., 1998). It has to be noted, however, that the effects can operate in both ways;

motor imagery can influence motor performance, and, in the reverse direction, the extent of physical practice can change the areas activated during motor imagery (see for example Takahashi et al., 2005).

Later in this chapter we will discuss a specific example on the role of motor imagery training and how it can influence perception. In the following sections we will review studies on imagined body rotations as they relevant to the example we will describe further below.

IMAGINED ROTATIONS OF THE BODY

Yet another area of mental imagery concerns imagined rotation of one's own body. Unlike the examples discussed above, the object of the mental operation is then no longer a single part of the body but rather the body as a whole. The typical paradigm used to investigate this category of mental transformation comes from experiments designed to study cognitive development in children. Piaget and Inhelder (1956) showed children a model of three mountains and asked them what this scene would look like if viewed from a different perspective. In a modified version of the same task, children viewed arrays of abstract objects. The children were then asked what the objects would look like after having imagined moving themselves around the array versus having imagined the array rotated in front of them while they remained stationary. The former case requires the children to perform a mental viewer rotation, whereas the latter requires a mental array rotation (Huttenlocher and Presson, 1979). Results showed that the relative difficulty of object and viewer transformation varied systematically depending on the questions the experimenter asked. The children performed the array rotation task faster and made fewer errors than in the viewer rotation task when they had to match the transformed array to one of several pictures (appearance questions). This pattern reversed when the children had to report which item was at a particular location (e.g., what's on your right?) after performing the mental transformation (item questions).

Several studies have then demonstrated the viewer advantage in tasks with item questions (Wraga et al., 1999; 2000). The viewer advantage has been explained by the fact that we actively move and orient ourselves in an otherwise perfectly stationary environment. Moreover, it disappeared when the rotation had to be performed in a plane other than the familiar horizontal plane (Carpenter and Proffitt, 2001). Subsequent research, however, has shown that it is the orthogonality of the relationship between the viewer's body orientation and the imagined rotation that determines the viewer advantage rather than the orientation of the viewer with respect to gravity (Creem et al., 2001a).

Taken together, a wealth of behavioral research conducted over the last fifty years provides evidence for the existence of two distinct cognitive mechanisms underlying the mental operation of imagined array and viewer rotations. With the use of neuroimaging technology, however, we are now able to further explore the neuronal correlates associated with these two different types of mental rotation.

VIEWER BASED MENTAL ROTATION

Several neuroimaging studies have investigated the neural mechanisms underlying mental rotations in imagery and have added to our understanding of how different frames of reference are involved in cognitive tasks (e.g. Zähle et al., 2007).

Regarding the mental object or array rotation tasks, the results converge on consistent involvement of parietal areas. Using the task developed by Shepard and Metzler (1971), Cohen and colleagues (1996) reported that the mental rotation process is computed mainly in the superior parietal lobe. Some neuroimaging studies reported bilateral activation in the posterior parietal lobe (Jordan et al., 2001; Kosslyn et al., 1998; Cohen et al., 1996; Tagaris et al., 1996), whereas other studies revealed subdominant right hemispheric activation when people are engaged in mental rotation tasks (Harris et al., 2000; Carpenter et al., 1999; Zacks et al., 2003). Yet other studies report left sided activation in the posterior parietal lobe (Alivisatos and Petrides, 1997; Just et al., 2001; Vingerhoets et al., 2001). Evidently, there is a wide-ranging agreement on the parietal involvement in mental rotation, but the question about laterality is still debated. In this context, several factors were discussed such as the role of task demands (Tagaris et al., 1996), complexity of the stimuli used (Corballis, 1997), and the influence of extensive practice (Voyer, 1995).

What about the neuronal mechanisms involved in the viewer rotation task? It is of interest to see whether the differences observed at the behavioral level correspond to different neuronal mechanisms. Two recent fMRI studies investigated the neural mechanisms associated with mental viewer rotations (Creem et al., 2001b; Zacks et al., 2003). Both studies revealed a subdominant left sided activation of the parietal cortex during imagined viewer rotation tasks. In the study by Creem and colleagues (2001b) subjects were instructed to imagine themselves positioned in the middle of an array, and then to imagine a rotation of the array around their main body-axis. Results showed hemodynamic increases in the posterior superior parietal cortex, with a peak of activation in the left hemisphere. In the study by Zacks and colleagues (2003), subjects performed an imagined viewer rotation task with visually presented arrays. In this task, participants imagined themselves moving around the array. The results showed increased activation in the left parieto-temporo-occipital junction for the mental viewer rotation task when compared to the condition in which the subjects mentally rotated the object. In comparison to the study by Creem and colleagues (2001b), this peak of activation is located relatively inferior. The reason for this difference remains an open question. It is possible that there are again differences in task demands between the two studies. For example, the rotation axis of the mental self rotation was different. Furthermore, there were also differences in how the data were analyzed. While Creem and colleagues (2001b) analyzed the activation during the imagined viewer rotation task relative to a non-rotational control condition, Zacks and colleagues (2003) reported activations associated with mental viewer rotation relative to an array rotation task.

Taken together, several neuroimaging studies of mental object rotation reveal converging evidence for an essential role of the parietal lobe. Only a small number of studies have been carried out using mental viewer rotation tasks. These studies consistently show activations of the left posterior parietal lobe associated with mental viewer rotation. Due to the lack of studies in which both spatial rotation processes were compared directly, neurofunctional differences between mental object rotation and mental viewer rotation are still hypothetical. Based on the current knowledge, however, we favor the hypothesis that mental spatial transformations are not performed by one single mechanism and engage – at least partly – separate neurofunctional mechanisms.

But what is the nature of these two different mechanisms? What is it that makes a viewer-based rotation different from an object-based rotation? Shared mechanisms involved in perception and mental imagery have been proposed already but only sporadic attempts have been made to apply this idea to viewer-based rotations.

IS VESTIBULAR INFORMATION INVOLVED IN MENTAL IMAGERY?

Whole body rotations involve a rigid rotation of one's own body with respect to an external frame of reference. Unlike movements of a body part, physical whole body rotations inevitably involve a stimulation of the vestibular end organ. When body movements are imagined, however, participants are completely still, which precludes any vestibular stimulation. Still, mechanisms that are normally associated with the processing of vestibular information during real body rotation may also be involved when the body movement is physically absent but instead vividly imagined. To the present, there is still a lack of knowledge on the role of the vestibular system (in particular its cortical projection areas) in cognitive tasks. The principal roles of the vestibular system include maintenance of balance, reflex contractions of the trunk and limbs, reflex control of eye movements, and detection and constant perception of the position and movement of the head. In consequence, the vestibular system connects with numerous brain areas. It is known to project to the thalamus, and is thought to project to the hippocampus to aid with formation of spatial memory. Some anatomical studies have suggested potential pathways for the thalamus to carry vestibular information to the hippocampus, although direct pathways may also exist (Smith, 1997).

Primate studies have identified several cortical areas as being vestibular in nature (Guldin and Grüsser, 1998). Vestibular responses have been reported physiologically in area 2v at the tip of the intraparietal sulcus (Schwarz and Fredrickson, 1971; Büttner and Buettner, 1978; but see Grüsser et al., 1990), area 3av in the central sulcus (Akbarian et al., 1994), the parietoinsular vestibular cortex next to the posterior insula (PIVC) (Grüsser et al., 1990), and area 7 in the inferior parietal lobule (Faugier-Grimaud et al., 1997; Akbarian et al., 1994). These areas have been shown to receive converging information from the vestibular, visual, and somatosensory systems and are thought to be involved in the analysis of self-motion, and/or modifying the reference frame in which visual and other sensory inputs are represented (Bremmer et al., 1997a, 1997b).

These cortical connections have been confirmed in human studies primarily through the use of fMRI and PET during caloric stimulation of the semicircular canals or galvanic stimulation of the vestibular system (Bottini et al., 1994; Lobel et al., 1998; Bense et al., 2001). Some authors contend that vestibular thalamo-cortical projections end in the parietoinsular cortex and the parieto-occipital cortex: Areas primarily involved with perception of self-motion and judgments of verticality. Dieterich and Brandt (2001) have shown that patients with parietal lesions have difficulties aligning a disk with a line on it to the earth vertical when any surrounding visual cues were absent. Interestingly, Akbarian and colleagues (1994) have also demonstrated cortical vestibular projections back to the vestibular nuclei, which is possibly important to understand the influence of top-down control of vestibular information. The influence in the reverse direction is presumably just as strong and it is rather surprising that still so little is known about how cognitive functions are influenced by vestibular information (e.g., Mast et al., 2006).

AN APPLIED EXAMPLE: ARTIFICIAL GRAVITY

As described above motor imagery can be used to improve athletic ability such as the timing and execution of a golf swing (Ross et al., 2003). Another application comes from a unique area: adaptation to artificial gravity (AG), which has been proposed as a potential countermeasure to the debilitating effects of long-duration space flight (Young, 1999). This artificial form of gravity is produced by 'standing' at the rim of a rotating object and comes in two basic varieties: 1) continuous stimulation on a large-radius, slowly spinning vehicle as it is shown in Stanley Kubrick's movie 2001: A Space Odyssey, or 2) intermittent stimulation aboard a short-radius, fast spinning device. The latter has the advantage of being more cost-effective and technologically easier to implement. The disadvantage arises from the cross-coupled stimulus (CCS) that is produced whenever a head turn is made out of the plane of the device's rotation. This CCS leads to inappropriate vestibular driven eye movements, sensations of whole-body tumbling, and nausea in many subjects.

Recent studies suggest that subjects can adapt to the rotating environment through a series of repeated CCS exposures conducted over multiple days (Hecht et al., 2002). However, roughly 25% of the participants in these studies drop out due to severe motion sickness symptoms, and therefore never acquire adaptation. Clearly, an alternative adaptation approach is needed for this segment of the population if AG is ever to attain universal acceptance. Mast and colleagues (2003) have demonstrated that the disorienting sensations, such as perception of whole-body tumbling, arising from CCS adapt more quickly and effectively than lower level central nervous system responses such as the vestibulo-ocular reflex. Meliga and colleagues (2005) have further shown that the adaptation process to CCS is cognitively malleable. Based on these findings we (Newby and Mast) designed a study in which we substituted motor imagery for actual CCS exposures, on the supposition that there is a shared neural substrate between imagining one's body tumbling and actually sensing one's body tumbling (Newby et al., 2006).

The experiment looked at adaptation trajectories over three days among three different groups: 1) a group physically adapted through repeated exposures to 42 CCS per day; 2) a group receiving 6 CCS exposures, followed by 30 imagined CCS exposures, and a last set of 6 CCS exposures per day; and 3) a control group receiving 6 CCS exposures, followed by 30 non-CCS exposures, and then a last set of 6 CCS exposures per day. The result was that the group using imagery as a pseudo-stimulus attained significantly better adaptation than the control group, and achieved similar adaptation as the physically adapted group for some response variables. One subject was able to completely adapt away the perceptual responses to CCS (i.e. head turns felt like normal head turns in a stationary environment); something rarely even observed in subjects physically adapted to CCS.

The fact that motor imagery of a whole body movement can be used to adapt to a stimulus that induces a sensation of whole-body motion suggests two very interesting possibilities: 1) neurofunctional areas that perceive vestibular sensations and whole-body motions overlap with neurofunctional areas subserving imagery of whole-body motion, and 2) motor imagery may be used to achieve some level of neuroplasticity. The former is not too surprising given what is known from single body-part motor imagery. The latter, while highly speculative, is given some support by the fact that subjects can completely adapt to a strong vestibular stimulus over just three days of mental practice. The end organs, in this case the semicircular canals, presumably still detect the stimulus and respond by sending an afferent signal upstream. From there, however, normal cortical processes that deal with this information must have undergone some type of modification to yield such a remarkable result.

FUTURE PERSPECTIVES

Spatial cognition is an important field of research and it has implications, which go far beyond the spatial nature of the experimental tasks. One of these abilities is perspective taking. It is central in the context of reasoning about others, and future research will help to further investigate how far spatial processing is involved in yet other seemingly unrelated domains. It is by all means striking how many mechanisms related to early perception or motor execution are also involved in higher cognitive tasks. As discussed above, this raises several questions regarding the underlying neuronal mechanisms as for example the possible motor inhibition when we vividly imagine a movement. The existence of shared mechanisms opens a huge amount of possibilities such as mental training, the benefits of which are by far not yet exploited. Future research needs to define more precisely the conditions under which what kind of mental training is appropriate. In terms of neuro-rehabilitation the kind of lesion may already confine whether at all a mental training can be taken into consideration. Yet another field concerns inter-individual differences, which is – unfortunately – almost entirely divorced from the study of cognitive functions. Even though we are sometimes told that anybody can achieve the most difficult goals if only she or he is determined enough to go for it, we know at the same time that this is by far too good to be true. Evidently, individuals differ widely in their cognitive abilities and taking the differences into account could in fact open up new possibilities to create innovative teaching tools tailored to the student's cognitive abilities. In this chapter we emphasized the role of vestibular information, which is not yet clearly understood in terms of its involvement in cognitive tasks. This is surprising given the fact that the cortical projections are so numerous. Future research will need to develop more precisely what role the vestibular system plays in higher cognitive tasks, and with what other cognitive functions vestibular cortical areas overlap.

ACKNOWLEDGMENTS

This research was supported by Swiss NSF grant no. 611-066052 and PDFM1-114406 and by NSBRI grant no. NCC 9-58 and NASA's Artificial Gravity Project.

REFERENCES

Akbarian, S., Grusser, O. J. and Guldin, W. O., 1994, Corticofugal connections between the cerebral cortex and brainstem vestibular nuclei in the macaque monkey, *J Comp Neurol* **339**: 421-437.

Alivisatos, B. and Petrides, M., 1997, Functional activation of the human brain during mental rotation, *Neuropsychologia* **35**: 111-118.

Alkadhi, H., Brugger, P., Hotz Boendermaker, S., Crelier, G., Curt, A., Hepp-Reymond M.-C. and Kollias, S. S., 2005, What disconnection tells about motor imagery: evidence from paraplegic patients, *Cereb Cortex* **15**: 131-140.

Barnes, J., Howard, R. J., Senior, C., Brammer, M., Bullmore, E. T., Simmons, A., Woodruff, P. and David, A. S., 2000, Cortical activity during rotational and linear transformations, *Neuropsychologia* **38**: 1148-1156.

Bense, S., Stephan, T., Yousry, T. A., Brandt, T. and Dieterich, M., 2001, Multisensory cortical signal increases and decreases during vestibular galvanic stimulation (fMRI), *J Neurophysiol* **85**: 886-899.

Bottini, G., Sterzi, R., Paulesu, E., Vallar, G., Cappa, S. F., Erminio, F., Passingham, R. E., Frith, C. D. and Frackowiak, R. S., 1994, Identification of the central vestibular projections in man: a positron emission tomography activation study, *Exp Brain Res* **99**: 164-169.

Bremmer, F., Distler, C. and Hoffmann, K. P., 1997a, Eye position effects in monkey cortex. II. Pursuit- and fixation-related activity in posterior parietal areas LIP and 7A, *J Neurophysiol* **77**: 962-977.

Bremmer, F., Ilg, U. J., Thiele, A., Distler, C. and Hoffmann, K. P., 1997b, Eye position effects in monkey cortex. I. Visual and pursuit-related activity in extrastriate areas MT and MST, *J Neurophysiol* **77**: 944-961.

Büttner, U. and Buettner, U. W., 1978, Parietal cortex (2v) neuronal activity in the alert monkey during natural vestibular and optokinetic stimulation, *Brain Res* **153**: 392-397.

Carpenter, M. and Proffitt, D. R., 2001, Comparing viewer and array mental rotations in different planes, *Mem Cognit* **29**: 441-448.

Carpenter, P. A., Just, M. A., Keller, T. A., Eddy, W. and Thulborn, K., 1999, Graded functional activation in the visuospatial system with the amount of task demand, *J Cogn Neurosci* **11**: 9-24.

Cohen, M. S., Kosslyn, S. M., Breiter, H. C., DiGirolamo, G. J., Thompson, W. L., Anderson, A. K., Bookheimer, S. Y., Rosen B. R. and Belliveau, J. W., 1996, Changes in cortical activity during mental rotation. A mapping study using functional MRI, *Brain* **119**: 89-100.

Corballis, M. C., 1997, Mental rotation and the right hemisphere, *Brain Lang* **57**: 100-121.

Creem, S. H., Downs, T. H., Wraga, M., Harrington, G. S., Proffitt, D. R. and Downs, J. H., 3rd., 2001a, An fMRI study of imagined self-rotation, *Cogn Affect Behav Neurosci* **1**: 239-249.

Creem, S. H., Wraga, M. and Proffitt, D. R., 2001b, Imagining physically impossible self-rotations: geometry is more important than gravity, *Cognition* **81**: 41-64.

Deiber, M. P., Ibanez, V., Honda, M., Sadato, N., Raman, R. and Hallett, M., 1998, Cerebral processes related to visuomotor imagery and generation of simple finger movements studied with positron emission tomography, *Neuroimage* **7**: 73-85.

Dieterich, M. and Brandt, T., 2001, Vestibular system: anatomy and functional magnetic resonance imaging, *Neuroimaging Clin N Am* **11**: 263-273, ix.

Dominey, P., Decety, J., Broussolle, E., Chazot, G. and Jeannerod, M., 1995, Motor imagery of a lateralized sequential task is asymmetrically slowed in hemi-Parkinson's patients, *Neuropsychologia* **33**: 727-741.

Ehrsson, H. H., Geyer, S. and Naito, E., 2003, Imagery of voluntary movement of fingers, toes, and tongue activates corresponding body-part-specific motor representations, *J Neurophysiol* **90**: 3304-3316.

Gerardin, E., Sirigu, A., Lehéricy, S., Poline, J.-B., Gaymard, B., Marsault, C., Agid, Y. and Le Bihan, D., 2000, Partially overlapping neural networks for real and imagined hand movements, *Cereb Cortex* **10**: 1093-1104.

Grabherr, L., Jola, C., Berra, G., Möckli, R., Theiler, R. and Mast, F. W., 2006, Motor imagery training improves hand movement in hemiplegic patients. Paper presented at the Meeting of the International Neuropsychological Society (INS), Zürich, Switzerland, 26-30 July 2006, Abstract book, p. 86.

Grusser, O.-J., Pause, M., Schreiter, U., 1990, Localization and responses of neurones in the parieto-insular vestibular cortex of awake monkeys (Macaca fascicularis), *J Physiol* **430**: 537-557.

Guldin, W. O. and Grusser, O. J., 1998, Is there a vestibular cortex? *Trends Neurosci* **21**: 254-259.

Hanakawa, T., Immisch, I., Toma, K., Dimyan, M. A., Van Gelderen, P. and Hallett, M., 2003, Functional properties of brain areas associated with motor execution and imagery, *J Neurophysiol* **89**: 989-1002.

Harris, I. M., Egan, G. F., Sonkkila, C., Tochon-Danguy, H. J., Paxinos, G. and Watson, J. D., 2000, Selective right parietal lobe activation during mental rotation: a parametric PET study, *Brain* **123:** 65-73.

Hecht, H., Brown, E. L. and Young, L. R., 2002, Adapting to artificial gravity (AG) at high rotational speeds, *J Gravit Physiol* **9**: P1-5.

Huttenlocher, J. and Presson, C. C., 1979, The coding and transformation of spatial information, *Cognit Psychol* **11**: 375-394.

Jeannerod, M., 2001, Neural Simulation of action: a unifying mechanism for motor cognition, *Neuroimage* **14**: 103-109.

Jeannerod, M. and Decety, J., 1995, Mental motor imagery: a window into the representational stages of action, *Curr Opin Neurobiol* **5**: 727-732.

Jeannerod, M. and Frak, V., 1999, Mental imaging of motor activity in humans, *Curr Opin Neurobiol* **9**: 735-739.

Jordan, K., Heinze, H. J., Lutz, K., Kanowski, M. and Jancke, L., 2001, Cortical activations during the mental rotation of different visual objects, *Neuroimage* **13**: 143-152.

Just, M. A., Carpenter, P. A., Maguire, M., Diwadkar, V. and McMains, S., 2001, Mental rotation of objects retrieved from memory: a functional MRI study of spatial processing, *J Exp Psychol Gen* **130**: 493-504.

Kosslyn, S. M., DiGirolamo, G. J., Thompson, W. L. and Alpert, N. M., 1998, Mental rotation of objects versus hands: neural mechanisms revealed by positron emission tomography. *Psychophysiology* **35**: 151-161.

Kosslyn, S. M., Thompson, W. L., Wraga, M. and Alpert, N. M., 2001, Imagining rotation by endogenous versus exogenous forces: distinct neural mechanisms, *Neuroreport* **12**: 2519-2525.

Lamm, C., Windischberger, C., Leodolter, U., Moser, E. and Bauer, H., 2001, Evidence for premotor cortex activity during dynamic visuospatial imagery from single-trial functional magnetic resonance imaging and event-related slow cortical potentials, *Neuroimage* **14**: 268-283.

Lobel, E., Kleine, J. F., Bihan, D. L., Leroy-Willig, A. and Berthoz, A., 1998, Functional MRI of galvanic vestibular stimulation, *J Neurophysiol* **80**: 2699-2709.

Lotze, M., Montoya, P., Erb, M., Hulsmann, E., Flor, H., Klose, U., Birbaumer, N. and Grodd, W., 1999, Activation of cortical and cerebellar motor areas during executed and imagined hand movements: an fMRI study, *J Cogn Neurosci* **11**: 491-501.

Mast, F. W., Merfeld, D. M. and Kosslyn, S. M., 2006, Visual mental imagery during caloric vestibular stimulation, *Neuropsychologia* **44**: 101-119.

Mast, F. W., Newby, N. J. and Young, L. R., 2003, Sensorimotor aspects of high-speed artificial gravity: II. The effect of head position on illusory self motion, *J Vestib Res* **12**: 283-289.

Meliga, P., Hecht, H., Young, L. R. and Mast, F. W., 2005, Artificial gravity head movements during short-radius centrifugation: influence of cognitive effects, *Acta Astronaut* **56**: 859-866.

Mulder, T., Zijlstra, S., Zijlstra, W. and Hochstenbach, J. (2004). The role of motor imagery in learning a totally novel movement, *Exp Brain Res* **154**: 211-217.

Naito, E., Kochiyama, T., Kitada, R., Nakamura, S., Matsumura, M., Yonekura, Y. and Sadato, N., 2002, Internally simulated movement sensations during motor imagery activate cortical motor areas and the cerebellum, *J Neurosci* **22**: 3683-3691.

Newby, N., Mast, F. W., Natapoff, A. and Paloski, W. H., 2006, Motor imagery as an adaptation method for head turns on a short-arm centrifuge. Paper presented at the 7th Symposium on the Role of Vestibular Organs in Space Exploration, Nordwjik, The Netherlands, 7 - 9 June. http://www.congrex.nl/06A07/

Nico, D., Daprati, E., Rigal, F., Parsons, L. and Sirigu, A., 2004, Left and right hand recognition in upper limb amputees, *Brain* **127**: 120-132.

Nyberg, L., Eriksson, J., Larsson, A., Marklund, P., 2006, Learning by doing versus learning by thinking: An fMRI study of motor and mental training, *Neuropsychologia* **44**: 711-717.

Piaget, J. and Inhelder, B., 1956, *The child's conception of space*, London: Routledge.

Porro, C. A., Cettolo, V., Francescato, M. P. and Baraldi, P., 2000, Ipsilateral involvement of primary motor cortex during motor imagery, *Eur J Neurosci* **12**: 3059-3063.

Richter, W., Somorjai, R., Summers, R., Jarmasz, M., Menon, R. S., Gati, J. S., Georgopoulos, A. P., Tegeler, C., Ugurbil, K. and Kim, S. G., 2000, Motor area activity during mental rotation studied by time-resolved single-trial fMRI, *J Cogn Neurosci* **12**: 310-320.

Ross, J. S., Tkach, J., Ruggieri, P. M., Lieber, M. and Lapresto, E., 2003, The mind's eye: functional MR imaging evaluation of golf motor imagery, *AJNR Am J Neuroradiol* **24**: 1036-1044.

Schwarz, D. W. and Fredrickson, J. M., 1971a, Rhesus monkey vestibular cortex: a bimodal primary projection field, *Science* **172**: 280-281.

Schwarz, D. W. and Fredrickson, J. M., 1971b, Tactile direction sensitivity of area 2 oral neurons in the rhesus monkey cortex, *Brain Res* **27**: 397-401.

Smith, P. F., 1997, Vestibular-hippocampal interactions, *Hippocampus* **7**: 465-71.

Schwoebel, J., Boronat, C. B. and Branch Coslett, H., 2002, The man who executed "imagined" movements: evidence for dissociable components of the body schema, *Brain Cogn* **50**: 1-16.

Shepard, R. N. and Metzler, J., 1971, Mental rotation of three-dimensional objects, *Science* **171**: 701-703.

Sirigu, A., Daprati, E., Ciancia, S., Giraux, P., Nighoghossian, N., Posada, A. and Haggard, P., 2004, Altered awareness of voluntary action after damage to the parietal cortex, *Nat Neurosci* **7**: 80-84.

Sirigu, A., Duhamel, J. R., Cohen, L., Pillon, B., Dubois, B. and Agid, Y., 1996, The mental representation of hand movements after parietal cortex damage, *Science* **273**: 1564-1568.

Tagaris, G. A., Kim, S. G., Strupp, J. P., Andersen, P., Ugurbil, K., and Georgopoulos, A. P., 1996, Quantitative relations between parietal activation and performance in mental rotation, *Neuroreport* **7**: 773 776.

Takahashi, M., Hayashi, S., Ni, Z., Yahagi, S., Favilla, M. and Kasai, T., 2005, Physical practice induces excitability changes in human hand motor area during motor imagery. *Exp Brain Res* **163**: 132-136.

Thobois, S., Dominey, P. F., Decety, P. J., Pollak, P. P., Gregoire, M. C., Le Bars, P. D. and Broussolle, E., 2000, Motor imagery in normal subjects and in asymmetrical Parkinson's disease: a PET study, *Neurology* **55**: 996-1002.

Tomasino, B., Budai, R., Mondani, M., Skrap, M. and Rumiati, R. I., 2005a, Mental rotation in a patient with an implanted electrode grid in the motor cortex, *Neuroreport* **16**: 1795-1800.

Tomasino, B., Borroni, P., Isaja, A. and Rumiati, R. I., 2005b, The role of the primary motor cortex in mental rotation: A TMS study, *Neuropsychology* **22**: 348-363.

Vingerhoets, G., Santens, P., Van Laere, K., Lahorte, P., Dierckx, R. A. and De Reuck, J., 2001, Regional brain activity during different paradigms of mental rotation in healthy volunteers: a positron emission tomography study, *Neuroimage* **13**: 381-391.

Voyer, D., 1995, Effect of practice on laterality in a mental rotation task, *Brain Cogn* **29**: 326-335.

Wolbers, T., Weiller, C. and Buchel, C., 2003, Contralateral coding of imagined body parts in the superior parietal lobe, *Cereb Cortex* **13**: 392-399.

Wraga, M., Creem, S. H. and Proffitt, D. R., 1999, The influence of spatial reference frames on imagined object- and viewer rotations, *Acta Psychol (Amst)* **102**: 247-264.

Wraga, M., Creem, S. H. and Proffitt, D. R., 2000, Updating displays after imagined object and viewer rotations, *J Exp Psychol Learn Mem Cogn* **26**: 151-168.

Wraga, M., Thompson, W. L., Alpert, N. M. and Kosslyn, S. M., 2003, Implicit transfer of motor strategies in mental rotation, *Brain Cogn* **52**: 135-143.

Yaguez, L., Nagel, D., Hoffman, H., Canavan, A.G., Wist, E., Homberg, V., 1998, A mental route to motor learning: improving trajectorial kinematics through imagery training, *Behav Brain Res* **90**: 95-106.

Young, L. R., 1999, Artificial gravity considerations for a mars exploration mission, *Ann N Y Acad Sci* **871**: 367-378.

Yue, G. and Cole, K. J., 1992, Strength increases from the motor program: comparison of training with maximal voluntary and imagined muscle contractions, *J Neurophysiol* **67**: 1114-1123.

Zacks, J. M., Vettel, J. M. and Michelon, P., 2003, Imagined viewer and object rotations dissociated with event-related FMRI, *J Cogn Neurosci* **15**: 1002-1018.

Zaehle, T., Jordan, K., Wüstenberg, T., Baudewig, J., Dechent, P., & Mast, F. W., 2007, The neural basis of the egocentric and allocentric spatial frame of reference, *Brain Res* **1137**,92-103.

Chapter 19

BOTTOM-UP EFFECTS OF SENSORY CONFLICT AND ADAPTATION ON MENTAL IMAGERY: SENSORIMOTOR GROUNDS FOR HIGH LEVEL COGNITION?

Gilles Rode[1,3], Sophie Jacquin-Courtois[1,3], Patrice Revol[1,2,3], Laure Pisella[1], Anne Sylvie Sacri[1], Dominique Boisson[1,2,3] and Yves Rossetti[1,2,3]

[1]Espace et Action, UMR-S 534, Institut National de la Santé et de la Recherche Médicale, Université Claude Bernard, and Hospices Civils de Lyon (HCL), Bron, France; [2]'Mouvement et Handicap' Plateforme IFNL-HCL, Institut Fédératif des Neurosciences de Lyon (IFNL) and HCL, France; [3]Hôpital Henry Gabrielle, HCL, St Genis Laval, France

Abstract: A large proportion of right-brain damaged patients show unilateral neglect, a neurological deficit of perception, attention, representation, and/or performing actions within their left-sided space. The intriguing symptom is a spontaneous orientation bias toward the right leading to neglect of objects or persons on the left. This ipsilesional behaviour orientation bias may also affect the representational space. This particular aspect of neglect was seen as a failure to generate or maintain a normal representation of the left side of the mental image. Affected mental images can be of spatial or numerical nature. Representational neglect represents a cognitive disorder reflecting the selective damage of structures located in the right hemisphere and involved in spatial cognition. Surprisingly, these cognitive deficits may be positively modulated by passive physiological stimulation as caloric vestibular stimulation or stimulation of sensorimotor plasticity by prism adaptation procedure. These findings suggest that using low-level sensorimotor transformation may act on higher cognitive levels of space representation and consciousness according a bottom-up track.

Key words: neglect; visual mental imagery; number bisection; vestibular stimulation; prism adaption.

1. INTRODUCTION

This chapter aims at exploring the cross-road between two areas of investigation. First, visual imagery deficits have been repeatedly described in patients with unilateral neglect, a neurological condition characterised by a deficit for orienting, perceiving, attending and acting to the left side of space. Second, visuo-manual adaptation to distorting lenses has been extensively explored over the last century. On one side it has been suggested that the visuo-motor effects of such adaptation can be extremely specific to the exposure conditions but on the other side recent research has shown that visuo-manual adaptation can alleviate numerous symptoms of unilateral neglect. Here we review evidence for the effects of visuo-manual adaptation on mental imagery and discuss possible hypotheses about their mechanisms.

2. REPRESENTATIONAL NEGLECT, A SPATIAL IMAGERY DEFICIT

Unilateral neglect is defined as the patient's failure to report, respond to, or orient toward novel and/or meaningful stimuli presented to the side opposite to the brain lesion (Heilman et al., 1985; Vallar and Bisiach, 1997). For example, neglect patient can forget to read the left part of a journal or a book; omit to eat the left half of a plate or forget to sheave the left hemi-face. This syndrome is frequently consecutive to the damage of the right brain hemisphere. It often is in association with contralesional hemiplegia, hemi-anesthesia and hemi-anopia. It also worsen the severity of motor or sensory associated deficit inducing many functionally debilitating effects on everyday life, and responsible for poor functional recovery and ability to benefit from treatment (Halligan et al. 1989; Denes et al., 1982; Fullerton et al., 1986). For these reasons, many attempts have been made in the last 20 years to rehabilitate neglect. Different approaches have been proposed relying mainly on passive physiological stimulations or active training (see review in: Rossetti and Rode, 2002; Rode et al., 2003). The main goal of these methods is to favour the re-orientation of the motor behaviour toward the neglect side and the first difficulty to obtain a generalization of effects at a functional level.

Unilateral neglect may also affect the representational domain. This particular aspect of neglect was seen as a failure to generate or maintain a normal representation of the left side of the mental image (Bisiach et al., 1979; Bisiach and Berti, 1987; Berti, 2004). In their pioneering study, Bisiach and Luzzatti (1978) described 'representational neglect' in two patients who not only neglected the left side of their visual space, but also failed to report the left side of mental images. When asked to describe a familiar scene, the Piazza del Duomo in Milan, the patients reported a larger number of landmarks from the right side of the imagined square. Moreover, the bias to report fewer landmarks from the left side was apparent both when the patients imagined themselves standing with their back turned on the cathedral and when they were facing the cathedral from the opposite end of the square. This impairment was ascribed to the brain damage on the side opposite to the mentally neglected side of space (Bisiach et al., 1979; Bisiach and Berti, 1987). This finding was replicated by Bisiach et al. (1981) in 28 neglect patients and the authors proposed that neglect patients suffer from 'a representational map reduced to one half' (p.549).

Representational neglect is also commonly assessed by requiring patients to draw objects from memory (Chokron et al., 2004) or to name the towns or the countries on an imagined map (Bartolomeo et al., 1994; Rode and Perenin, 1994). For example when neglect patients were asked to evoke mentally the map of France a representational deficit of the left half of map was evidenced in several conditions: when the patient had to imagine the map in either a canonical or an inverted condition - with a vertical axis splitting the map in two approximately equivalent halves (Rode et al., 1995); when no mental rotation was required and the patient had to name as many towns as possible either on the right of the vertical axis then on the left or in the reverse order, i.e. by starting on the left side (Rode and Perenin, 1994) and finally when the mental exploration was performed without any directional instruction or mental rotation (Rode et al., 1999, 2001).

Furthermore, this specific representational deficit occurs only when patients are asked to evoke visually the map of France, and not when they have to establish a list of French geographical locations without imagining those (Rode et al., 2004). In this case-report, a patient continued to show imaginal neglect even when, six months after the first testing, extra-personal neglect had resolved (for similar patterns of selective recovery, see Bartolomeo et al., 1994; Bartolomeo and Chokron, 2001). The retrieval and generation from visual long-term memory of an inner image of the map of France did not succeed in providing topographic information about towns on the western part of the map (see Fig. 19-1). However, that part of geographical knowledge could be recalled from a linguistic description, clearly indicating that the representational neglect did not result from a memory or verbal fluency impairment but was rather due to a disturbance in mental processes (Rode et al., 2004).

A

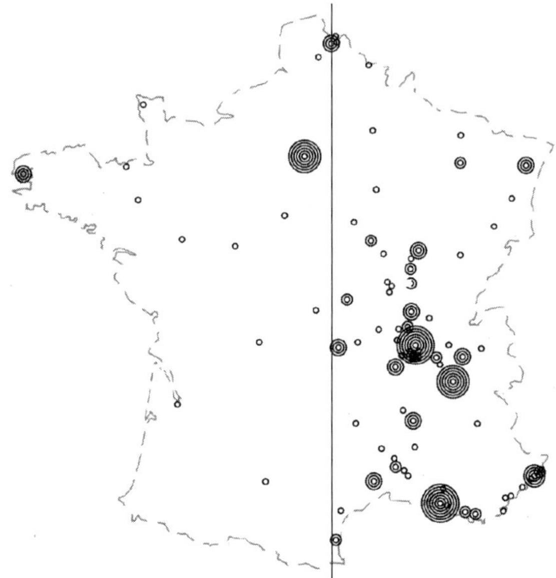

B

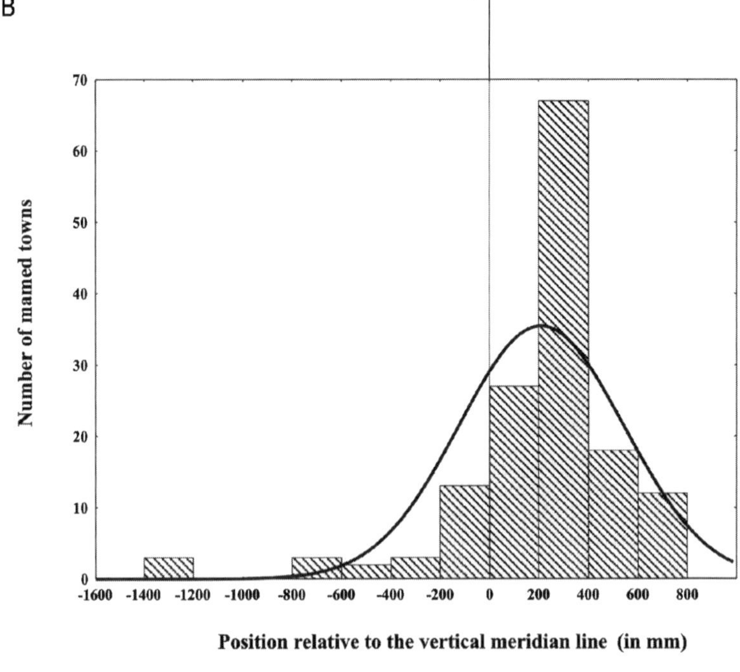

Figure 19-1. Mental evocation of map of France in 8 neglect patients with closed eyes. Location of named town by a circle on a tracing of map. For each town, the size of circle reflects the number of repetitions for all neglect patients. Distribution of named towns according to their position (in mm) relative to the vertical meridian line, measured on a map.

Taken as a whole, these different experiments demonstrated a similar representational deficit in accordance with the analogical model proposed by Bisiach et al. (1979). Nevertheless in order to have a better estimate of the strategy used by the patient and the healthy subject, Rode et al. (2004) measured on the map the distance between two successive responses (towns). In the visual imagery task, the mean geographic distances were shorter both in the patient and in the healthy subjects, suggesting a strategy relying on an inner visual scanning, which would be limited to the rightward part of the map in the patient's case (see Fig. 19-2).

An alternative view could be that the mental representation of controlesional space was not lacking, but rather failed to be activated or explored. This explanation is consistent with the hypothesis postulating that visual mental imagery involves some of the attentional-exploratory mechanisms that are employed in visual behaviour (Bartolomeo and Chokron, 2001; Thomas, 1999), in particular an inability to direct attention to areas of imagined space (Bisiach et al., 1979; Meador et al., 1987). The positive influence of head position (Meador et al., 1987) on representational neglect in a pure imaging task fits well with this explanation. Recently, Chokron et al. (2004) asked six neglect patients to perform two different tasks involving spatial representations: a clock-drawing task and a drawing from memory task. In different test conditions, patients performed the tasks in free vision or while blindfolded. Results showed that blindfolding was able to decrease and even suppress left neglect.

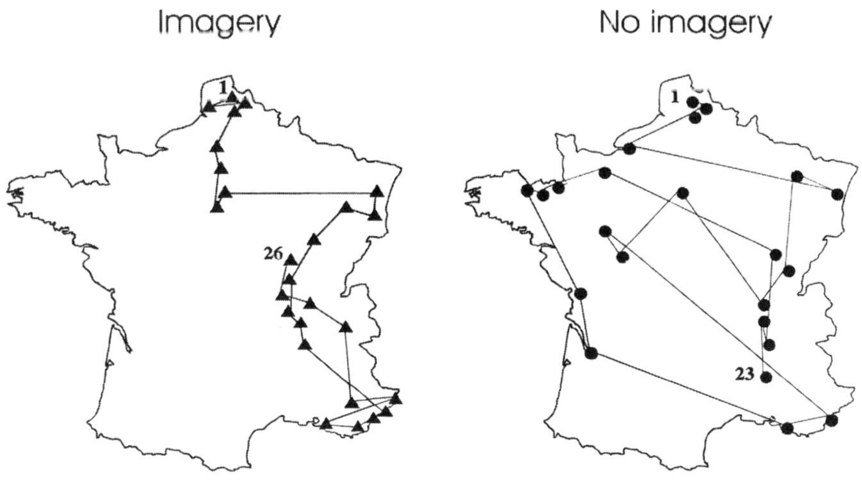

Figure 19-2. Evocation of French towns in a visual imagery task and a related task not employing imagery by a right-brain-damaged with a representational neglect.

These results support the hypothesis that orienting attention through visual control can also influence space-related imagery (Bartolomeo and Chokron, 2002). Chedru (1976) was the first to mention the negative effect of vision on the severity of left neglect. Similar findings were replicated in a clock drawing task (Mesulam, 1985; Anderson, 1993), in a geometric shape drawing and a letter cancellation task (Marshall and Halligan, 1993), showing that visual feedback may exacerbate left neglect. Taken as a whole, these results suggest a strong link between vision and spatial representation. In a recent study, this question was asked in eight healthy subjects and eight brain-damaged patients with left imaginal neglect to imagine the map of France and to name as many towns as possible within two minutes. Two different conditions were tested, either in free vision or while blindfolded. The results displayed no appreciable effect of the presence or absence of a visual context. Imaginal neglect may thus be completely independent of visual attention. Alternative possibilities are that stimuli must be task-relevant to engender an attentional bias in neglect, or that vision might influence spatial representations only in tasks requiring a manual response (Rode et al., 2006).

3. THE BOTTOM-UP TRACK: A COGNITIVE DEFICIT ALLEVIATED BY PHYSIOLOGICAL STIMULATIONS

The attentional-exploratory mechanisms that are employed in visual behaviour (Bartolomeo and Chokron, 2001; Thomas, 1999), in particular an inability to direct attention to areas of imagined space (Bisiach et al., 1979; Meador et al., 1987) may be positively influenced by sensory manipulations (Rode and Perenin, 1994; Geminiani and Bottini, 1992; Vallar, 1997; Rossetti and Rode, 2002) and prismatic visuo-motor adaptation (Rode et al., 2001) according a bottom-up track. Indeed many manifestations of neglect have been shown to be alleviated by sensory stimulation (vestibular, optokinetic, transcutaneous electrical, transcutaneous mechanical vibration, auditory) (reviews: Vallar et al., 1997; Kerkhoff, 2000; Rossetti and Rode, 2002). The improvement has been mainly reported for extrapersonal neglect (classical neuropsychology testing), but many other aspects have been investigated including personal neglect and sensory and motor deficits of left hemibody associated with neglect and extinction. More cognitive symptoms such as anosognosia, somatoparaphrenia or representational neglect were also improved after sensory stimulation. For example, a vestibular caloric stimulation with iced water in the left ear (Geminiani and Bottini, 1992;

Rode and Perenin, 1994) or transcuatenous electrical nervous stimulation (TENS) of the left side of the neck (Guariglia et al., 1998) produced significant improvement of performance only on the left side of mental representation of mental images of space.

3.1 Vestibular stimulations

We have specifically studied the effects of vestibular stimulation on representational neglect assessed by the mental evocation of the map of France in a group of 14 right-brain damaged patients in different conditions: before and after mental rotation of the map (Rode et al., 1995), in a cued condition i.e according a rightward or leftward direction of exploration and in a non cued condition (Rode and Perenin, 1994; Rode et al., 1996). Vestibular stimulation was performed by a cold caloric irrigation of the left ear in order to stimulate the horizontal semicircular canal. The cold stimulation provokes a lateral bias of eye position toward the left and a nystagmus with a slow phase toward the same side. Before stimulation the mental exploration of the image was limited to the right half of the map. After stimulation the patient was also able to evoke town's names located into the left part of the map (see Fig. 19-3).

Results of this patient are similar to those reported in others experimental conditions. In all cases vestibular activation did not change the right side performances. This means that it does not act through a non-specific mechanism of increased arousal which would also increase the scores on the right side. The possibility that it would produce a global activation of the right hemisphere and decrease an imbalance between the two hemispheres has to be rejected too, in the absence of a decrease of the right side scores. Thus the reversibility of representational neglect through vestibular stimulation does not support the hypothesis that unilateral neglect could be due to an imbalance between two concurrent systems of lateral attention (Kinsbourne, 1987). Moreover the lack of repeated items during mental evocation on both conditions suggests also that patients did not mentally "revisit" the same locations and that the geographic representational deficit does not rely on a trans-saccadic working memory deficit (Husain et al., 2001).

Under stimulation, the patients were again able to build a symmetrical representation of space without distortion of the left part or to orient easier attentional-exploratory mechanisms toward the left side of image. This cognitive effect is not symmetric; no improvement of performance on the right side was noted, as also shown by the positive effect of sensory stimulations on sensory or motor deficit of the left hemibody (Vallar et al., 1993, 1997; Rode et al., 1998).

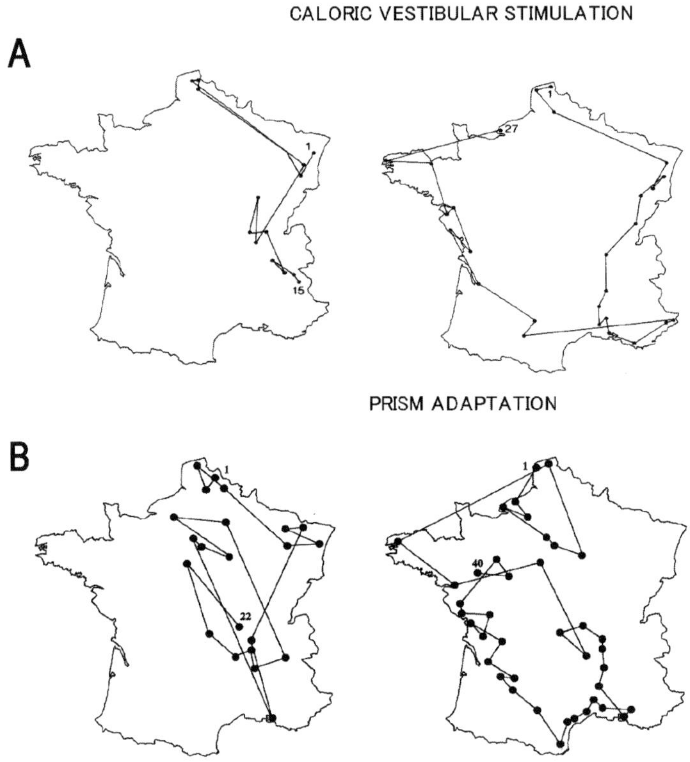

Figure 19-3. Improvement of attentional-exploratory mechanisms in a pure imaging task after caloric vestibular stimulation **(A)** and visuo-motor prismatic adaptation **(B)** in two right-brain-damaged patients with representational neglect.

As shown by previous studies, vestibular stimulation appears to act on most, if not all neglect phenomena and related disorders. This reinforces the original view, argued by Bisiach and Berti (1987) of a unitary conception of the syndrome, whose multiple manifestations, although often dissociated, would rely on a common basic disturbance. A disordered awareness or representation of one side of space (corporeal and/or extracorporeal) would best account for both the negative (neglect) and positive (misrepresentation, e.g. somatoparaphrenia) manifestations of the syndrome. The fact that a sensory manipulation can suppress or reduce unilateral neglect of mental space and also false beliefs on one side of the body, a disorder which strikingly resists to any tentative of persuasion (Cappa et al., 1987; Rode et al., 1992), provides interesting insight on the neuroarchitecture of cognition. Due to the topological relations of misrepresentation and neglect with brain lesions (the disorders affect the side contralateral to the lesion), it has been argued that mental representation of space would be subserved by an analogue mode of processing, much similar to that involved in visuo-spatial perception (Bisiach et al., 1981; Geminiani and Bottini, 1992). The possibility to restore mental space through vestibular stimulation means that the representation is not definitively lost following brain lesion producing neglect (Rode and Perenin, 1994). One way to reconcile this result with the analogue hypothesis is to regard mental representation of space as resulting from dynamic and distributed processes, which does not seem to be the case for the more peripheral representation of a visual image on the primary visual cortex. Although anatomo-clinical studies suggest that the right posterior parietal lobe would be the main center of space representation in man, other association cortex areas and subcortical structures participate and may account for spontaneous recovery of neglect (Vallar and Perani 1986; Vallar, 2001; Mort et al., 2003; Thiebaut de Schotten et al., 2005). The idea of a distributed system underlying mental space has been strongly supported by studies in the monkey (see review in Guldin and Grüsser, 1996). Interestingly, the structures involved are all the site of polysensory and/or sensorimotor integrations, which would provide a continuous updating of space representation. Vestibular stimulation in neglect patients, by acting on these integrative mechanisms, would modify the representation, while preserving its analogue format.

Functional imaging studies with vestibular stimulation have shown that vestibular information was thus integrated into a complex network of areas located in the posterior insula and retroinsular regions, the STG, the parts of the IPL, the post-central and pre-central gyrus, the anterior cingulated gyrus, the precuneus and hippocampus (Bottini et al., 1994, 2001; Dieterich et al., 2003). Activation of this cortical network is not symmetric in both hemispheres. Activation was stronger in the non-dominant hemisphere, in the hemisphere ipsilateral to the stimulated ear (Dieterich et al., 2003). These structures are not strictly 'vestibular' and can also integrate other sensory stimulation as somatosensory, visual optokinetic suggesting thus their multimodal character represents a significant site for the neural transformation of converging vestibular, auditory, neck proprioceptive and visual input into higher order spatial representations. The right hemisphere shows thus a dominance for vestibular processing and also for the occurrence of severe and persistent neglect in favour of a close relationship between vestibular function on the one hand, spatial neglect and mental representation on the other hand (Karnath and Dieterich, 2006). The physiological stimulations, particularly vestibular stimulation, may modulate, through a bottom-up track, the higher-order spatial representations, which are more frequently disrupted by lesions in this hemisphere (Vallar et al., 1997; Bottini et al., 1994, 2001).

3.2 Prism adaptation

The influence of sensori-motor levels on higher-order spatial representations may be also documented by the cognitive effects of a visuomotor adaptation task on representational neglect. In contrast to more complex visual reorganisation which requires several days of exposure, adaptation to shifting wedge prisms has been known to quickly develop over the course of a five minute simple pointing session (Rossetti et al., 1998). Despite over 100 years of studies, only visuo-motor after-effects had been described until the demonstration that prism adaptation can improve unilateral neglect (Redding et al., 2005). Many of the therapeutical effects described following prism adaptation involve a visual or a manual component, which maybe directly affected by the visuo-manual adaptation procedure. It was therefore of prime interest to investigate the possibility that prism adaptation can also improve symptoms of unilateral neglect that may not be directly affected by the adaptation. The level of space representation assessed by mental imagery tasks clearly differs from the sensory-motor level which is directly stimulated by prism adaptation.

The effect of prism adaptation on representational neglect was explored in two right-brain-damaged patients who exhibited a representational neglect. The prism exposure procedures consisted of 60 pointing trials made toward 2 visual targets in a random order (as in Rossetti et al., 1998). The goggles were fitted with glass prisms sustaining a wide visual field of 105° (each monocular visual field was 75° and the binocular visual field was 45°). Patients were asked to point, as fast and as accurately as possible, with sagittal movements of the right arm to each of two visual targets (located on the table 10° on each side of the subject's body midline). The sham goggles were made of two pairs of 5° prisms producing opposite shifts, i.e. a total shift of 0° (same weight and same opacity as the 10° prisms). The prisms were fitted inside glacier goggles (Cébé®) in order to prevent any access to unshifted vision (Optique Peter, Lyon). The adaptation procedure lasted for about 8 minutes while subjects wore goggles producing a 10-degree rightward lateral visual shift in the adaptation session and a 0° shift for the control session. Vision of the starting position of the hand was occluded to ensure the optimal development of the adaptation (Redding and Wallace, 1997). A pointing task without visual feedback (open loop) was performed before and after the adaptation procedure to check the development of a visuo-manual adaptation to the visual shift.

Before prism exposure, both patients showed a left visual neglect on free drawing of daisy and a left neglect of imagined space on a task requiring the patients to report from an imagined representation of the map of France. In two patients, the same side of space (left) was affected by representational and visuospatial neglect. In two patients, immediately after prism adaptation, modifications of the mental evocation were evidenced: the total number of towns named by the two patients within 2 minutes was increased, suggesting that the prism adaptation facilitated the mental evocation of the map; this increase concerned mainly the evocation of towns located on the left half of map, suggesting a reduction of left representational neglect and no town was evoked on the east-most part of map, in one patient, suggesting a significant shift of mental evocation toward the left side of the image and a paradoxical 'neglect' of the right side (see Fig. 19-3 B). Moreover, following adaptation, drawing also improved in both patients, with reduced asymmetry of the daisy. This improvement was also paralleled by the significant increase of the straight-ahead shift toward the left side, which again suggests that these two symptoms can be dissociated. The reversal effect observed after prism adaptation strongly supports the idea that the effect of this manipulation on neglect is attributable to a specific directional mechanism, and not just to a general improvement of cognitive functions.

After a delay of about 24 hours, following the prism exposure, the previous modifications of mental evocation had disappeared and the patients again showed left neglect of imaginal space, comparable with performance prior to adaptation. However the total number of towns named remained still larger than before prism exposure, probably reflecting learning of the mental task. On the other hand, the improvement in daisy drawing was partially maintained. This difference may be explained by the involvement of a manual response in drawing task. In this task, the subject had to generate an inner image of a map from long-term memory and explore it. It should be noticed that this level of space representation clearly differs from the sensory-motor level, directly stimulated by prism adaptation. The reduction of imaginal neglect shown by both patients after the prism adaptation suggests that the stimulation of active processes involved in the plasticity of sensori-motor correspondences can also influence cognitive processes at the level of mental representation. This result further supports the idea that the process of prism adaptation may activate brain functions related to multisensory integration and higher spatial representations (Rossetti et al., 1998; Rode et al., 2003).

Another interesting type of imagery deficit is worth addressing here. Recently an unexpected feature of neglect has been elegantly uncovered. Zorzi et al. (2002) reported that the mental bisection between two numbers was systematically shifted to the right (i.e. towards bigger numbers) showing the same properties as when neglect patients are asked to mark the centre of a physical line (Marshall and Halligan 1989). This observation elegantly confirmed that there is anatomical vicinity between the areas representing number and space. In our clinical practice we have frequently observed number bisection problems in patients with unilateral neglect (e.g. Lacour et al., 2004). Not only patients are frequently biased towards higher numbers, but several of them exhibit a peculiar difficulty when the number interval is presented in the decreasing order. For example patients may exhibit no apparent difficulty finding out where the middle between 1 and 9 is (i.e. responding "7" or "8" without hesitation), but remain blocked by the same interval presented in the other order. Some of them widely increase their reaction times in such condition, but others simply fail to provide a response. Asked about the middle between say, 9 and 1, they may reply that "there is nothing in between!", or that "it is impossible!", or that they "don't know...". This new feature of neglect allowed us to show that the therapeutic effects of prism on unilateral neglect do in fact generalise to tasks with no visual nor manual component and with no explicit spatial aspect and thereby to directly investigate the *functional* link between number and space representation.

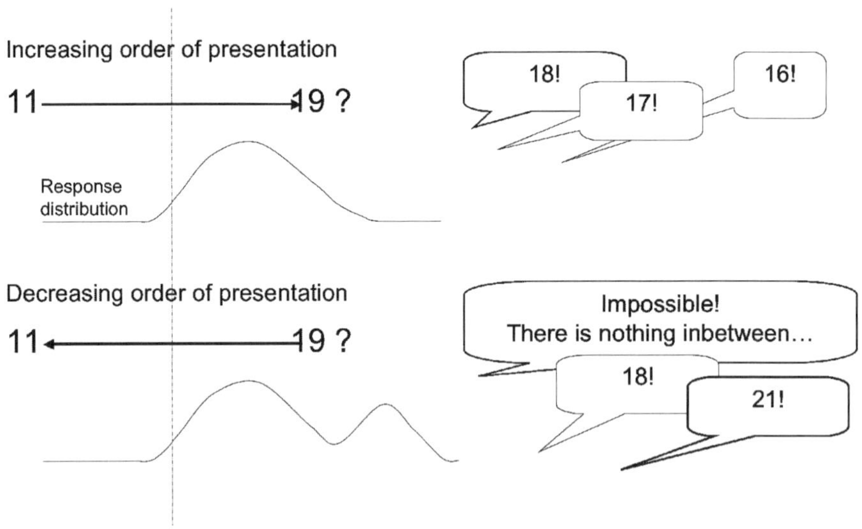

Figure 19-4. Number bisection in unilateral neglect. Asked about the middle between "11 and 19", patients typically respond "16", "17" or "18" without hesitation. But when the question is about the middle between "19 and 11", they may even find it impossible to provide a response.

In a first study, we described the number bisection bias of two patients before and after prism adaptation sessions (Rossetti et al., 2004). Despite a reliable bias was observed for the two patients over the first two sessions a clear improvement was observed after the prism adaptation session. Interestingly, the range of improvement was within the largest effects described after prism adaptation in neglect patients. One of the two patients exhibited a normal-range performance following the true adaptation session, even though he had not been improved by the sham adaptation. The magnitude of this effect suggested that prism adaptation may affect non-visual and non-manual tasks as much as visuo-manual tasks like drawing. In contrast to the average response bias, the total number of error was not modified by the adaptation, implying that the patients mainly shifted their distribution of errors towards zero. This is similar to one of the control patients described by Zorzi et al. (2002) who performed as many errors as the neglect patients but exhibited no response bias. Our result thus confirms that the global error rate is not linked to the unilateral neglect bias. The specific improvement of the response bias indicates that this improvement does not result from a non specific factor such as an increase in arousal. As a matter of facts, the two patients did not notice any difference between the sham and prism adaptation sessions, as is often noted with unilateral neglect patients. This effect appears to be the first report of an effect of a visuo-manual procedure on a highly cognitive task. Although visuo-motor adaptation has been found to improve many aspects of neglect including mental imagery (Rode et al., 1999, 2001), visuo-verbal tasks (e.g. Farné et al., 2002), haptic tasks (e.g. McIntosh et al., 2002) and postural control (Tilikete et al., 2001), mental number bisection is the first non-spatial task to be improved after prism adaptation. Another patient was followed-up over several days and was administered two sessions of prism adaptation. This patient exhibited a very strong bias on the first test session performed prior to prism adaptation, such that she could sometimes 'bisect' the number interval beyond its higher edge! Following the first prism adaptation session her performance improved dramatically. As her bisecting bias deteriorated over the following days, a second adaptation session was performed on day 4. Although the bias was less severe than at day 1, a new improvement was clearly noted in the bisection (Fig. 19-5).

The neural substrates of prism adaptation remain under investigation. This phenomenon appears to be more complex than vestibular activation and its underlying neural networks are not yet clear. The cerebellum is clearly involved in visuo-manual adaptation to wedge prisms (Weiner et al., 1983; Pisella et al., 2005), whereas the posterior parietal cortex is not necessary for adapting (Pisella et al., 2004). It is even possible that the posterior parietal cortex contributes to error reduction in such a way that it reduces the amount of prism adaptation reflected in the compensatory after-effects (Pisella et al., 2004). On the other hand, when it is lesioned as is the case in most unilateral neglect patients, the magnitude of the aftereffects is increased (Rossetti et al., 1998; review: Rode et al., 2003). The contribution of the cerebellum has been confirmed recently by a functional imaging study in unilateral neglect patients (Luauté et al., 2006). It was additionally suggested that the cerebellum activation by prism adaptation could indirectly interfere with cognitive functions via an inhibition of cortical structures.

The fact that prism adaptation has been shown to be effective on line bisection (e.g. Rossetti et al., 1998) and on mental imagery (Rode et al., 1999; 2001) requires further considerations about the potential links between line bisection and number bisection. There is evidence that the mental number line is represented with number growing from left to right (e.g. Hubbard et al., 2005) but it must be emphasised that the verbal presentation of the task implies that there is no spatialised access to the information and no intrinsic spatial constraints in this task. With this respect it is interesting to note again that mental representation of places are biased in unilateral neglect patients (Bisiach and Luzatti, 1978) only when the topographic information must be spatialised (Rode et al., 2003). When French patients are requested to name cities of France without being previously explicitly instructed to visualise the map of France, no bias can be observed in their mental representation (Rode et al., 2003).

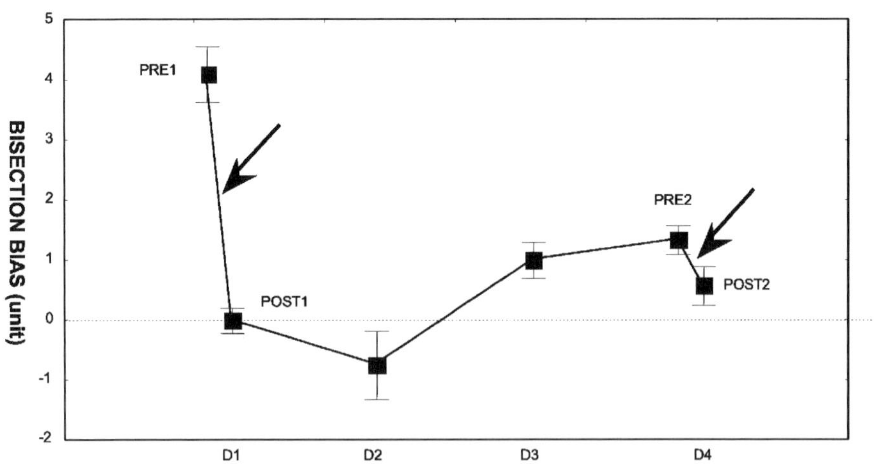

Figure 19-5. Improvement of number bisection following prism exposure. The bisection bias exhibited by a unilateral neglect patient was improved twice by two sessions of visuo-manual prism adaptation.

In addition, the difference found between physical line bisection and number bisection shows that the number bisection cannot be considered as a simple mental version of line bisection. The finding that a simple visuo-manual task can affect such higher-level of cognition as number bisection is therefore of prism interest and cannot be viewed as a simple addition of line bisection and imagery effects.

Until recently experiments on prism adaptation had reported only visuo-motor after-effects in normals and an improvement of neglect for tasks that were explicitly spatial. The striking effect of a single visuo-manual adaptation session on a numerical estimation task shows that a simple visuo-manual exercise can be capable of altering the mental number line. The central representation of space that governs our actions can be updated by a short visuo-manual task and is likely to play a central role in cognitive representations including visual imagery and even imagery that is not explicitly spatial in nature. This finding provides direct evidence for the role of sensori-motor interaction in the organisation of cognitive abilities in adults, extending to the organisation of number representation.

4. CONCLUSION

The observations reviewed here have raised the question of the link between sensori-motor plasticity and cognitive space representation, extending on Piaget theories on child development. Although unilateral neglect is no longer considered a unitary syndrome (Halligan and Marshall, 1998), it is striking that prism adaptation, as well as other physiological manipulations (review: Rossetti and Rode, 2002; Rode et al., 2003; Revol et al., in press), have been shown to improve many aspects of neglect including line bisection (Rossetti et al., 1998; Farnè et al., 2002). In addition adaptation to prisms was found to affect line midpoint perception in normals (Colent et al., 2000; Berberovic and Mattingley, 2003; Michel et al., 2003a,b; Girardi et al., 2004). Hence series of results collected both in healthy subjects and unilateral neglect patients suggest that visuo-manual adaptation can modify spatial tasks on a cognitive level. However all tasks performed to date involved either a manual or a visual component or both, i.e., factors that may be directly modified by the prism adaptation procedure. The question also arises as to whether the effects of vestibular stimulation may result from the intersensory conflict generated during the stimulation. It is likely that the acute stimulation of the internal ear produces an activation of not only the vestibular cortex but also of the associative cortices. The only difference between vestibular stimulation and prism adaptation would then lie in the presence of active movements performed during prism exposure. This active behaviour is responsible for the adaptive reactions that are responsible for the much longer lasting effects described with prism adaptation.

REFERENCES

Anderson, B., 1993, Spared awareness for the left side of internal visual images in patients with left sided extrapersonal neglect, *Neurology* **43**:213-216.

Bartolomeo, P., and Chokron, S., 2001, Levels of impairment in unilateral neglect, in: Handbook of Neuropsychology, F. Boller and J. Grafman, eds., vol. 4, 2nd ed., Elsevier Science Publishers, Amsterdam, pp 67-98.

Bartolomeo, P., and Chokron, S., 2002, Can we change our vantage point to explore imaginal neglect? Commentary on Pylyshyn: Mental imagery: In search of a theory, *Behav and Brain Sci* **25**:184-185.

Bartolomeo, P., D'Erme, P., and Gainotti, G., 1994, The relationship between visuospatial and representational neglect, *Neurology*, **44**:1710-1714.

Berberovic, N., and Mattingley, J. B., 2003, Effects of prismatic adaption on judgements of spatial extent in peripersonal and extrapersonal space. *Neuropsychologia* **41**(4):493-503.

Berti, A., 2004, Cognition in dyschiria: Edoardo Bisiach's theory on misconception of space and consciousness, *Cortex* **24**:275-280.

Bisiach, E., and Berti, A., 1987, Dyschiria. An attempt at its systemic explanation, in: *Neurophysiological and Neuropsychological Aspects of Spatial Neglect*, M. Jeannerod, ed., North Holland, Amsterdam, pp.183-201.

Bisiach, E., Capitani, E., Luzzatti, C., and Perani, D., 1981, Brain and conscious representation of outside reality, *Neuropsychologia* **19**:543-551.

Bisiach, E., and Luzzatti, C., 1978, Unilateral neglect of representational space, *Cortex* **14**:129-133.

Bisiach, E., Luzzatti, C., and Perani, D., 1979, Unilateral neglect, representational schema and consciousness, *Brain* **102**:609-618.

Bottini, G., Karnath, H. O., Vallar, G., Sterzi, R., Frith, C. D., Frackowiak, R. S., and Paulesu, E., 2001, Cerebral representations for egocentric space: functional-anatomical evidence from caloric vestibular stimulation and neck vibration, *Brain* **124**:1182-1196.

Bottini, G., Sterzi, R., Paulesu, E., Vallard, G., Cappa, S. F., Erminio, F., Passingham, R. E., Frith, C. D., and Frackowiak, R. S., 1994, Identification of the central vestibular projections in man: a positron emission tomography activation study, *Exp Brain Res* **99**:164-169.

Cappa, S. F., Sterzi, R., Vallar, G., and Bisiach, E., 1987, Remission of hemineglect and anosognosia during vestibular stimulation, *Neuropsychologia* **25**:775-782.

Chedru, F., Space representation in unilateral neglect, 1976, *J Neurol Neurosurg Ps* **39**: 1057-1061.

Chokron, S., Colliot, P., and Bartolomeo, P., 2004, The role of vision on spatial representations, *Cortex* **40**:281-290.

Colent, C., Pisella, L., Bernieri, C., Rode, G., and Rossetti, Y., 2000, Cognitive bias induced by visuo-motor adaptation to prisms: a simulation of unilateral neglect in normal individuals? *Neuroreport* **9**:1899-1902.

Denes, G., Semenza, C., Stoppa, E., and Lis, A., 1982, Unilateral spatial neglect and recovery from hemiplegia: a follow-up study, *Brain* **105**:543-552.

Dietrich, M., Bense, S., Lutz, S., Drzezga, A., Stephan, T., and Brandt, T., 2003, Dominance for vestibular cortical function in the non-dominant hemisphere, *Cereb Cortex* **13**: 994-1007.

Farnè, A., Rosetti, Y., Toniolo, S., and Làdavas, E., 2002, Ameliorating neglect with prism adaption: Visuo-manual vs. visuo-verbal measures. *Neuropsychologia* **40**:1069-80.

Fullerton, K. J., McSherry, D., and Stout, R. W., 1986, Albert's test: a neglected test of perceptual neglect. *Lancet* **327**:430-432.

Geminiani, G., and Bottini, G., 1992, Mental representation and temporary recovery from unilateral neglect after vestibular stimulation, *J Neurol Neurosurg Psychiatry* **55**: 332-333.

Girardi, M., McIntosh, R. D., Michel, C., Vallar, G., and Rossetti, Y., 2004, Sensorimotor effects on central space representation: Prism adaptation influences haptic and visual representations in normal subjects, *Neuropsychologia* **42**:1477-1487.

Guariglia, C., Lippolis, G., and Pizzamiglio, L., 1998, Somatosensory stimulation improves imagery disorders in neglect, *Cortex* **34**:233-41.

Guldin, W. O., and Grüsser, O. J., 1996, The anatomy of the vestibular cortices of primates, in: *Le Cortex Vestibulaire*, M. Collard, M. Jeannerod, Y. Christen, eds., Editions IRVINN, Paris, pp. 17-26.

Halligan, P. W., and Marshall, J. C., 1998, Visuospatial neglect: the ultimate deconstruction? *Brain Cognition* **37**:419-438.

Halligan, P. W., Marshall J. C., and Wade D. T., 1989, Visuospatial neglect: underlying factors and test sensitivity. Lancet **2**(8668):908-11.

Heilman, K. M., Watson, R., and Valenstein, E., 1985, in: *Clinical Neuropsychology*, K. M. Heilman, E. Valenstein, eds., Oxford University Press, New York, pp. 243-93.

Hubbard, E. M., Piazza, M., Pinel, P., and Dehaene, S., 2005, Interactions between number and space in parietal cortex. Nat Rev Neurosci **6**(6):435-48.

Husain, M., Mannan, S., Hodgson, T., Wojciulik, E., Driver, J., and Kennard, C., 2001, Impaired spatial working memory across saccades contributes to abnormal search in parietal neglect, *Brain* **124**:941-952.

Karnath, H. O., and Dietrich, M., 2006, Spatial neglect – a vestibular disorder? *Brain* **129**:293-305.

Kinsbourne, M., 1987, Mechanisms of unilateral neglect, in: *Neurophysiological and neuropsychological aspects of spatial neglect*, M. Jeannerod, ed., Elsevier Science, Amsterdam, pp. 69-86.

Kerkhoff, G., 2000, Multiple perceptual distortions and their modulation in left sided visual neglect, *Neuropsychologia* **38**:1073-86.

Lacour, S., Jacquin, S., and Rosetti, Y., 2004, Liens entre représentations spatiales et représentations numériques: apport de la négligence spatiale unilatérale et de la plasticité visuo-motrice, in: *Vision, espace et cognition: fonctionnement normal et pathologique*, Y. Coello, S. Casalis, C. Moroni, eds., Septentrion, Sillery (Québec), pp. 61-72.

Luaute, J., Michel, C., Rode, G., Pisella, L., Jacquin-Courtois, S., Costes, N., Cotton, F., le Bars, D., Boisson, D., Halligan, P., and Rosetti, Y., 2006, Functinal anatomy of the therapeutic effects of prism adaption on left neglect, *Neurology* **66**(12):1859-67.

Marshall, J. C., and Halligan, P. W., 1989, When right goes left: an investigation of line bisection in a case of visual neglect, *Cortex* **25**(3):503-15.

Marshall, J. C., and Halligan, P. W., 1993, Visuo-spatial neglect: a new copying test to assess perceptual parsing, *J Neurol* **240**:37-40.

McIntosh, R. M., Rosetti, Y., and Milner, A. D., 2002, Prism adaption improves chronic visual and haptic neglect, *Cortex* **38**:309-20.

Meador, K. J., Loring, D. W., Bowers, D., and Heilman, K. M., 1987, Remote memory and neglect syndrome, *Neurology* **37**:522-526.

Mesulam, M. M., 1985, *Principles of Behavioral Neurology*, F. A. Davis, Philadelphia.

Michel, C., Pisella, L., Halligan, P. W., Luauté, J., Rode, G., Boisson, D., and Rossetti, Y., 2003, Simulating unilateral neglect using prism adaptation: Implications for theory, *Neuropsychologia* **41**:25-39.

Michel, C., Rossetti, Y., Rode, G., and Tilikete, C., 2003, After-affects of visuo-manual adaptation to prisms on body posture in normal subjects, *Exp Brain Res* **148**:219-226.

Mort, D. J., Malhotra, P., Mannan, S. K., Rorden, C., Pambakian, A., Kennard, C., and Husain, M., 2003, The anatomy of visual neglect, *Brain* **126**:1986-97.

Pisella, L., Michel, C., Tilikete, C., Vighetto, A., and Rossetti, Y., 2004, Preserved prism adaptation in bilateral optic ataxia: strategic versus adaptive reaction to prisms, *Exp Brain Res* **156**:399-408.

Pisella, L., Rossetti, Y., Michel, C., Rode, G., Boisson, D., Pélisson, D., and Tilikete, C., 2005, Ipsidirectional impairment of prism adaptation after unilateral lesion of anterior cerebellum, *Neurology* **65**:150-152.

Redding, G., Rossetti, Y., and Wallace, B., 2005, Application of prism adaption: a tutorial in theory and method, *Neurosci Biobehav Rev* **29**:431-444.

Redding, G., and Wallace, B., 1997, *Adaptive spatial alignment*, Lawrence Erlbaum Associates, Mahwah, New York.

Revol, P., Pisella, L., Luauté, J., Jacquin-Courtois, S., Farne, A., Ota, H., Rode, G., Boisson, D., and Rossetti, Y., Prism adaption therapy for unilateral neglect: from the facts to a bottom-up explanatory model, in: *Complex medical engineering*, J. Wu, H. Fukuyama, M. Mitsuishi, K. Ito, S. Tobimatsu, T. Nishida, , eds., Springer, Norwell/Heidelberg, in press.

Rode, G., Charles, N., Perenin, M. T., Vighetto, A., Trillet, M., and Aimard, G., 1992, Partial remission of hemiplegia and somatoparaphrenia through vestibular stimulation in a case of unilateral neglect, *Cortex* **28**: 203-208.

Rode, G., and Perenin, M. T., 1994, Temporary remission of representational hemineglect through vestibular stimulation, *Neuroreport* **5**:869-872.

Rode, G., Perenin, M. T., and Boisson, D., 1995, Négligence de l'espace représenté: mise en évidence par l'évocation mentale de la carte de France, *Rev Neurol* **151**:161-164.

Rode, G., Perenin, M. T., and Boisson, D., 1996, Effets de la stimulation vestibulaire sur la réversibilité de la négligence de l'espace représenté, in: *Le Cortex Vestibulaire*, M. Collard M. Jeannerod, Y. Christen, eds., Editions IRVINN, Paris, pp. 109-116.

Rode, G., Pisella, L., Rossetti, Y., Farne, A., and Boisson, D., 2003, Bottom-up transfer of sensory-motor plasticity to recovery of spatial cognition: visuomotor adaptation and spatial neglect, *Prog Brain Res* **142**:273-287.

Rode, G., Revol, P., Rossetti, Y., Boisson, D., and Bartolomeo, P., Looking while imagining: The influence of visual input on representational neglect, *Neurology*, in revision.

Rode, G., Rossetti, Y., Li, L., and Boisson, D., 1998, The effect of prism adaption on neglect for visual imagery, *Behav Neurol* **11**:251-258.

Rode, G., Rossetti, Y., Li, L., and Boisson, D., 1999, Improvement of mental imagery after prisms exposure in neglect: a case study, *Behav Neurol* **11**:251-258.

Rode, G., Rossetti, Y., and Boisson, D., 2001, Prism adaptation improves representational neglect. *Neuropsychologia* **39**:1250-1254.

Rode, G., Rossetti, Y., Perenin, M. T., and Boisson, D., 2004, Geographic information has to be spatialised to be neglected: a representational neglect case, *Cortex* **40**:391-397.

Rossetti, Y., Jacquin-Courtois, S., Rode, G., Ota, H., Michel, C., and D. Boisson, 2004, Does action make the link between number and space representation? Visuo-manual adaptation improves number bisection in unilateral neglect, *Psychol Sci* **15:**426-430.

Rossetti, Y., Koga, K., and Mano, T., 1993, Prismatic displacement of vision induces transient changes in the timing of eye-hand coordination, *Percept Psychophys* **54**:355-364.

Rossetti, Y., Rode, G., Pisella, L., Farné, A., Li, L., Boisson, D., and Perenin, M. T., 1998, Prism adaptation to a rightward optical deviation rehabilitates left hemispatial neglect, *Nature* **395**:166-169.

Rossetti, Y., and Rode, G., 2002, Reducing spatial neglect by visual and other sensory manipulations: non-cognitive (physiological) routes to the rehabilitation of a cognitive disorder, in: *The Cognitive and Neural Bases of Spatial Neglect*, H.-O. Karnath, A. D. Milner, Vallar, G., eds., Oxford University Press, New York, NY, pp. 375-396.

Thiebaut de Schotten, M., Urbanski, M., Duffau, H., Volle, E., Levy, R., Dubois, B., and Bartolomeo, P., 2005, Direct evidence for a parietal-frontal pathway subserving spatial awareness in humans, *Science* **309**(5744):2172-2173.

Thomas, N. J. T., 1999, Are theories of imagery theories of imagination? An active perception approach to conscious mental content, *Cognitive Science* **23**:207-245.

Tilikete, C., Rode, G., Rossetti, Y., Pichon, J., Li, L., and Boisson, D., 2001, Prism adaption to rightward optical deviation improves postural imbalance in left-hemiparetic patients, *Curr Biol* **11**:524-528.

Vallar, G., 2001, Extrapersonal visual unilateral spatial neglect and its neuroanatomy, *Neuroimage* **14:**S52-58.

Vallar, G., Bottini, G., Rusconi, M. L., and Sterzi, R., 1993, Exploring somatosensory hemineglect by vestibular stimulation, *Brain* **116:**71-78.

Vallar, G., Guariglia, C., Nico, D., and Pizzamiglio, L., 1997, Motor deficits and optokinetic stimulation in patients with left hemigeglect, *Neurology* **49:**1364-1370.

Vallar, G., Guariglia, C., and Rusconi, M. L., 1997, Modulation of the neglect syndrome by sensory stimulation, in: *Parietal Lobe Contribution to Orientation in 3D Space,* P. Thier and H.-O. Karnath, eds., Springer Verlag, Heidelberg, pp 555-579.

Vallar, G., and Perani, D., 1986, The anatomy of unilateral neglect after right-hemisphere stroke lesions. A clinical/CT-scan correlation study in man, *Neuropsychologia* **24**(5): 609-22.

Weiner, M. J., Hallett, M., and Funkenstein, H. H.,1983, Adaptation to lateral displacement of vision in patients with lesions of the central nervous system, *Neurology* **33**:766-772.

Zorzi, M., Priftis, K., and Umilta, C., 2002, Neglect disrupts the mental number line, *Nature* **417**:138-139.

Chapter 20

CORTICAL PROCESSING OF AUDITORY SPACE: PATHWAYS AND PLASTICITY

Josef P. Rauschecker
Department of Physiology and Biophysics, Georgetown University School of Medicine, Washington, DC, USA

Abstract: Contrary to popular belief, which places auditory space processing wholly in the brainstem, several lines of evidence suggest that auditory cortex plays an important role in spatial perception. Lesion studies in animals and humans demonstrate severe deficits in sound localization after damage to auditory cortex. Single-unit recording studies find neurons tuned to spatial location in auditory cortical areas. While these neurons exist already in primary auditory cortex, their prevalence and sharpness of spatial tuning increases in nonprimary areas of the caudal belt, as defined in nonhuman primates. The firing of neurons in these latter areas also shows a tighter correlation with the behavioral performance of alert monkeys engaged in sound localization behavior. Caudal belt and parabelt project to posterior parietal cortex and to areas of prefrontal cortex, such as the frontal eye and pinna fields, known to be involved in spatial perception. This has led to the notion that a posterior-dorsal processing stream is intimately involved in aspects of auditory spatial perception. The existence of such an auditory "where"-stream is also suggested by functional neuroimaging studies in humans in which subjects process stationary or moving sounds in space. Consistently, posterior aspects of the superior temporal cortex and adjoining inferior parietal cortex are activated during these tasks. Thus, while brainstem nuclei perform an important service by computing some of the basic parameters necessary for spatial processing, such as interaural time and intensity differences, these parameters are integrated (together with monaural spectral cues that depend on head and pinnae) at the cortical level. Auditory space perception, including perception of motion in space, is, therefore, ultimately accomplished at the cortical level. Animals and humans that grow up blind use their auditory modality for localization in far space. Areas in parietal and occipital cortex that are ordinarily used for vision become activated by auditory input. This leads to an expansion of auditory areas in the dorsal stream into visual territory and to a simultaneous sharpening of auditory spatial tuning in these neurons. Together, this massive cross-modal reorganization leads to superior performance of blind as compared to sighted individuals in auditory spatial tasks.

Key words: auditory cortex; primates, neurophysiology; functional magnetic resonance imaging (fMRI); parallel processing; dual-pathway hypothesis; parietal cortex; prefrontal cortex; blindness; cortical reorganization.

1. BRAINSTEM PROCESSING OF AUDITORY SPACE

The superior colliculus (SC) is often considered the seat of auditory spatial perception, because it contains a map of auditory space. This has been shown in various species (owls: (Knudsen and Konishi, 1978); guinea pigs: (King and Hutchings, 1987); cats: (Middlebrooks and Knudsen, 1984)). The SC receives its input from the inferior colliculus (IC), in particular the "external" nuclei of the IC, which in turn receive their input from the dorsal part of the cochlear nuclei (DCN). The medial superior olive (MSO) and the lateral superior olive (LSO) are responsible for encoding interaural time and level differences (ITD and ILD), respectively. (For a more complete review of brainstem mechanisms of auditory space processing see Irvine, 1992).

2. SPATIAL SELECTIVITY IN PRIMARY AND NONPRIMARY AUDITORY CORTEX

Early studies have suggested a role for auditory cortex in sound localization (Diamond et al., 1956; Heffner and Masterton, 1975; Ravizza and Masterton, 1972). The first study, however, to unequivocally demonstrate that a lesion of primary auditory cortex (A1) in cats causes a deficit in sound localization was performed by Jenkins and Merzenich (1984). Particularly convincing was the fact that the sound localization deficits after small A1 lesions were frequency-specific. These findings were confirmed in later studies using different tasks (Beitel and Kaas, 1993; Heffner and Heffner, 1990). In these studies, A1 appeared to be the only region of auditory cortex whose ablation caused a localization deficit. However, cats have an auditory cortical region that is hidden deep in the anterior ectosylvian sulcus (AES) which was later implicated in sound localization: the anterior ectosylvian auditory area (AEA; (Korte and Rauschecker, 1993; Middlebrooks et al., 1994; Rauschecker and Korte,

1993)) or field AES ((Meredith and Clemo, 1989)). It is the main source of auditory cortical input to the SC. In addition, on the basis of cortical cooling studies, the posterior auditory field (PAF) has also been claimed to play a role in sound localization (Malhotra et al., 2004).

Spatial tuning of single cortical neurons in A1 was measured by a number of groups (Imig et al., 1990; Rajan Aitkin and Irvine, 1990; Rajan Aitkin Irvine et al., 1990). They all found two types of spatial tuning: single-peak and hemi-field. In single-peak neurons, the best response is found at a particular azimuth location; in hemi-field neurons, the response is largely restricted to speaker locations in the contra- or ipsi-lateral hemifield.

In rhesus monkeys, spatially tuned neurons are also found in A1, but at an even higher rate they are found in the caudo-medial field (CM) (Rauschecker et al., 1997; Recanzone, 2000). When monkeys are trained in an auditory localization task, the firing rate of neurons in CM correlates more tightly with behavioral performance than that of neurons in A1, which is a strong indication that CM plays an important role in sound localization (Recanzone et al., 2000).

3. EARLY PARALLEL PROCESSING IN THE AUDITORY CORTEX

Parallel processing streams in the auditory cortex start as early as the core areas: Areas A1 and R are both koniocortical areas with neurons sharply tuned for frequency and with tonotopic maps that are mirror symmetric. Combined lesion and tracer studies (Rauschecker et al., 1997) have shown that both cortical core areas receive input from the principal relay nucleus of the auditory thalamus, the ventral nucleus of the medial geniculate (MGv). By contrast, area CM, the other prominent area on the supratemporal plane mentioned above, does not receive input from MGv but only from the medial and dorsal subnuclei of the medial geniculate (MGd and MGm). As a consequence, lesions of A1 lead to unresponsiveness of neurons in CM to tonal stimulation, but not of neurons in area R, which receive independent input from MGv.

The parallel input to areas of the supratemporal plane may start even more peripherally than the thalamus. Studies of the auditory brainstem indicate that the ventral and dorsal cochlear nuclei have very different response characteristics and may subserve different functions of hearing, including auditory object and space processing, respectively (Yu and Young, 2000). Area CM, which contains large numbers of spatially tuned neurons (Rauschecker et al., 1997; Recanzone, 2000) (see above), could receive at least some of its input from the dorsal cochlear nucleus via the external nuclei of the inferior colliculus and the MGd (Rauschecker, 1997).

4. SPATIAL AND PATTERN SELECTIVITY IN THE LATERAL BELT

In order to compare the spatial selectivity of neurons in the rostral and caudal lateral belt (LB) directly in the same animals, broad-band species-specific communication calls were presented from different locations (Tian et al., 2001). In order to quantify the response selectivity to different monkey calls (MC), a monkey call preference index (MCPI) was calculated depending on the number of calls the neuron responds to. The LB areas differed in their degree of MC selectivity, as quantified on this basis. The anterolateral area (AL) had the greatest percentage of highly selective neurons (MCPI ≤ 2), followed by the middle lateral area (ML), whereas the caudolateral area (CL) had the smallest percentage of highly MC-selective neurons. Naturally, for the most non-selective neurons (MCPI$\geq$6), the opposite was found: CL had the greatest percentage of such non-selective neurons, AL the least, with ML somewhere between those two extremes.

Spatial tuning in neurons of the LB showed the opposite areal distribution: the highest selectivity was found in CL and the lowest in AL. This has led to the hypothesis that these two areas, which lie on opposite ends of the LB along its rostro-caudal extent, form the beginning of two processing streams for the processing of auditory space and pattern information (see Fig. 20-1) (Rauschecker and Tian, 2000; Tian et al., 2001). The anterior 'what'-stream extends all the way to the temporal pole, which has recently been demonstrated unequivocally to be activated by auditory stimuli (Poremba et al., 2003) using the 2-deoxyglucose technique in monkeys. By contrast, the posterior 'where'-stream projects to parietal cortex and dorsolateral prefrontal cortex (see below).

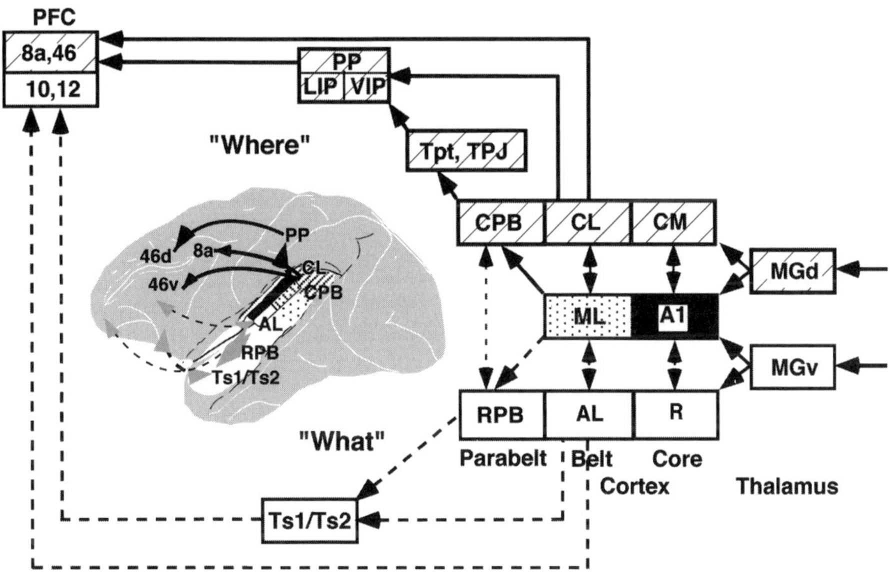

Figure 20-1. Schematic diagram of dual auditory cortical pathways in primates representing auditory object/pattern ("what") processing in an antero-ventral projection and auditory space ("where") processing in a postero-dorsal projection (modified and expanded from (Rauschecker, 1998) and (Rauschecker and Tian, 2000)). The auditory "where"-pathway is the main topic of the present chapter. Its projections are highlighted in solid lines; participating cortical areas are marked with oblique lines. The antero-ventral pathway is shown in dashed lines. Areas that are not uniquely participating in either pathway are shown in dark blocks (primary auditory cortex, A1) or stippled (middle lateral belt area, ML). Prefrontal connections of the lateral belt areas are also shown directly on a lateral view of a rhesus monkey brain (from (Romanski et al., 1999)). Abbreviations: MGd = medial geniculate nucleus, dorsal division; MGv = medial geniculate nucleus, ventral division; CM = caudomedial area; R = rostral area; CL = caudolateral area; CPB = caudal parabelt area; RPB = rostral parabelt area; Tpt = temporoparietal area; TPJ = temporoparietal junction; PP = posterior parietal cortex; LIP = lateral intraparietal area; VIP = ventral intraparietal area; Ts1, Ts2 = rostral temporal areas of (Pandya and Sanides, 1972); PFC = prefrontal cortex. Brodmann areas are abbreviated with their respective numbers.

5. AUDITORY BELT PROJECTIONS TO PARIETAL AND PREFRONTAL CORTEX

5.1 Auditory projections to prefrontal cortex

Anatomical studies in rhesus monkeys have demonstrated the existence of largely separate pathways that originate in the LB and project to different target regions in the prefrontal cortex (Romanski et al., 1999) (see Fig. 20-1). In these studies, three different fluorescent tracers were injected into matched frequency regions of the three belt areas after these had been physiologically mapped. Injections into area AL produced label in ventrolateral and orbital regions of prefrontal cortex (areas 10, 12), whereas CL injections led to labeling of dorsolateral prefrontal cortex (areas 8a, 46). The latter is known for its involvement in spatial working memory, whereas the former regions are assumed to participate in object working memory (Goldman-Rakic, 1996).

These projection patterns conform to the physiological response properties found in the aforementioned study of Tian et al. (2001), which assigned superior selectivity for auditory patterns and space to areas AL and CL, respectively. The studies by Tian et al. (2001) and Romanski et al. (1999), therefore, form the cornerstones of a recent theory according to which dual processing streams in nonprimary auditory cortex underlie the perception of auditory objects and auditory space, respectively (Fig. 20-1 and (Rauschecker and Tian, 2000)): One pathway projecting antero-ventrally from A1 through AL and the rostral STG and STS into orbitofrontal cortex forms the main substrate for auditory pattern recognition and object identification. Indeed, an auditory domain is found in ventrolateral prefrontal cortex, in which neurons show responses to complex sounds, including animal and human vocalizations (Romanski and Goldman-Rakic, 2002). Another pathway projecting caudo-dorsally from A1 to the caudal belt (areas CM and CL) and parabelt (areas Tpt and CPB) into posterior parietal (PP) and dorsolateral prefrontal cortex is thought to be involved in auditory spatial processing.

5.2 Auditory projections to parietal cortex

A projection from posterior STG to posterior parietal cortex in monkeys has been found independently by Lewis and Van Essen (Lewis and Van Essen, 2000). Specifically, the ventral intraparietal area (VIP) in the PP has been identified as the primary recipient of auditory input to PP. The lateral intraparietal area (LIP) has been found to contain auditory neurons as well, but only after training monkeys on auditory saccades (Andersen, 1997; Mazzoni et al., 1996; Stricanne et al., 1996). Auditory activation of inferior PP has also been demonstrated in human imaging studies (Bushara et al., 1999; Weeks et al., 1999) (Bremmer et al., 2001). By testing the subjects in a visual as well as in an auditory task during the same imaging session it was shown that PP does contain a unimodal auditory spatial representation before multisensory convergence occurs in superior parietal cortex (Bushara et al., 1999). A similar conclusion was reached on the basis of clinical and psychophysical studies by Griffiths and co-workers (Griffiths et al., 1997; Griffiths et al., 1996; Griffiths et al., 1998).

6. HUMAN IMAGING STUDIES OF AUDITORY CORTICAL PROCESSING

6.1 Core and belt areas

Human neuroimaging studies have confirmed the organization of auditory cortex into core and belt areas by using the same types of stimuli as in the present study (Wessinger et al., 2001). Two core areas robustly activated by pure-tone stimuli and mirror-symmetric tonotopic organization were found along Heschl's gyri. A third such area was sometimes seen more laterally. While the first two areas quite obviously correspond to areas A1 and R, the third area may be homologous to area ML, which (like the core areas A1 and R (Rauschecker et al., 1997) has been shown to receive direct input from the MGv (Kaas and Hackett, 2000; Morel et al., 1993). These three pure-tone responsive areas were surrounded by belt regions both medially and laterally, which were activated only by BPN bursts. An exploration of the medial belt region in the monkey with BPN bursts is therefore indicated.

7. DUAL STREAMS IN HUMAN AUDITORY CORTICAL PROCESSING

Various findings from human neuroimaging support the dual-stream hypothesis of auditory processing: Antero-lateral areas of the superior temporal cortex are activated by intelligible speech or speech-like sounds (Alain et al., 2001; Barrett and Hall, 2006; Binder et al., 2000; Binder et al., 2004; Maeder et al., 2001; Obleser et al., 2005; Scott et al., 2000). Thus it becomes more and more obvious that behaviorally relevant auditory patterns, including speech sounds, are discriminated selectively within an anterior auditory 'what'-stream and not in the "planum temporale" which is located posterior to Heschl's gyrus. Auditory areas in the planum temporale are still quite unspecific and involved in a variety of auditory functions.

Further posterior in the STG and STS are regions of the caudal belt and parabelt (projecting up dorsally into inferior posterior parietal cortex) that are specifically active during spatial tasks, such as auditory spatial discrimination or tasks involving auditory motion in space (Arnott et al., 2004; Brunetti et al., 2005; Degerman et al., 2006; Jääskeläinen et al., 2004; Krumbholz Schonwiesner Rubsamen et al., 2005; Krumbholz Schonwiesner von Cramon et al., 2005; Maeder et al., 2001; Tata and Ward, 2005a, 2005b; Warren et al., 2002; Zatorre et al., 2002; Zimmer and Macaluso, 2005).

In a meta-analysis, Arnott et al. (2004) reviewed evidence from auditory functional magnetic resonance imaging (fMRI) and positron emission tomography (PET) studies to determine the reliability of the auditory dual-pathway model in humans. Activation coordinates from 11 "spatial" studies (i.e., listeners made localization judgments on sounds that could occur at two or more perceptually different positions) and 27 "nonspatial" studies (i.e., listeners completed nonspatial tasks involving sounds presented from the same location) were entered into the analysis. Almost all temporal lobe activity observed during spatial tasks was confined to posterior areas. In addition, all but one of the spatial studies reported activation within the inferior parietal lobule as opposed to only 41% of the nonspatial studies. Finally, inferior frontal activity (Brodmann's areas 45 and 47) was reported in only 9% of the spatial studies, but in 56% of the nonspatial studies. These results support an auditory dual-pathway model in humans in which nonspatial sound information (e.g., sound identity) is processed primarily along an antero-ventral stream whereas sound location is processed along a postero-dorsal stream, i.e. within areas posterior to primary auditory cortex.

In a PET study by Zatorre et al. (2002) posterior auditory cortex responded to sounds that varied in their spatial distribution, but only when multiple complex stimuli were presented simultaneously. Consistent with other studies, these authors also found that the right inferior parietal cortex was specifically recruited in localization tasks.

An fMRI study by Krumbholz and co-workers (2005) found that interaural time differences were represented along a posterior pathway comprising the planum temporale (PT) and inferior parietal lobe (IPL) of the respective contralateral hemisphere. The response was stronger and extended further into adjacent regions of the IPL when the sound was moving than when it was stationary, a finding that confirmed earlier results by Warren et al. (2002). In contrast to Zatorre et al., the study by Krumbholz et al. (2005) found that stationary lateralized sounds did produce a significant activation increase in the PT of the respective contralateral hemisphere compared to a centrally presented sound. This discrepancy may be due to the inferior sensitivity of PET relative to fMRI, or to the fact that the spatial ranges of the sounds used by Zatorre et al. were centered around the midline, and thus always comprised equal parts of both hemifields. This suggests that Zatorre et al. were unable to detect the contralateral tuning that was observed in the study of Krumbholz et al.

Timing differences between the two ears can be used to localize sounds in space only when the inputs to the two ears have similar spectro-temporal profiles (high binaural coherence). Zimmer and Macaluso (2005) used fMRI to investigate any modulation of auditory responses by binaural coherence. They assessed how processing of these cues depends on spatial information being task-relevant and whether brain activity correlates with subjects' localization performance. They found that activity in Heschl's Gyrus increased with increasing coherence, irrespective of localization being task-relevant. Posterior auditory regions also showed increased activity for high coherence, but only when sound localization was required and subjects successfully localized sounds. The authors concluded that binaural coherence cues are processed throughout the auditory cortex, but that these cues are used in posterior regions of the STG for successful auditory localization (Zimmer and Macaluso, 2005).

Tata and Ward (Tata and Ward, 2005a, 2005b) used auditory evoked potentials to explore the putative auditory "where"-pathway in humans. The mismatch negativity (MMN) elicited by deviations in sound location is comprised of two temporally and anatomically distinct phases: an early phase with a generator posterior to primary auditory cortex and contralateral to the deviant stimulus, and a later phase with generators that are more frontal and bilaterally symmetric. The posterior location of the early-phase generator suggests the engagement of neurons within a posterior "where"-pathway for processing spatial auditory information (Tata and Ward, 2005a). Transient attention oriented in cue-target paradigms results in several modulations of the auditory event-related potential. Its earliest component (the Nd1) also reflects modulation of neurons posterior to primary auditory cortex within or near the temporo-parietal junction (TPJ) (Tata and Ward, 2005b).

Selective attention was also used to differentiate the effects of sound location and pitch of an auditory stimulus in an fMRI study (Degerman et al., 2006). Attention to either sound feature produced activation in areas of the superior temporal cortex and in prefrontal and inferior parietal regions. However, during attention to location these activations were located more posterior on the STG than during attention to pitch.

Finally, in a superb study combining fMRI and MEG, Brunetti and co-workers found that the processing of sound coming from different locations activates a neural circuit similar to the auditory "where" pathway described in monkeys (Brunetti et al., 2005). This system included Heschl's gyrus, the posterior STG, and the inferior parietal lobule. Their MEG analysis allowed assessment of the timing of this circuit: activation of Heschl's gyrus was observed 139 ms after the auditory stimulus, the peak latency of the source located in the posterior STG was at 156 ms, and the inferior parietal lobule and the supramarginal gyrus peaked at 162 ms. Both hemispheres were found to be involved in the processing of sounds coming from different locations, but a stronger activation was observed in the right hemisphere (Brunetti et al., 2005).

In conclusion, it appears that, as in the visual system, studies of nonhuman primates can serve as excellent models for human studies. Conversely, human imaging studies can provide useful guidance for microelectrode studies in nonhuman primates, which permit analyses at much higher spatial and temporal resolution than would be possible in most human studies, with some exceptions (Howard et al., 1996; Howard et al., 2000).

8. PLASTICITY OF SOUND LOCALIZATION IN THE EARLY BLIND

8.1 Behavioral evidence in animals and humans

Behavioral data from auditory spatial testing after visual deprivation exist in two mammalian species, cats (Rauschecker and Kniepert, 1994) and ferrets (King and Parsons, 1999). Both sets of data demonstrate that auditory spatial acuity increases (and sound localization error decreases) in early blind mammals. The most pronounced effects were found in lateral and rear positions of azimuth, where the differences to sighted controls were highly significant.

All tests in visually deprived cats were performed with very brief tones (40 ms), so that the animals could not orient towards the sound source. However, if given the opportunity, visually deprived animals (just like blind humans) will use other strategies for the localization of sounds. Very frequently, binocularly lid-sutured cats can be observed to orient towards the azimuth position of a novel sound and then perform vertical scanning movements (in elevation) within that same azimuthal plane (Rauschecker and Henning, 2001). The frequency of these near-sinusoidal scanning movements is about 1 Hz. The scanning movements utilize the directional characteristics of the pinnae and may help blind cats gain not only more refined information about the elevational position of objects in space, but also about their shape and surface texture.

Early studies of sound localization in blind humans, e.g. Fisher (1964), were problematic in that they often studied small patient populations with mixed etiology. Only very recently studies with large blind cohorts have been undertaken that tested subjects with similar history under stringent conditions (Lessard et al., 1998; Muchnik et al., 1991; Röder, B et al., 1999; Zwiers et al., 2001). None of these studies found a disadvantage of the blind in their sound localization abilities and all except one (Zwiers et al., 2001) showed them to be superior. Most interestingly, the study by Lessard et al.

(1998) found patients with partial vision to be the worst of all three groups (fully sighted, completely blind, and partially sighted). The same study provided valuable hints for the neural basis of spatial tuning improvements in the blind: when the blind subjects were forced to use monaural spectral cues for sound localization, the biggest improvement was found. It has been argued previously that the evaluation of spectral cues must be particularly dependent on experience (and therefore based in the auditory cortex), as these cues change systematically during the growth of head and outer ears (Hofman et al., 1998; Rauschecker, 1999).

A study by Röder et al. (1999) found the most significant improvements of sound localization in blind humans for lateral and rear positions of azimuth, precisely as it had been demonstrated in visually deprived cats by Rauschecker and Kniepert (1994). Earlier studies by Rice and co-workers (1970; , 1965) also found improvements in the ability of the blind to judge the spatial direction of echoes, especially those coming from lateral and rear positions.

8.2 Neural plasticity of auditory space in visually deprived cats

Neurophysiologically, the most extensive studies on effects of visual deprivation have been performed in cats. This is true both for intra- as well as intermodal plasticity. Crossmodal changes have first been observed in the superior colliculus (Rauschecker and Harris, 1983), where an increased number of auditory neurons was found in visually deprived cats compared to normal controls. These auditory neurons were situated at high density in intermediate and deep layers of the SC, but also occasionally in superficial layers, where normally only visual cells exist.

Injections of the retrograde anatomical tracer horseradish peroxidase (HRP) into the SC of BD cats revealed a vastly impoverished population of neurons in layer V of striate cortex compared to normal controls (Rauschecker, 2005)(Rauschecker and Aschoff, unpublished). On the other hand, projections from association areas in the region around the anterior ectosylvian sulcus (AES) were well preserved or even strengthened in BD cats. The increased number of auditory responses in the SC may thus be conveyed by these corticotectal projections, and a study of the AES region appeared warranted.

Indeed, in the cortex of visually deprived cats, profound crossmodal changes were found in the AES region (Rauschecker and Korte, 1993). This cortical region normally contains visual, auditory, and somatosensory areas in close vicinity to each other, with some overlap between them (Benedek et al., 1988; Clarey and Irvine, 1990; Clemo and Stein, 1983; Jiang et al., 1994;

Mucke et al., 1982; Olson and Graybiel, 1987). In cats that were visually deprived from birth by means of binocular lid suture (and lids reopened for testing), the visual area (AEV) in the fundus of the AES had all but disappeared. However, neurons in this region did not simply become unresponsive to sensory stimulation altogether. Instead, they were found to be briskly responsive to auditory and (to some extent) tactile stimulation. In other words, the neighboring auditory and somatosensory fields had expanded into the formerly visual territory, at the expense of area AEV (Rauschecker and Korte, 1993).

The response properties of the expanded anterior ectosylvian auditory area (AEA) in the AES region of blind cats were homogeneous with neighboring auditory fields. In particular, the auditory spatial tuning (the tuning for the location of a sound source in free field) was significantly sharper in the whole AES region (including the anterior auditory field, AAF, on the anterior ectosylvian gyrus) compared to sighted control animals (Korte and Rauschecker, 1993). Whereas the control group comprised roughly 50% spatially tuned cells (with a spatial tuning ratio of better than 2:1), the blind animals had close to 90 % spatially specific auditory neurons in the AES region. In addition, neurons with spatial tuning ratios of 10:1 or better were more abundant in blind cats.

The increased number of auditory neurons, together with their sharpened spatial filtering characteristics, is likely to improve the sampling density of auditory space and is thought to underlie the improved spatial abilities of early blind animals (Rauschecker, 1995). Sharper tuning does increase the efficiency of a population code, in the sense that fewer neurons are required to achieve a given acuity (Fitzpatrick et al., 1997). If the number of neurons stays the same or even increases (as in our case), the resulting acuity increases too. Related theoretical considerations lead to the conclusion that a tuning optimum for best performance can be found (Baldi and Heiligenberg, 1988), and it appears that the tuning values found in blind cats come closer to this optimum. The reason why this optimum is not reached in sighted cats lies in the limit on the number of auditory-responsive neurons imposed by the competing visual input.

Subsequent studies found improvements of similar magnitude in the spatial tuning of neurons in primary auditory cortex (A1`; Rauschecker and Henning, 2001).

8.3 Cortical reorganization of auditory space in blind humans

With the advent of modern imaging techniques it became possible to map the distribution of neural activity during auditory stimulation in blind and sighted subjects directly in the human brain. Early studies of that kind had demonstrated metabolic rates in the occipital cortex of blind subjects that were as high as in sighted controls during visual stimulation (Wanet-Defalque et al., 1988). However, it did not become clear whether this increased rate was actually due to specific sensory activity until it could be shown that increased regional cerebral blood flow and metabolic rates were correlated with auditory stimulation (Arno et al., 2001; De Volder et al., 1999; Uhl et al., 1993).

Studies using event-related potentials (ERP) then demonstrated that the extent of cortical activation by changes in the frequency, intensity and location of a sound was expanded in blind people and shifted posteriorly into occipital areas (Kujala et al., 1995; Kujala et al., 1992; Liotti et al., 1998; Röder et al., 1996). Comparing patients blind from birth with those who became blind later in life, one study found a posteriorly-directed expansion also in late blind, which was intermediate in extent to the early blind (Kujala et al., 1997). The latter finding confirms the existence of at least partial crossmodal plasticity in the adult, which is consistent with behavioral findings in two visually deprived adult cats (Rauschecker and Kniepert, 1994). Overall, however, crossmodal plasticity does seem to follow the usual pattern of heightened cortical susceptibility early in life (Büchel et al., 1998; Cohen et al., 1999).

Finally, a recent study using positron emission tomography (PET) in congenitally blind and sighted subjects showed massive activation of occipital cortex during a sound localization task in virtual auditory space (Weeks et al., 2000). The sounds were presented via headphones and the spatial cues were programmed on the basis of standardized head-related transfer functions (HRTF`; Wightman and Kistler, 1989), taking into account monaural spectral cues in addition to binaural cues. The behavioral performance of the blind in localizing these sounds was just as good as that of the normal controls.

Localization of sounds in virtual auditory space by sighted subjects leads to activation in specific foci in the inferior parietal lobules (IPL) and in dorsolateral prefrontal cortex (Bushara et al., 1999; Weeks et al., 1999). The same foci light up in blind subjects but are vastly expanded towards parieto-occipital (BA 7) and occipital locations (BA 18 and 19), and a pronounced right-hemisphere bias becomes apparent. Inter-regional correlation analysis, using the right IPL as a reference region, reveals a functional network of connections involving inferior and posterior parietal and occipital areas of the right hemisphere (Weeks et al., 2000).

Comparison of blind and sighted subjects shows significantly greater inter-regional correlations in the blind between the right IPL and the right parieto-occipital (areas 7/19), the right peristriate (BA 18) and right superior temporal cortices (area 22). This suggests that auditory signals from the temporal areas reach the occipital cortex via parietal and parieto-occipital areas. Whether this reassignment involves the formation of new connections or the strengthening of existing ones remains to be elucidated. The animal work from binocularly deprived cats would suggest that the original crossmodal expansion happens in a competitive fashion in parietal cortex. In this case, the back-projections from parietal to occipital cortex would already carry an enhanced auditory signal (Fig. 20-2).

The question how the auditory input and the somatosensory input observed in the related studies mentioned earlier (Büchel et al., 1998; Cohen et al., 1997; Sadato et al., 1996) co-exist in occipital cortex and share this territory remains to be elucidated by subsequent studies. Under the hypothesis of a top-down input from parietal cortex, the parcellation into different modalities would, in all likelihood, be achieved at that level. This would render occipital cortex in the blind effectively an extension of parietal areas.

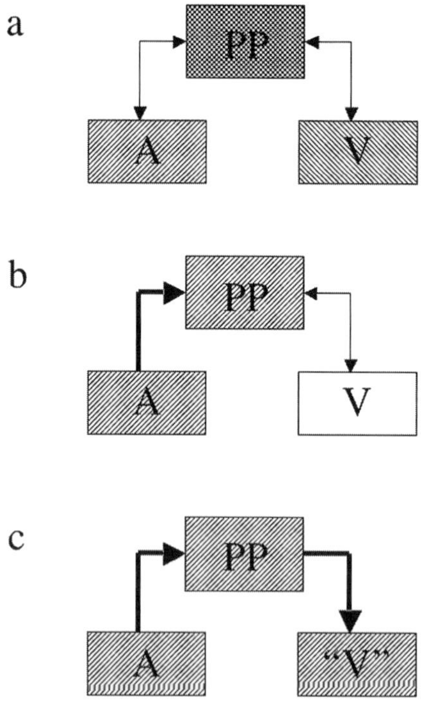

Figure 20-2. Cross-modal plasticity in early blind humans. The reassignment of ordinarily visual areas in occipital cortex to the auditory modality is illustrated schematically. In this

hypothetical scheme, posterior parietal areas of the dorsal pathway play a pivotal role for the reorganization: In newborns (a), parietal cortex receives input from both visual and auditory sensory regions. In visually deprived individuals (b), the visual input remains largely silent during development, so the auditory input obtains a competitive advantage and takes over large portions of posterior parietal cortex. After a period of blindness (c), "visual" occipital areas become activated by auditory input during a sound localization task. It has been shown that this nonvisual input is indeed functionally and behaviorally relevant. Abbreviations: A = auditory; V = visual; PP = posterior parietal.

8.4 Comparison of human and animal data

The areas of expansion in blind humans are probably homologous, in part, to the higher visual areas that were activated by auditory or somatosensory stimuli in visually deprived cats (Rauschecker and Korte, 1993) and monkeys (Hyvärinen Carlson et al., 1981; Hyvärinen Hyvärinen et al., 1981), but clearly include more primary areas in the occipital cortex, such as BA18, as well. This vast expansion of auditory activation into occipital cortex, corresponding to primary and secondary visual areas, may come as a surprise when compared to the animal data. However, there is a simple explanation for this seeming discrepancy. Auditory responses were never tested in primary occipital areas of blind cats and monkeys, because it seemed unlikely that auditory input could expand so far into normally visual territory. Crossmodal expansion was thought to be limited to neighboring areas with multimodal overlap, where competition could occur between overlapping input from different modalities. In light of the recent PET data by Weeks et al. (2000), a reexamination of this view appears now warranted.

On the other hand, multimodal overlap and competition between different modalities may indeed occur even in occipital cortex during early postnatal stages, when occipital areas are still wired to receive (transitory) input from auditory cortex (Innocenti and Clarke, 1984). Recent anatomical data with more sensitive tracers have demonstrated that a direct projection from auditory to visual cortex, though sparse, does exist in adult rhesus monkeys (Falchier et al., 2002; Rockland and Ojima, 2003). The continuance of a multipotential function even for primary sensory cortices has been suggested (Pascual-Leone and Hamilton, 2001).

9. GENERAL CONCLUSIONS

In this chapter, I have summarized the evidence for the existence of dual pathways in the auditory cortex for the processing of object/pattern information and for the processing of space and motion. The latter pathway follows a posterior-dorsal route, originating in primary auditory cortex and projecting through caudal belt and parabelt into inferior posterior parietal cortex. Evidence for these pathways comes from anatomical and physiological studies of nonhuman primates as well as neuroimaging studies in humans.

Plasticity of the dorsal pathway is at the core of cross-modal reorganization in the cerebral cortex of animals and humans that grow up blind. Multisensory competition between different modalities creates an advantage for the nondeprived senses. Auditory cortical regions expand at the expense of formerly visual regions within the anterior ectosylvian cortex of visually deprived cats. Auditory cortical neurons sharpen their tuning for the spatial location of a sound. Behaviorally, visually deprived cats are superior to their sighted littermates. In blind humans, auditory activation of occipital (normally visual) cortical regions is found in PET and ERP studies. It is assumed that the auditory input to occipital cortex is mediated by parietal regions that have undergone cross-modal reassignment.

ACKNOWLEDGEMENTS

The present chapter draws from the following prior publications: Rauschecker, 2002; Rauschecker and Tian, 2005. Comments should be addressed to Josef Rauschecker (E-Mail: rauschej@georgetown.edu; Tel.: ++1-202-687-1580; Fax: ++1-202-687-0617).

REFERENCES

Alain, C., Arnott, S. R., Hevenor, S., Graham, S., and Grady, C. L., 2001, "What" and "where" in the human auditory system, *Proc Natl Acad Sci U S A* **98**(21):12301-12306.

Andersen, R. A., 1997, Multimodal integration for the representation of space in the posterior parietal cortex, *Philos Trans R Soc Lond B Biol Sci* **352**(1360):1421-1428.

Arno, P., De Volder, A. G., Vanlierde, A., Wanet-Defalque, M. C., Streel, E., Robert, A., Sanabria-Bohorquez, S., and Veraart, C., 2001, Occipital activation by pattern recognition in the early blind using auditory substitution for vision, *Neuroimage* **13**(4):632-645.

Arnott, S. R., Binns, M. A., Grady, C. L., and Alain, C., 2004, Assessing the auditory dual-pathway model in humans., *Neuroimage* **22**(1):401-408.

Baldi, P., and Heiligenberg, W., 1988, How sensory maps could enhance resolution through ordered arrangements of broadly tuned receivers, *Biological Cybernetics* **59**(4-5):313-318.

Barrett, D. J., Hall, D. A., 2006, Response preferences for "what" and "where" in human non-primary auditory cortex, *Neuroimage* **32**(2):968-977

Beitel, R. E., and Kaas, J. H., 1993, Effects of bilateral and unilateral ablation of auditory cortex in cats on the unconditioned head orienting response to acoustic stimuli, *J Neurophysiol* **70**(1):351-369.

Benedek, G., Mucke, L., Norita, M., Albowitz, B., and Creutzfeldt, O. D., 1988, Anterior ectosylvian visual area (AEV) of the cat: physiological properties, *Prog Brain Res* **75**: 245-255.

Binder, J. R., Frost, J. A., Hammeke, T. A., Bellgowan, P. S., Springer, J. A., Kaufman, J. N., and Possing, E. T., 2000, Human temporal lobe activation by speech and nonspeech sounds, *Cereb Cortex* **10**(5):512-528.

Binder, J. R., Liebenthal, E., Possing, E. T., Medler, D. A., and Ward, B. D., 2004, Neural correlates of sensory and decision processes in auditory object identification, *Nat Neurosci* **7**(3):295-301.

Bremmer, F., Schlack, A., Shah, N. J., Zafiris, O., Kubischik, M., Hoffmann, K., Zilles, K., and Fink, G. R., 2001, Polymodal motion processing in posterior parietal and premotor cortex: a human fMRI study strongly implies equivalencies between humans and monkeys, *Neuron* **29**(1):287-296.

Brunetti, M., Belardinelli, P., Caulo, M., Del Gratta, C., Della Penna, S., Ferretti, A., Lucci, G., Moretti, A., Pizzella, V., Tartaro, A., Torquati, K., Olivetti Belardinelli, M., and Romani, G. L., 2005, Human brain activation during passive listening to sounds from different locations: An fMRI and MEG study, *Hum Brain Mapp* **26**(4):251-261.

Büchel, C., Price, C., Frackowiak, R. S., and Friston, K., 1998, Different activation patterns in the visual cortex of late and congenitally blind subjects, *Brain* **121**(Pt 3):409-419.

Bushara, K. O., Weeks, R. A., Ishii, K., Catalan, M. J., Tian, B., Rauschecker, J. P., and Hallett, M., 1999, Modality-specific frontal and parietal areas for auditory and visual spatial localization in humans, *Nat Neurosci* **2**(8):759-766.

Clarey, J. C., and Irvine, D. R. F., 1990, The anterior ectosylvian sulcal auditory field in the cat: I. An electrophysiological study of its relationship to surrounding auditory cortical fields, *J Comp Neurol* **301**:289-303.

Clemo, H. R., and Stein, B. E., 1983, Organization of a fourth somatosensory area of cortex in cat, *J Neurophysiol* **50**(4):910-925.

Cohen, L. G., Celnik, P., Pascual-Leone, A., Corwell, B., Falz, L., Dambrosia, J., Honda, M., Sadato, N., Gerloff, C., Catala, M. D., and Hallett, M., 1997, Functional relevance of cross-modal plasticity in blind humans, *Nature* **389**(6647):180-183.

Cohen, L. G., Weeks, R. A., Sadato, N., Celnik, P., Ishii, K., and Hallett, M., 1999, Period of susceptibility for cross-modal plasticity in the blind, *Ann Neurol* **45**(4):451-460.

De Volder, A. G., Catalan-Ahumada, M., Robert, A., Bol, A., Labar, D., Coppens, A., Michel, C., and Veraart, C., 1999, Changes in occipital cortex activity in early blind humans using a sensory substitution device, *Brain Res* **826**(1):128-134.

Degerman, A., Rinne, T., Salmi, J., Salonen, O., and Alho, K., 2006, Selective attention to sound location or pitch studied with fMRI, *Brain Res 1077(1):123-134.*

Diamond, I. T., Fisher, J. F., Neff, W. D., and Yela, M., 1956, Role of auditory cortex in discrimination requiring localization of sound in space, *J Neurophysiol* **19**(6):500-512.

Falchier, A., Clavagnier, S., Barone, P., and Kennedy, H., 2002, Anatomical evidence of multimodal integration in primate striate cortex, *J Neurosci* **22**(13):5749-5759.

Fisher, G. H., 1964, Spatial localization by the blind, *Am J Psychol* **77**:2-133.

Fitzpatrick, D. C., Batra, R., Stanford, T. R., and Kuwada, S., 1997, A neuronal population code for sound localization, *Nature* **388**(6645):871-874.

Goldman-Rakic, P. S., 1996, The prefrontal landscape: implications of functional architecture for understanding human mentation and the central executive, *Phil Trans R Soc Lond B* **351**(1346):1445-1453.

Griffiths, T. D., Rees, A., Witton, C., Cross, P. M., Shakir, R. A., and Green, G. G., 1997, Spatial and temporal auditory processing deficits following right hemisphere infarction. A psychophysical study, *Brain* **120**(Pt 5):785-794.

Griffiths, T. D., Rees, A., Witton, C., Shakir, R. A., Henning, G. B., and Green, G. G., 1996, Evidence for a sound movement area in the human cerebral cortex, *Nature* **383**(6599): 425-427.

Griffiths, T. D., Rees, G., Rees, A., Green, G. G., Witton, C., Rowe, D., Buchel, C., Turner, R., and Frackowiak, R. S., 1998, Right parietal cortex is involved in the perception of sound movement in humans, *Nat Neurosci* **1**(1):74-79.

Heffner, H., and Masterton, B., 1975, Contribution of auditory cortex to sound localization in the monkey (Macaca mulatta), *J Neurophysiol* **38**(6):1340-1358.

Heffner, H. E., and Heffner, R. S., 1990, Effect of bilateral auditory cortex lesions on sound localization in Japanese macaques, *J Neurophysiol* **64**(3):915-931.

Hofman, P. M., Van Riswick, J. G., and Van Opstal, A. J., 1998, Relearning sound localization with new ears, *Nat Neurosci* **1**(5):417-421.

Howard, M. A., Volkov, I. O., Abbas, P. J., Damasio, H., Ollendieck, M. C., and Granner, M., 1996, A chronic microelectrode investigation of the tonotopic organization of human auditory cortex, *Brain Res* **724**:260-264.

Howard, M. A., Volkov, I. O., Mirsky, R., Garell, P. C., Noh, M. D., Granner, M., Damasio, H., Steinschneider, M., Reale, R. A., Hind, J. E., and Brugge, J. H., 2000, Auditory cortex on the human posterior superior temporal gyrus, *J Comp Neurol* **416**:79-92.

Hyvärinen, J., Carlson, S., and Hyvärinen, L., 1981, Early visual deprivation alters modality of neuronal responses in area 19 of monkey cortex, *Neurosci Lett* **4**:239-243.

Hyvärinen, J., Hyvärinen, L., and Linnankoski, I., 1981, Modification of parietal association cortex and functional blindness after binocular deprivation in young monkeys, *Exp Brain Res* **42**:1-8.

Imig, T. J., Irons, W. A., and Samson, F. R., 1990, Single-unit selectivity to azimuthal direction and sound pressure level of noise bursts in cat high-frequency primary auditory cortex, *J Neurophysiol* **63**(6):1448-1466.

Innocenti, G. M., and Clarke, S., 1984, Bilateral transitory projection from auditory cortex in kittens, *Dev Brain Res* **14**:143-148.

Irvine, D. R. F., 1992, Physiology of auditory brainstem pathways., in: *Springer Handbook of Auditory Research. Vol 2, The Mammalian Auditory Pathway: Neurophysiology*, R. R. Fay and A. A. Popper, eds., Springer-Verlag, Berlin, pp. 153-231.

Jääskeläinen, I. P., Ahveninen, J., Bonmassar, G., Dale, A. M., Ilmoniemi, R. J. L., Lin, F. H., May, P., Melcher, J., Stufflebeam, S., Tiitinen, H., and Belliveau, J. W., 2004, Human posterior auditory cortex gates novel sounds to consciousness, *Proc Natl Acad Sci U S A* **101**(17):6809-6814.

Jenkins, W. M., and Merzenich, M. M., 1984, Role of cat primary auditory cortex for sound-localization behavior, *J Neurophysiol* **52**(5):819-847.

Jiang, H., Leporé, F., Ptito, M., and Guillemot, J. P., 1994, Sensory modality distribution in the anterior ectosylvian cortex (AEC) of cats, *Exp Brain Res* **97**(3):404-414.

Kaas, J. H., and Hackett, T. A., 2000, Subdivisions of auditory cortex and processing streams in primates, *Proc Natl Acad Sci U S A* **97**(22):11793-11799.

King, A. J., and Hutchings, M. E., 1987, Spatial response properties of acoustically responsive neurons in the superior colliculus of the ferret: a map of auditory space, *J Neurophysiol* **57**:596-624.

King, A. J., and Parsons, C., 1999, Improved auditory spatial acuity in visually deprived ferrets, *Eur J Neurosci* **11**(11):3945-3956.

Knudsen, E. I., and Konishi, M., 1978, A neural map of auditory space in the owl, *Science* **200**(4343):795-797.

Korte, M., and Rauschecker, J. P., 1993, Auditory spatial tuning of cortical neurons is sharpened in cats with early blindness, *J Neurophysiol* **70**(4):1717-1721.

Krumbholz, K., Schonwiesner, M., Rubsamen, R., Zilles, K., Fink, G. R., and von Cramon, D. Y., 2005, Hierarchical processing of sound location and motion in the human brainstem and planum temporale, *Eur J Neurosci* **21**(1):230-238.

Krumbholz, K., Schonwiesner, M., von Cramon, D. Y., Rubsamen, R., Shah, N. J., Zilles, K., and Fink, G. R., 2005, Representation of interaural temporal information from left and right auditory space in the human planum temporale and inferior parietal lobe, *Cereb Cortex* **15**(3):317-324.

Kujala, T., Alho, K., Huotilainen, M., Ilmoniemi, R., Lehtokoski, A., Leinonen, A., Rinne, T., Salonen, O., Sinkkonen, J., Standertskjold-Nordenstam, C., and Naatanen, R., 1997, Electrophysiological evidence for cross-modal plasticity in humans with early- and late-onset blindness, *Psychophysiology* **32**(2):213-216.

Kujala, T., Alho, K., Kekoni, J., Hamalainen, H., Reinikainen, K., Salonen, O., Standertskjold-Nordenstam, C. G., and Näätänen, R., 1995, Auditory and somatosensory event-related brain potentials in early blind humans., *Exp Brain Res* **104**(3):519-526.

Kujala, T., Alho, K., Paavilainen, P., Summala, H., and Näätänen, R., 1992, Neural plasticity in processing sound location by the early blind: an event-related potential study, *Electroencephalogr Clin Neurophysiol* **84**:469-472.

Lessard, N., Pare, M., Lepore, F., and Lassonde, M., 1998, Early-blind human subjects localize sound sources better than sighted subjects, *Nature* **395**(6699):278-280.

Lewis, J. W., and Van Essen, D. C., 2000, Corticocortical connections of visual, sensorimotor, and multimodal processing areas in the parietal lobe of the macaque monkey, *J Comp Neurol* **428**(1):112-137.

Liotti, M., Ryder, K., and Woldorf, M. G., 1998, Auditory attention in the congenitally blind: where, when and what gets, *Neuroreport* **9**(6):1007-1012.

Maeder, P. P., Meuli, R. A., Adriani, M., Bellmann, A., Fornari, E., Thiran, J. P., Pittet, A., and Clarke, S., 2001, Distinct Pathways Involved in Sound Recognition and Localization: A Human fMRI Study, *Neuroimage* **14**(4):802-816.

Malhotra, S., Hall, A. J., and Lomber, S. G., 2004, Cortical control of sound localization in the cat: unilateral cooling deactivation of 19 cerebral areas, *J Neurophysiol* **92**(3): 1625-1643.

Mazzoni, P., Bracewell, R. M., Barash, S., and Andersen, R. A., 1996, Spatially tuned auditory responses in area LIP of macaques performing delayed memory saccades to acoustic targets, *J Neurophysiol* **75**(3):1233-1241.

Meredith, M. A., and Clemo, H. R., 1989, Auditory cortical projection from the anterior ectosylvian sulcus (Field AES) to the superior colliculus in the cat: an anatomical and electrophysiological study, *J Comp Neurol* **289**(4):687-707.

Middlebrooks, J. C., Clock, A. E., Xu, L., and Green, D. M., 1994, A panoramic code for sound location by cortical neurons, *Science* **264**(5160):842-844.

Middlebrooks, J. C., and Knudsen, E. I., 1984, A neural code for auditory space in the cat's superior colliculus, *J Neurosci* **4**:2621-2634.

Morel, A., Garraghty, P. E., and Kaas, J. H., 1993, Tonotopic organization, architectonic fields, and connections of auditory cortex in macaque monkeys, *J Comp Neurol* **335**(3):437-459.

Muchnik, C., Efrati, M., Nemeth, E., Malin, M., and Hildesheimer, M., 1991, Central auditory skills in blind and sighted subjects, *Scand Audiol* **20**:19-23.

Mucke, L., Norita, M., Benedek, G., and Creutzfeldt, O., 1982, Physiologic and anatomic investigation of a visual cortical area situated in the ventral bank of the anterior ectosylvian sulcus of the cat, *Exp Brain Res* **46**(1):1-11.

Obleser, J., Boecker, H., Drzezga, A., Haslinger, B., Hennenlotter, A., Roettinger, M., Eulitz, C., and Rauschecker, J. P., 2005, Vowel sound extraction in anterior superior temporal cortex, *Hum Brain Mapp* **27**(7):562-571.

Olson, C. R., and Graybiel, A. M., 1987, Ectosylvian visual area of the cat: location, retinotopic organization, and connections, *J Comp Neurol* **261**(2):277-294.

Pandya, D. N., and Sanides, F., 1972, Architectonic parcellation of the temporal operculum in rhesus monkey and its projection pattern, *Z Anat Entw-Gesch* **139**:127-161.

Pascual-Leone, A., and Hamilton, R., 2001, The metamodal organization of the brain, *Prog Brain Res* **134**:427-445.

Poremba, A., Saunders, R. C., Crane, A. M., Cook, M., Sokoloff, L., and Mishkin, M., 2003, Functional mapping of the primate auditory system, *Science* **299**(5606):568-572.

Rajan, R., Aitkin, L. M., and Irvine, D. R., 1990, Azimuthal sensitivity of neurons in primary auditory cortex of cats. II. Organization along frequency-band strips, *J Neurophysiol* **64**(3):888-902.

Rajan, R., Aitkin, L. M., Irvine, D. R., and McKay, J., 1990, Azimuthal sensitivity of neurons in primary auditory cortex of cats. I. Types of sensitivity and the effects of variations in stimulus parameters, *J Neurophysiol* **64**(3):872-887.

Rauschecker, J. P., 1995, Compensatory plasticity and sensory substitution in the cerebral cortex, *TINS* **18**:36-43.

Rauschecker, J. P., 1997, Processing of complex sounds in the auditory cortex of cat, monkey and man, *Acta Otolaryngol* **532**:34-38.

Rauschecker, J. P., 1998, Cortical processing of complex sounds, *COIN* **8**:516-521.

Rauschecker, J. P., 1999, Auditory cortical plasticity: a comparison with other sensory systems, *Trends Neurosci* **22**(2):74-80.

Rauschecker, J. P., 2002, Cortical map plasticity in animals and humans, *Prog Brain Res* **138**:73-88.

Rauschecker, J. P., 2005, Plasticity in auditory functions., in: *Textbook of Neural Repair and Rehabilitation, Vol. 1, Neural Repair and Plasticity*, M. Selzer, S. Clarke, L. Cohen, P. Duncan and F. Gage, eds., Cambridge University Press, Cambridge, England, pp. 162-197

Rauschecker, J. P., and Harris, L. R., 1983, Auditory compensation of the effects of visual deprivation in the cat's superior colliculus, *Exp Brain Res* **50**:69-83.

Rauschecker, J. P., and Henning, P., 2001, Crossmodal expansion of cortical maps in early blindness, in: *The Mutable Brain*, J. Kaas, ed., Harwood Academic Publishers, Singapore, pp. 243- 259.

Rauschecker, J. P., and Kniepert, U., 1994, Enhanced precision of auditory localization behavior in visually deprived cats, *Eur J Neurosci* **6**:149-160.

Rauschecker, J. P., and Korte, M., 1993, Auditory compensation for early blindness in cat cerebral cortex, *J. Neurosci* **13**:4538-4548.

Rauschecker, J. P., and Tian, B., 2000, Mechanisms and streams for processing of "what" and "where" in auditory cortex, *Proc Natl Acad Sci U S A* **97**(22):11800-11806.

Rauschecker, J. P., and Tian, B., 2005, Hierarchic processing of communication sounds in primates, in: *Behavior and Neurodynamics for Auditory Communication*, J. S. Kanwal and G. Ehret, eds., Cambridge University Press, Cambridge, pp.

Rauschecker, J. P., Tian, B., Pons, T., and Mishkin, M., 1997, Serial and parallel processing in rhesus monkey auditory cortex, *J Comp Neurol* **382**(1):89-103.

Ravizza, R. J., and Masterton, B., 1972, Contribution of neocortex to sound localization in opossum (Didelphis virginiana), *J Neurophysiol* **35**(3):344-356.

Recanzone, G. H., 2000, Spatial processing in the auditory cortex of the macaque monkey, *Proc Natl Acad Sci U S A* **97**(22):11829-11835.

Recanzone, G. H., Guard, D. C., Phan, M. L., and Su, T. K., 2000, Correlation between the activity of single auditory cortical neurons and sound-localization behavior in the macaque monkey, *J Neurophysiol* **83**(5):2723-2739.

Rice, C. E., 1970, Early blindness, early experience, and perceptual enhancement, *Res Bull Am Found Blind* **22**:1-22.

Rice, C. E., Feinstein, S. H., and Schusterman, R. J., 1965, Echo-detection ability of the blind: size and distance factor, *J Exp Psychol* **70**:246-251.

Rockland, K. S., and Ojima, H., 2003, Multisensory convergence in calcarine visual areas in macaque monkey, *Int J Psychophysiol* **50**(1-2):19-26.

Röder, B., Rosler, F., Henninghausen, E., and Nacker, F., 1996, Event-related potentials during auditory and somatosensory discrimination in sighted and blind human subjects, *Brain Res Cog Brain Res* **4**(2):77-93.

Röder, B., Rosler, F., and Neville, H. J., 1999, Effects of interstimulus interval on auditory event-related potentials in congenitally blind and normally sighted humans, *Neurosci Lett* **264**(1-3):53-56.

Röder, B., Teder-Salejarvi, W., Sterr, A., Rosler, F., Hillyard, S. A., and Neville, H. J., 1999, Improved auditory spatial tuning in blind humans, *Nature* **400**(6740):162-166.

Romanski, L. M., and Goldman-Rakic, P. S., 2002, An auditory domain in primate prefrontal cortex, *Nat Neurosci* **5**:15-16.

Romanski, L. M., Tian, B., Fritz, J., Mishkin, M., Goldman-Rakic, P. S., and Rauschecker, J. P., 1999, Dual streams of auditory afferents target multiple domains in the primate prefrontal cortex, *Nat Neurosci* **2**(12):1131-1136.

Sadato, N., Pascual-Leone, A., Grafman, J., Ibanez, V., Deiber, M.-P., Dold, G., and Hallett, M., 1996, Activation of the primary visual cortex by Braille reading in blind subjects, *Nature* **380**(526-528).

Scott, S. K., Blank, C. C., Rosen, S., and Wise, R. J., 2000, Identification of a pathway for intelligible speech in the left temporal lobe, *Brain* **123(**Pt 12):2400-2406.

Stricanne, B., Andersen, R. A., and Mazzoni, P., 1996, Eye-centered, head-centered, and intermediate coding of remembered sound locations in area LIP, *J Neurophysiol* **76**(3):2071-2076.

Tata, M. S., and Ward, L. M., 2005a, Early phase of spatial mismatch negativity is localized to a posterior "where" auditory pathway, *Exp Brain Res* **167**(3):481-486.

Tata, M. S., and Ward, L. M., 2005b, Spatial attention modulates activity in a posterior "where" auditory pathway, *Neuropsychologia* **43**(4):509-516.

Tian, B., Reser, D., Durham, A., Kustov, A., and Rauschecker, J. P., 2001, Functional specialization in rhesus monkey auditory cortex, *Science* **292**(5515):290-293.

Uhl, F., Franzen, P., Podreka, I., Steiner, M., and Deecke, L., 1993, Increased regional cerebral blood flow in inferior occipital cortex and cerebellum of early blind humans, *Neurosci Lett* **150**:162-164.

Wanet-Defalque, M. C., Veraart, C., De Volder, A., Metz, R., Michel, C., Dooms, G., and Goffinet, A., 1988, High metabolic activity in the visual cortex of early blind human subjects, *Brain Res* **446**(2):369-373.

Warren, J. D., Zielinski, B. A., Green, G. G. R., Rauschecker, J. P., and Griffiths, T. D., 2002, Analysis of sound source motion by the human brain, *Neuron* **34**:1-20.

Weeks, R., Horwitz, B., Aziz-Sultan, A., Tian, B., Wessinger, C. M., Cohen, L. G., Hallett, M., and Rauschecker, J. P., 2000, A positron emission tomographic study of auditory localization in the congenitally blind, *J Neurosci* **20**(7):2664-2672.

Weeks, R. A., Aziz-Sultan, A., Bushara, K. O., Tian, B., Wessinger, C. M., Dang, N., Rauschecker, J. P., and Hallett, M., 1999, A PET study of human auditory spatial processing, *Neurosci Lett* **262**(3):155-158.

Wessinger, C. M., VanMeter, J., Tian, B., Van Lare, J., Pekar, J., and Rauschecker, J. P., 2001, Hierarchical organization of the human auditory cortex revealed by functional magnetic resonance imaging, *J Cogn Neurosci* **13**(1):1-7.

Wightman, F. L., and Kistler, D. J., 1989, Headphone simulation of free-field listening. I: Stimulus synthesis, *J Acoust Soc Am* **85**(2):858-867.

Yu, J. J., and Young, E. D., 2000, Linear and nonlinear pathways of spectral information transmission in the cochlear nucleus, *Proc Natl Acad Sci U S A* **97**(22):11780-11786.

Zatorre, R. J., Bouffard, M., Ahad, P., and Belin, P., 2002, Where is 'where' in the human auditory cortex?, *Nat Neurosci* **5**(9):905-909.

Zimmer, U., and Macaluso, E., 2005, High binaural coherence determines successful sound localization and increased activity in posterior auditory areas, *Neuron* **47**(6):893-905.

Zwiers, M. P., Van Opstal, A. J., and Cruysberg, J. R., 2001, Two-dimensional sound-localization behavior of early-blind humans, *Exp Brain Res* **140**(2):206-222.

Chapter 21

NETWORKS FOR ATTENTIONAL CONTROL AND SELECTION IN SPATIAL VISION

George R. Mangun and Sean P. Fannon
Center for Mind and Brain, University of California, Davis, CA, USA

Abstract: Visual selective attention is a powerful cognitive ability that involves the interactions of widespread cortical networks, resulting in modulations of sensory processing at early stages of visual information processing. Visual attention can be directed voluntarily based on the goals and intentions of the individual or captured reflexively by bottom-up processes that are driven by the nature of sensory inputs. Attention can be focused based on spatial (i.e., location) or non-spatial (i.e., color or form, etc.) information, or their conjunction. In this chapter, the focus is on voluntary visual spatial attention, and in particular on the neural mechanisms involved in the control of the focus of attention for spatial versus non-spatial information. Although many models posit that voluntary attention relies on supramodal attentional control networks, evidence presented here will challenge this view, and will argue that attentional control for location and motion information involves partly specialized neural mechanisms in frontal and parietal cortex. The result of activity in these specialized networks is to tune visual cortex to select some information over other competing inputs, facilitating actions in response to relevant sensory information.

Key words: selective attention; visual attention; spatial attention; dorsal stream.

1. INTRODUCTION

The sensory world is rich in information, but we have long understood from both our subjective experience (e.g., James, 1981, 1890; Helmholtz, 1962, 1924-5) and experimental demonstrations (e.g., Broadbent, 1962; Posner, 1978; Treisman, 1964) that the nervous system is limited in how much of this flurry of information it can process effectively at any moment in time. Selective attention provides a means by which organisms regulate the flow of sensory information by "selecting" relevant information for continued processing and/or by inhibiting irrelevant or distracting events.

Selective attention involves interacting neural processes including top-down and bottom-up mechanisms. *Top-down* attentional mechanisms are typically thought of as those that are based on the goals and intentions of the individual. In general, these top-down attentional functions are referred to as attentional control processes (e.g., Corbetta and Shulman, 2002; Hopfinger et al., 2000; Posner and Petersen, 1990). In this context, "control" refers to the top-down mechanisms involved in establishing selective sensory processing, presumably by biasing neural processing in favor of some stimulus inputs over others. Attentional biasing is the idea that attention influences the outcome of neural competition in neural networks at a variety of stages of information processing (Desimone and Duncan, 1995). Neural competition arises from the local and global network properties within populations of neurons involved in sensory, cognitive, or motor function, and models of attentional regulation propose that this competition can be biased in favor of attended inputs. The resultant biased processes and their outputs can be referred to as attentional selection; that is, the effective selection of behaviorally relevant inputs.

Because the term "selection" has been used in a variety of ways in the attention literature, it is useful to define how it will be used here. The term *selection* is a concept borrowed from cognitive psychology (e.g., Deutsch and Deutsch, 1963) that refers in the strongest sense to a mechanism by which attended information is tagged (selected) for further processing while competing information is rejected from further processing (e.g., Broadbent, 1962). We will use the term selection in a more general sense as shorthand to refer to the outcome of the biasing of information processing that results from attentional regulation, and we do *not* assume that unattended information is completely rejected from further influence on neural activity or behavior (e.g., Treisman, 1964, 1969; Vogel, Luck and Shapiro, 1998).

In contrast to top-down processes, bottom-up or "stimulus-driven" mechanisms are triggered by the properties of sensory inputs. There are two general types of bottom-up influences in information processing. The first is the sensory influence of a stimulus itself, which, depending on its properties such as stimulus saliency, may bias neural competition (e.g., Robinson and Petersen, 1992) and therefore interact with top-down attentional control processes (e.g., Wright, 2005). A second aspect of bottom-up processing is that which has itself been described as attentional in nature, such as when bottom-up sensory signals trigger an automatic or "reflexive" orienting of attention (e.g., Hopfinger and Mangun, 1998; Klein, 2000; Posner and

Cohen, 1984). Bottom-up factors and top-down influences interact, and can generally be considered to be in competition for the control of neural processing and therefore behavior. Indeed, top-down attentional regulation is most relevant when goals and strategies are in competition with bottom-up stimulus-driven influences or prepotent response tendencies (e.g., Beck and Kastner, 2005). In this chapter, reflexive attention will not be discussed further, and the focus will be on voluntary visual spatial selective attention.

Directing attention selectively to a location is known as "spatial attention". When one directs spatial attention in the visual field this facilitates psychophysical performance for stimuli appearing at the attended location. Reaction times are faster and discrimination accuracy is enhanced for events at attended versus unattended locations (e.g., Downing, 1988; Handy et al., 1996; Heinze and Mangun, 1995; Posner et al., 1978). Spatial attention can also alter stimulus appearance by, for example, altering apparent stimulus contrast (Carrasco et al., 2004), spatial frequency, and gap size (Gobell and Carrasco, 2005). Such findings are consistent with the idea that voluntary spatial selective attention influences information processing, and much research has been dedicated to identifying the stage(s) of information processing affected by attention, as well as to unraveling the underlying neural mechanisms. Nearly a half century ago, psychologists and physiologists proposed that spatial attention might involve modulations of neural processing in the early visual pathways (e.g., Hernandez-Peon et al., 1961) based on the mapping of visual space in subcortical and early cortical visual regions.

Visual space is coded at multiple levels of the ascending visual pathways. At early stages, spatiotopic maps are represented in a point-for-point fashion across visual structures beginning with the retina and lateral geniculate nucleus (LGN) of the thalamus, and continuing in early visual cortical areas. As information ascends in the visual hierarchy, precise visual spatial coding progressively gives way to spatially invariant visual representations in inferotemporal and parietal cortex (Marr, 1982). This organization of early visual processing provides a substrate upon which attentional processes might act to select information at some locations over that at other locations in the visual field (i.e., spatial attention), and much research has been directed at understanding the properties and neural mechanisms of spatial attention.

Two principal questions about the operation of spatial attention have dominated investigation. The first asks where and by what mechanisms spatial attention influences information processing. This has led to a long debate, the early- versus late-selection debate, about whether attention could influence sensory- and perceptual-level visual processes or would act only later on post-perceptual representations (e.g., Deutsch and Deutsch, 1963;

Luck et al., 1994; Pashler, 1998). Although purely behavioral analyses have provided conflicting evidence about the level of selection in spatial attention, physiological evidence has provided conclusive proof that selective attention can influence early visual processing.

Evidence in humans and animals has demonstrated that focused attention influences sensory analyses at early stages of visual processing. Almost forty years ago, the first properly controlled studies of visual selective attention were conducted using the ERP methods in humans. Prior to this period, physiological studies in humans and animals (e.g., Hernandez-Peon et al., 1961) failed to control for a variety of nuisance variables that prevented definitive statements from being made about whether selective attention modulated stimulus processing (see Näätänen, 1975, for a review). Robert Eason and his colleagues, and Steve Hillyard and his colleagues (both groups then in San Diego, California, but working separately) devised the first controlled studies of selective attention in studies of spatial attention. Prior work by Hillyard and colleagues in the auditory system is generally regarded as the first definitive evidence in tightly controlled studies for selective sensory processing in humans (Hillyard et al., 1973). We first summarize the relevant human ERP literature on selective attention effects in sensory processing (attentional selection), and then consider related studies in animals. This will be followed by a brief review of attentional control networks and then a discussion of the specialization in neural networks that may be involved in voluntary attentional control.

1.1 Event-Related Potentials and Selective Spatial Attention

ERP studies of voluntary spatial attention have demonstrated that sensory-evoked activity can be modulated by *selective* visual attention. Sensory ERP components are greater in amplitude when subjects are covertly attending the location at which the eliciting stimuli are presented than when the subjects' attention is directed elsewhere (e.g., Eason et al., 1969; Eason, 1981; Van Voorhis and Hillyard, 1977). Because these sensory ERPs begin within 70-80 ms of stimulus onset and arise from visual cortical areas V1-V4 (e.g., Heinze et al., 1994; Khoe et al., 2005; Martinez et al., 1999), the finding that spatial attention affects their amplitudes suggests that attention operates by modulating the transmission of information in the ascending visual pathways, and does so in a spatially specific manner (reviewed in Luck et al., 2000; Mangun, 1995).

ERP evidence for modulations of V1 with spatial attention has been inconsistent, but recent studies suggest that ERPs generated in V1 can also be modulated by attention under some circumstances. Most studies have failed to find evidence for modulation of the earliest cortical evoked potential, the C1, which is generated in V1 with an onset latency of 50-60 ms. However, more recent studies using dipole source localization have demonstrated attentional modulation of activity in V1 at longer post-stimulus latencies in the range of about 150-300 ms (e.g., Martinez et al., 1999, 2001; Noesselt et al., 2002). These findings suggest that attention may modulate activity in V1 via delayed feedback from higher visual areas, and help to reconcile fMRI evidence for attentional modulation in V1. There is also enticing recent evidence for modulation of early (70-80 ms) V1 activity in humans in response to movement of an exogenously cued surface (Khoe et al., 2005).

ERP evidence to suggest that subcortical transmission in the LGN is affected by spatial attention has been highly controversial, although several prominent models have proposed that gating of afferent activity in the LGN might form the basis for an early selection mechanism (e.g., Yingling and Skinner, 1976). Despite the fact that LGN responses are small when recorded at the scalp, there have been several reports of human ERPs (independent of attention) that were consistent with sources in the LGN (e.g., Cracco and Cracco, 1978; Hackley et al., 1990; Harding and Rubinstein, 1980; Pratt et al., 1982), and such reports are broadly consistent with evidence from human depth recordings in the thalamus (e.g., Krolak-Salmon et al., 2004) and findings in monkeys using both depth recordings and surface electrode ERPs (Schroeder et al., 1992). In support of a role for LGN in early attentional selection, Eason and colleagues (e.g., Oakley and Eason, 1990) provided evidence of short-latency (40-70 msec) human ERP changes with spatial attention that have been attributed to subcortical (presumably LGN) activity. However, as discussed above, other studies have generally failed to find modulations of very early ERPs with attention. As a result, the evidence from human ERPs for attentional modulation in the LGN has been, to say the least, modest. Nonetheless, given that it is possible to obtain ERP measures of human LGN activity using appropriately designed stimuli (e.g., Pratt et al., 1982) and recent evidence from human imaging studies (e.g., O'Connor et al., 2002) and animal research (reviewedbelow) that attention may affect processing in the LGN, it is important that the possible subcortical effects of selective attention in humans be investigated further. Therefore, for now, the question of whether selective visual attention involves subcortical gating of visual sensory transmission must, in our opinion, remain open.

1.2 Neuronal Recordings in Animals and Selective Spatial Attention

Attentional modulation of sensory processing has been observed in recordings of neuronal activity in awake behaving monkeys in several areas of visual cortex, including striate (V1) (e.g., McAdams and Maunsell, 1999; McAdams and Reid, 2005) and extrastriate visual areas V2-V4 (e.g., Chelazzi et al., 1993, 2001; Luck et al., 1997; Maunsell and Cook, 2002; Moran and Desimone, 1985; Reynolds and Chelazzi, 2004; Treue 2001, 2003). Selective attention has also been shown to influence higher-order perceptual processing in ventral-stream areas such as inferotemporal cortex (e.g., Chelazzi et al., 1993; Spitzer and Richmond, 1991), and in dorsal-stream areas such as the motion processing areas MT and MST (Recanzone et al., 1993; Treue and Trujillo, 1999). Animal studies of spatial attention have demonstrated that a cell's response to a stimulus presented in its receptive field is highly sensitive to where the animal is attending. For example, Moran and Desimone (1985), in the first well-controlled study of selective attention in animals, recorded from V4 of the macaque monkey and found that the response to a stimulus that was effective in driving a cell was dramatically reduced if the monkey was attending to a different stimulus within the receptive field, but only minimally reduced if the monkey was attending to a different stimulus outside the receptive field. In other words, spatial attention gated the response to the cell's preferred stimulus (see also Luck et al., 1997; Reynolds et al., 1999).

In animals, there has been a long history of interest in the role of subcortical "gating" in selective attention (e.g., Hernandez-Peon et al., 1961), and subcortical mechanisms have been described that could, in principle, provide a basis for attentional gating at the level of the LGN (e.g., Sherman and Guillery, 2002; Sherman, 2004; Yingling and Skinner, 1976). Given models of attention involving hypotheses about the role of key subcortical mechanisms in gating information flow to cortex, animal studies have revealed surprisingly little evidence for a role in LGN in selective attention. Recently, however, Vanduffel and colleagues (2000) used deoxyglucose mapping in macaques and found evidence for attention-related changes in metabolic activity in the magno-cellular layers of LGN.

1.3 Attentional Control Networks

Research in animals, patients with neurological dysfunction, and healthy human subjects using ERPs and functional neuroimaging suggests that in addition to changes in perceptual processing regions of cortex, visual spatial attention involves activity in a complex network of widely distributed brain areas, including dorsolateral-prefrontal cortex, anterior cingulate cortex, posterior parietal cortex, and thalamic and midbrain structures (e.g., Bushnell et al., 1981; Goldberg and Bruce, 1985; Harter et al., 1989; Knight et al., 1995; LaBerge, 1997; Mesulam, 1981; Miller, 2000). For example, the clinical syndrome of spatial neglect or hemi-inattention that results from lesions in association cortex, in particular the posterior parietal and posterior-superior temporal cortex, suggests that networks outside the sensory and motor cortex are involved in the ability to focus attention (e.g., Karnath et al., 2004; Mesulam, 1981; for a recent empirical report that contains a cogent review, see Rorden et al., 2005). Such observations, combined with studies of sensory attentional selection, have resulted in models of attention that have distinguished between attentional control processes and the influence of these processes at a site of action in the visual perceptual system (e.g., Harter et al., 1989; Posner and Petersen, 1990). It is essential to lay out the theoretical distinction between attentional control and attentional selection in greater detail so that the operation of the attentional system from control to selection can be discussed further. Although earlier in the Introduction we described a general view of attentional control and selection, it is useful to now describe how attentional control circuitry might be distinguished from the resultant attention selection: In some circumstances, this is not straightforward and requires definition.

Most models of attention consider the concepts of attentional control and attentional selection (the result of control). Posner and Petersen (1990) referred to the "source" of attentional control signals and the "site" of action of such signals, the sources being the cortical attentional control network and the sites being the loci within the sensory-perceptual system at which attentional selection was manifest as changes in sensory stimulus processing during focused attention. Recently, Serences and Yantis (2005) used similar terminology to distinguish between "sources" and "targets" of attention, referring to the same distinction as did Posner and Petersen (1990), although Serences and Yantis went on to argue for a weaker distinction. If we consider only voluntary processes, we can impose a temporal characteristic on these processes of control and selection such that attentional control activity must necessarily precede attentional selection as the former is by definition required to generate the latter. This provides one simple way to distinguish between control and selection networks which roughly maps onto sources and sites of attention: Attentional control (source of attention) is neural activity related in time to focusing attention based on internal goals or external cues that are related to internal goals, whereas activity related in time to stimulus processing and its attentional modulation clearly reflects attentional selection processes (site of attention). This assumes a model whereby attention acts prior to inputs and is not dependent on the creation of a sensory trace in order to be engaged. This has been referred to as preparatory attention (e.g., Corbetta et al., 2005).

Event-related fMRI has recently been used to identify superior frontal, inferior parietal and superior temporal regions as being selectively activated during attentional control processing (e.g., Corbetta et al., 2000; Hopfinger et al., 2000; Serences et al., 2005; Wilson et al., 2005; Woldorff et al., 2004). Current models hypothesize that these networks reflect top-down attentional influences that result in changes in excitability in multiple visual cortical areas in order to achieve selective sensory processing of relevant visual targets (reviewed in Kanwisher and Wojciulik, 2000; Kastner and Ungerleider, 2000; Serences and Yantis, 2005).

Our group has conducted several studies of attentional control circuitry in spatial and non-spatial attention using cuing paradigms that permit the separation of control from selection processes. Across these studies, we have consistently observed activity in the attentional control network involving superior frontal and parietal cortex when subjects prepare to selectively process upcoming stimuli based on their location (Giesbrecht et al., 2003; Hopfinger et al., 2000; Wilson et al., 2005), their color (Giesbrecht et al., 2003) or their global or local properties (Weissman et al., 2002a, 2002b, 2003). In general, this pattern of consistent activation in the frontoparietal network is consistent with the view of some that this attentional control network is task and modality non-specific (e.g., Wojciulik and Kanwisher, 1999; Serences and Yantis, 2005; see also Corbetta et al., 2005). However, in direct comparisons of spatial and non-spatial (feature) attention, we have provided evidence for specialization in the frontoparietal network for the control of spatial attention (Giesbrecht et al., 2003), which is reviewed in the next section.

2. SPECIALIZATIONS IN TOP-DOWN ATTENTIONAL CONTROL

2.1 Top-down Control During Spatial and Feature Attention

Most studies investigating attentional control networks have not addressed the question of specializations in attentional control, and those that have not typically included designs and conditions that permit strong statistical tests of differences in the pattern of neural activation for different forms of attentional control. In Giesbrecht et al. (2003) we conducted a direct test of whether the same cortical network supports top-down control of both spatial and non-spatial attention. We identified specializations in the dorsal frontoparietal attention network for spatial attention.

The study involved randomly intermingled trials in which either the location or the color of an upcoming target was cued. Subjects were told that the cues were instructions telling them what stimulus to prepare for such that when a stimulus was presented, the correct discrimination could be made. The cues themselves were letters located at fixation (in one condition) or above fixation in the periphery (second condition) (e.g., L = attend left, and B = attend blue). The target stimuli were outline rectangles that were either located in the left or right hemifields (both conditions) or overlapping outline rectangles in ten healthy right-handed subjects participated in the study. The task required them to prepare attention to select an upcoming stimulus based on the relevant feature (location or color) and to indicate its orientation. Event-related fMRI measures were obtained to the cues when followed by their respective targets at random intervals (between 1000 ms and 8000 ms). Gradient-echo, echo-planar images with a TR = 2 sec, and slice thickness of 5 mm were taken in the axial plane using a 1.5T GE scanner. SPM99 was used to perform a group analysis of the cue-related activity over the ten subjects.

2.2 Results and Discussion

Regions of the frontoparietal network were commonly activated by the spatial and non-spatial cues in this task (not shown). This similarity in the patterns of activity likely reflects those aspects of the task that the two orienting conditions had in common. This should include low-level sensory processing of the cues, extraction of the linguistic information represented in the cue letter, mapping that information to the relevant task instruction (i.e., letter "L" means attend left visual field, etc.), executing the task instruction (i.e., selectively preparing attention for processing location or color), maintaining relevant information during the cue-target interval, and preparing to respond. The notion that portions of the frontoparietal network generalize over multiple dimensions is consistent with recent models of top-down attentional control (Corbetta and Shulman, 2002; Shulman et al., 2002). However, overlap in common mental operations across two complex tasks that involve non-specific mental resources (e.g., working memory) does not address the issue of how the brain directs attention to spatial versus non-spatial stimulus features. To test whether any of the areas activated in response to the cues were selective for spatial or non-spatial orienting, it is necessary to directly quantitatively compare activity in response to location versus color cues.

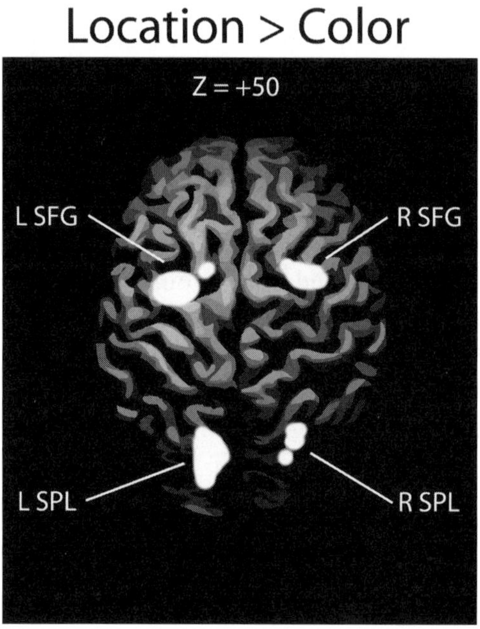

Figure 21-1. Results of the direct comparison between location and color cue conditions overlaid on an axial slice. Greater activity to location cues than color cues is seen in the superior frontal gyrus (SFG) and the superior parietal lobule (SPL) bilaterally.

In Giesbrecht et al., because the two conditions were designed to be as similar as possible except for the attended feature (i.e., location or color), the direct comparison of location and color cue conditions should reveal those brain areas that are selectively engaged (i.e., specialized) for controlling attention to those features.

The results of the direct comparison of neural activity related to location versus color attention cues are shown in Figure 21-1. Two statistical contrasts were conducted in the population maps. The brain regions more active for location versus color cues could be compared to the contrast for brain regions where color cues produced more activity than location cues. Figure 21-1 shows the former contrast, revealing non-overlapping regions of frontal and parietal cortex that were activated during orienting to location versus color. Orienting attention based on stimulus location activated regions of the dorsal frontal cortex, posterior parietal cortex, and supplementary motor cortex. For the complementary contrast, regions of ventral occipital cortex were more active to color than to location cues. No regions of the frontoparietal control network were more active for non-spatial (color) cues in comparison to spatial cues. This pattern of selectivity in the frontoparietal network is consistent with the idea that neural specializations exist for the control of orienting attention to locations. The regions of superior frontal and parietal cortex more active for spatial than non-spatial attention may reflect neural circuitry involved in mapping spatial coordinates for covert and overt orienting (e.g., Corbetta, 1998). In contrast, the pattern of color attention selectivity in ventral visual cortex likely reflects the enhancement of color-specific areas in preparation for the target display (Chawla et al., 1999), and therefore we argue that these do not necessarily reflect attentional control regions. To put it another way: Non-spatial attention (at least for color selective attention) is engaged by control circuitry that is shared with modality non-specific systems in the frontoparietal attention network, whereas spatial attention includes additional, specialized circuitry in dorsal frontal and parietal regions.

2.3 Top-down Control During Motion and Color Attention

There is substantial evidence that the visual system represents spatial information and perceptual information about objects in distinct pathways. The 'dorsal stream' pathway carries information about the movement and locations of objects from primary visual cortex to posterior parietal regions and plays a critical role in guiding actions, whereas the 'ventral stream' projects from striate cortex to inferotemporal cortex and plays a key role in the perceptual identification of objects (e.g., Ungerleider and Mishkin, 1982; Goodale and Milner, 1992). The study by Giesbrecht et al. (2003) that we described in the preceding section found that subregions of the frontoparietal network appear to be specialized for controlling attention to spatial locations. However, it is possible that these areas are recruited for directing attention not just to spatial locations *per se*, but rather to dorsal-stream features more generally. Indeed, the superior parietal lobule (SPL), which was selectively active in the location condition, is a key region within the dorsal pathway that contains a variety of visual areas involved in visual-spatial and visual-motor processes (Goodale and Milner, 1992). We tested this alternative hypothesis that specializations in superior frontal and parietal cortex for preparatory spatial selective attention also respond to other dorsal-stream features by directly comparing two different forms of preparatory attention. In this study we compared preparatory attention for stimulus motion, a dorsal stream process, with the same non-spatial ventral stream feature we used previously (color). One can question the extent to which motion processing is wholly distinct from spatial processing, but as a first approximation preparing to attend for the motion of a foveal stimulus is very different from orienting attention to a peripheral field location while ignoring other locations.

The paradigm was as follows. At the start of each trial, an auditory word cue (presented via headphones) instructed subjects to attend for a target within a given stream (i.e., color or motion) or to attend for a specific feature within one of these streams (e.g., red or leftward motion). The specific motion cues were the words ‘left’, ‘right’, ‘up’ or ‘down’ and the specific color cues were ‘orange’, ‘blue’, ‘green’, and ‘purple’. Participants were instructed to use the cues to prepare to detect faint stimuli in the cued feature. If cued to a color, then they were to prepare for and detect brief color flashes within a display of randomly moving dots presented during a subsequent test period. If cued to motion, then they were to prepare for and detect the brief coherent motion stimulus in the display of randomly moving dots. The test periods began 2500 ms after the cue onset. In each run, 44% of trials were cue plus test trials, 22% were cue-only trials (where no test stimulus appeared – see below), and 33% were “null trials”. Null trials consisted only of the unchanging fixation cross and in effect created a random inter-trial interval permitting deconvolution of hemodynamic responses to adjacent trials despite overlapping responses (e.g., Burock et al., 1998; Woldorff et al., 2004). Cue-only trials consisted of a cue followed 2500 ms later by a 1000 ms dimming of the fixation cross. This dimming occurred at the point in time at which the test stimulus would have occurred on a cue plus test trial and indicated to the subjects that they should cease preparing for a target and simply wait for the next cue. Cue-only trials were included to permit separation of the responses to cue and target events (Ollinger et al., 2001; Shulman et al., 1999; Woldorff et al., 2004). Of the cue plus test trials, 11% were catch trials in which the randomly moving dots were presented but no target appeared within the attended dimension. The remainder of cue plus test trials included a target. For trials in which a specific feature was cued, the target that appeared in the relevant stream (if it appeared) was always of the cued feature. These specific feature cues were most similar to those used in the location and color study of Giesbrecht et al. (2003) and the neural responses to these will therefore be the focus of our discussion.

Functional images were acquired from 16 right-handed subjects with a 1.5T whole-body neuro-optimized GE Signa Horizon LX NV/I CV/I MRI system scanner using a BOLD sensitive gradient-echo EPI sequence. Two hundred sixty MR frames were acquired per run, each frame consisting of 18 contiguous 6 mm axial slices (FOV = 220 mm, 64 x 64 matrix). Coplanar T2-weighted spin-echo and high-resolution T1-weighted SPGR images were also acquired. Each subject completed a training session within the two weeks prior to the imaging session.

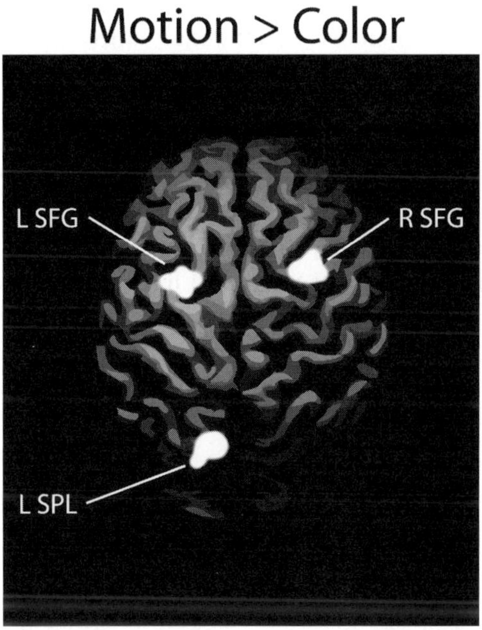

Figure 21-2. Results of the direct comparison between motion and color cue conditions overlaid on an axial slice. Greater activity to motion cues than color cues is seen in the superior frontal gyrus (SFG) bilaterally, but the superior parietal lobule (SPL) only on the left.

During this session the percentage of coherently moving target dots and the saturation of target colors were varied between runs to establish levels at which each subject's performance stabilized between 65% and 75% correct. During the imaging session, subjects completed 8 to 10 runs. The motion coherence and color saturation levels determined during the training session were used as the initial settings during the imaging session, but these levels were adjusted as needed between runs to maintain a stable level of performance.

2.4 Results and Discussion

As in the Giesbrecht et al. study, we observed a broad network of both cortical and subcortical brain areas that were active for both motion and color cues. This network closely matched that reported in previous studies of the attentional control network (e.g., Corbetta et al., 2000; Giesbrecht et al., 2003; Hopfinger et al., 2000; Kastner and Ungerleider, 2000; Kincade et al., 2005; Wilson et al., 2005). Significant overlap could be observed in the activity related to motion and color cues, suggesting that orienting attention to motion cues and color cues in the context of this cuing task involve overlapping mental operations engaged for task performance (e.g., sensory processing, working memory, attentional orienting, etc.). But as in Giesbrecht and colleagues' comparison of preparatory attention for location versus color processing, the question is whether some subregions of these widespread brain networks for attentional control may be specialized for motion versus color processing, in accord with the predominant dorsal- versus ventral-stream organization of these visual attributes. If spatial selective attention is special, engaging the regions shown in Figure 21-1 only when location is selectively attended, then no regions should be more active for motion versus color attention in direct statistical contrasts. However, if the activations in Figure 21-1 that were greater for spatial attention cues than color attention cues reflect differences between dorsal- and ventral-stream attention processes, then one would expect increased activity in dorsal-stream areas for motion versus color cues.

The result of the direct statistical contrast for motion versus color cues supported the idea that specializations in dorsal stream areas may be related to preparatory attention for dorsal-stream processes versus ventral-stream processes. Subregions of the attentional control network were more active for motion than for color, and vice versa. Subregions of the left and right SFG and the left SPL were selectively active to motion cues (see Fig. 21-2, Table 21-1). Importantly, these activations closely match 3 of the areas reported by Giesbrecht et al. to be selectively active for attention to locations. This suggests that these areas may control attention to dorsal-stream features generally.

Table 21-1. Activations for motion greater than color

Region	x	y	z	Volume (mm^3)	Max T value	Mean P value
Left SFG	-22	-11	60	1969	4.665167	0.003413
Right SFG	29	-6	58	1151	4.537501	0.003265
Left SPL	-17	-73	53	910	3.537341	0.005789

Centers of gravity for each region followed by the number of voxels in that region and the maximum T and P values.

However, unlike location cues, motion cues did not selectively activate the right SPL. Even analysis of an ROI placed at the Talairach coordinates reported for this region failed to reveal significant differences between the cue conditions (see Figure 21-3). Because it was selectively active for location cues, but not for motion cues, it is possible that this region may indeed be specialized for directing attention to locations. If true, this finding might shed light on a discrepancy between the neuropsychological literature of attention and functional imaging studies.

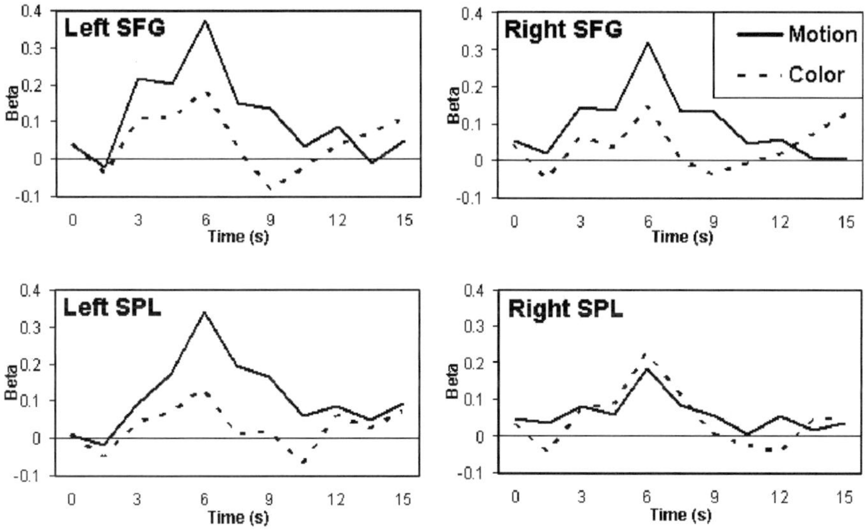

Figure 21-3. Hemodynamic responses to motion and color cues in four regions of interest within the frontoparietal network. The left SFG, right SFG, and left SPL regions are those more active in response to motion than color cues shown in Fig. 21-2. The right SPL region was located at the coordinates for that region reported by Giesbrecht et al. (2003), and confirms that there was no difference between motion and color cues at that site.

Spatial neglect is commonly observed in patients with damage to the right parietal lobe; however, neuroimaging studies typically find bilateral parietal activity during spatial attention. It may be that both parietal lobes are active during spatial attention simply because it is a dorsal stream feature, but that only the right side is critical for attending to locations.

A compelling argument has been made that the dorsal stream represents visual information required for generating actions (e.g., Cohen and Andersen, 2002; Goodale and Milner, 1992; Goodale and Westwood, 2004), rather than representing location *per se*, as originally conceived (Ungerleider and Mishkin, 1982). Following this reasoning, it may be no coincidence that the areas specific to dorsal stream attentional control have also been implicated in the control of voluntary eye movements (e.g., Corbetta et al., 1998). These areas may represent planned eye movements corresponding to the location or direction of motion that is attended (Rizzolatti et al., 1987; Shepherd et al., 1986; Chelazzi et al., 1995; Hoffman and Subramaniam, 1995). Indeed, several reports highlight the tight correspondence between frontal and parietal areas that control eye movements and those that control spatial attention, as assessed using fMRI (Corbetta et al., 1998, 2005; Moore and Fallah, 2004; Nobre et al., 2000). Given this, it is important to ask how attention to motion interacts with action systems such as those involved in eye movement generation, and therefore whether the present specialization for motion more than color attention may reflect processes closely related to those engaged during spatial attention that are involved in planning eye movements. Relevant to this question is the field of research on how anticipation of the direction of motion of an upcoming stimulus affects one's ability to subsequently make smooth-pursuit eye movements when a target appears.

Behavioral work on the role of cue information on the ability to make anticipatory pursuit eye movements has shown that top-down influences engaged by cues that predict the direction of motion of an upcoming target do affect the latency and amplitude of pursuit eye movements to motion stimuli (e.g., Chakraborti et al., 2002; de Hemptinne et al., 2006). Hence, one possible explanation for the close correspondence of the present findings and our past work (Giesbrecht et al., 2003) is that superior frontal and parietal activity present for location and motion selective attention more than for non-spatial (color) attention is related via the role of spatial attention in preparing actions, specifically those involved in oculomotor output. The present studies do not address this directly but raise the possibility that specializations in superior frontal and parietal cortex for preparatory spatial attention and preparatory motion attention may be linked via the role of the dorsal system for preparing actions. Such an idea predicts that preparatory attention for actions such as reaching should also selectively activate this network, and indeed, evidence in favor of this idea exists. Astafiev and colleagues (2003) investigated brain activations in cued covert attention, overt saccade and pointing tasks. They found overlapping regions in the superior frontal and parietal cortex that were activated in all three tasks. Such a pattern strongly suggests a dorsal-stream attentional process involved in programming actions and that generalizes across effectors.

3. CONCLUSIONS

The ability to selectively attend and act is dependent on highly evolved brain mechanisms that impart essential survival advantages in a complex and unforgiving world. The neural networks supporting these mechanisms are slowly revealing themselves as evidence from behavioral, neurological, and neurophysiological approaches converges to relate specific behaviors with specific brain structures. The brain's attention system is complex and has been described as consisting of widespread neural networks. Nonetheless, these widespread neural networks for attention are not wholly non-specific, but rather support a variety of attentional phenomena whose complementary functions are delicately balanced to permit the moment-to-moment choreography of our awareness and intention. In this chapter we have described the cortical system for voluntary covert visual attention, relaying how attentional control and attentional selection in sensory processing interact. We investigated the question of whether some forms of attention may be supported by specialized systems that form sub-networks in the cortical attentional control system. Our studies have revealed a highly specialized network of regions in the superior frontal cortex and superior parietal cortex that are more active during the preparatory period of spatial attention than for non-spatial attention. We asked whether this specialized dorsal network was selective for spatial attention, that is, location-selective attention, or whether it might rather reflect attentional processes related to the dorsal visual-action processing stream. Our studies show that direct statistical contrasts between activation maps for preparatory attention for detection of motion stimuli versus non-spatial color stimuli activate virtually the same regions of the dorsal frontal and parietal cortex. One possibility is that that this close correspondence between the specialized networks for spatial (location) and motion selective attention may reflect the activity of a common attentional system in the dorsal cortex that is involved in attention for action. The present data cannot distinguish this from an attention system that acts in the service of impending action via a mapping of spatial attention, and further research will be required to distinguish these related but different attentional models. Nonetheless, the present work reveals specializations for the control of visual attention that support visual spatial processing and visual motion processing, two characteristics of dorsal-stream neural function.

ACKNOWLEDGEMENTS

The authors are grateful to Barry Giesbrecht, Marty Woldorff, Allan Song, Michael Buonocore and Daniel Weissman for their contributions to portions of the work reviewed here, and to Santani Teng for assistance with preparation of this manuscript. Supported by NIMH grants MH55714 and MH02019 to G.R.M. and a NSF Graduate Fellowship to S.P.F..

REFERENCES

Astafiev S. V., Shulman, G. L., Stanley, C. M., Snyder, A. Z., Van Essen, D. C. and Corbetta, M., 2003, Functional organization of human intraparietal and frontal cortex for attending, looking, and pointing, *J Neurosci* **23**(11):4689-99.

Beck, D. M. and Kastner, S., 2005, Stimulus context modulates competition in human extrastriate cortex, *Nat Neurosci* **8**(8):1110-6.

Broadbent, D. E., 1962, *Attention and the Perception of Speech*, W.H. Freeman and Co., New York.

Burock, M. A., Buckner, R. L., Woldorff, M. G., Rosen, B. R., and Dale, A. M., 1998, Randomized event-related experimental designs allow for extremely rapid presentation rates using functional MRI, *Neuroreport* **9**:3735-3739.

Bushnell, M. C., Goldberg, M. E., and Robinson, D. L., 1981, Behavioral enhancement of visual responses in monkey cerebral cortex. I. Modulation in posterior parietal cortex related to selective visual attention, *J Neurophysiol* **46**:755-772.

Carrasco, M., Ling, S., and Read, S., 2004, Attention alters appearance, *Nat Neurosci* **7**: 308–313.

Chakraborti, S. R., Barnes, G. R., and Collins, C. J., 2002, Factors affecting the longevity of a short term velocity store for predictive oculomotor tracking, *Exp Brain Res* **144**:152-8.

Chawla, D., Rees, G. and Friston, K. J., 1999, The physiological basis of attentional modulation in extrastriate visual areas, *Nat Neurosci* **2**(7):671-6.

Chelazzi, L., Miller, E. K., Duncan, J., and Desimone, R., 1993, A neural basis for visual search in inferior temporal cortex, *Nature* **363**:345-347.

Chelazzi, L., Miller, E. K., Duncan, J., and Desimone, R., 2001, Responses of neurons in macaque area V4 during memory-guided visual search *Cereb Cortex* **11**:345-347.

Chelazzi, L., Biscaldi, M., Corbetta, M., Peru, A., Tassinari, G. and Berlucchi, G., 1995, Oculomotor activity and visual spatial attention, *Behav Brain Res* **71**(1-2):81-8.

Cohen, Y. E. and Andersen, R. A., 2002, A common reference frame for movement plans in the posterior parietal cortex, *Nature Rev Neurosci* **3**:553-562.

Corbetta, M., 1998, Frontoparietal cortical networks for directing attention and the eye to visual locations: Identical, independent, or overlapping neural systems?, *Proc Natl Acad Sci USA* **95**:831-838.

Corbetta, M., Akbudak, E., Conturo, T. E., Snyder, A. Z., Ollinger, J. M., Drury, H. A., Linenweber, M. R., Petersen, S. E., Raichle, M. E., Van Essen, D. C. and Shulman, G. L., 1998, A common network of functional areas for attention and eye movements, *Neuron* **21**(4):761-773.

Corbetta, M., Kincade, J. M., Ollinger, J. M., McAvoy, M. P., and Shulman, G. L., 2000, Voluntary orienting is dissociated from target detection in human posterior parietal cortex, *Nat Neurosci* **3**:292-297.

Corbetta, M. and Shulman, G., 2002, Control of goal-directed and stimulus-driven attention in the brain, *Nat Rev Neurosci* **3**(3):201-15.

Corbetta, M., Tansy, A. P., Stanley, C. M., Astafiev, S. V., Snyder, A. Z. and Shulman, G. L., 2005, A functional MRI study of preparatory signals for spatial location and objects, *Neuropsychologia* **43**(14):2041-56.

Cracco, R. Q., and Cracco, J. B., 1978, Visual evoked potential in man: early oscillatory potentials, *Electroencephalogr Clin Neurophysiol* **45**(6):731-9.

de Hemptinne, C., Lefevre, P. and Missal, M., 2006, Influence of cognitive expectation on the initiation of anticipatory and visual pursuit eye movements in the rhesus monkey, *J Neurophysiol*, in press.

Desimone, R., and Duncan, J., 1995, Neural mechanisms of selective visual attention, *Ann Rev Neurosci* **18**:193-222.

Deutsch, J. A. and Deutsch, D., 1963, Attention: Some theoretical considerations, *Psychological Rev* **70**(1):51-60.

Downing, C. J., 1988, Expectancy and visual-spatial attention: Effects on perceptual quality, *J Exp Psychol Human* **14**:188-202.

Eason, R. G., 1981, Visual evoked potential correlates of early neural filtering during selective attention, *Bull Psychon Soc* **18**:203-206.

Eason, R. G., Harter, M. R. and White, C. T., 1969, Effects of attention and arousal on visually evoked cortical potentials and reaction time in man, *Physiol Behav* **4**(3):283-289.

Giesbrecht, B., Woldorff, M. G., Song, A. W., and Mangun, G. R., 2003, Neural mechanisms of top-down control during spatial and feature attention, *Neuroimage* **19**:496-512.

Gobell, J. and Carrasco, M., 2005, Attention alters the appearance of spatial frequency and gap size, *Psychol Sci* **16**(8):644-51.

Goldberg, M. E., and Bruce, C. J., 1985, Cerebral cortical activity associated with the orientation of visual attention in the rhesus monkey, *Vision Res* **25**:471-481.

Goodale, M. and Milner, A., 1992, Separate visual pathways for perception and action, *Trends Neurosci* **15**(1):20-25.

Goodale, M. and Westwood, D., 2004, An evolving view of duplex vision: separate but interacting cortical pathways for perception and action, *Curr Opin Neurobiol* **14**(2): 203-211.

Hackley, S. A., Woldorff, M. and Hillyard, S. A., 1990, Cross-modal selective attention effects on retinal, myogenic, brainstem, and cerebral evoked potentials, *Psychophysiology* **27**(2):195-208.

Handy, T., Kingstone, A. and Mangun, G. R., 1996, Spatial distribution of visual attention: Perceptual sensitivity and response latency, *Percept Psychophys* **58**:613-627.

Harding, G. F. and Rubinstein, M. P., 1980, The scalp topography of the human visually evoked subcortical potential, *Invest Ophthalmol Vis Sci* **19**(3):318-21.

Harter, M. R., Miller, S. L., Price, N. J., LaLonde, M. E., and Keyes, A. L., 1989, Neural processes involved in directing attention, *J Cognitive Neurosci* **1**:223-237.

Heinze, H. J., and Mangun, G. R., 1995, Electrophysiological signs of sustained and transient attention to spatial locations, *Neuropsychologia* **33**:889-908.

Heinze, H. J., Mangun, G. R., Burchert, W., Hinrichs, H., Scholz, M., Münte, T. F., Gös, A., Johannes, S., Scherg, M., Hundeshagen, H., Gazzaniga, M. S., and Hillyard, S. A., 1994, Combined spatial and temporal imaging of spatial selective attention in humans, *Nature* **392**:543-546.

Helmholtz, H. von, 1962/1924-5, Treatise on Physiological Optics, translated from the 3d German ed., edited by James P. C. Southall, Dover Publications, New York.

Hernandez-Peon, R., Brust-Carmona, H., Penaloza-Rojas, J. and Bach-Y-Rita, G., 1961, The efferent control of afferent signals entering the central nervous system, *Ann N Y Acad Sci* **89**:866-82.

Hillyard, S. A., Hink, R. F., Schwent, V. L. and Picton, T. W., 1973, Electrical signs of selective attention in the human brain, *Science* **182**(108):177-80.

Hoffman, J. E. and Subramaniam, B., 1995, The role of visual attention in saccadic eye movements, *Percept Psychophys* **57**(6):787-795.

Hopfinger, J. B. and Mangun, G. R., 1998, Reflexive attention modulates processing of visual stimuli in human extrastriate cortex, *Psychol Sci* **9**:441-447.

Hopfinger, J. B., Buonocore, M. H. and Mangun, G. R., 2000, The neural mechanisms of top-down attentional control, *Nat Neurosci* **3**:284-291.

James, W., 1981/1890, *The Principles of Psychology*, Harvard University Press, Cambridge, MA.

Kanwisher, N. and Wojciulik, E., 2000, Visual Attention: Insights from Brain Imaging, *Nat Rev Neurosci* **1**:91-100.

Karnath, H., Berger, M. F., Küker, W. and Rorden, C., 2004, The anatomy of spatial neglect based on voxelwise statistical analysis: A study of 140 patients, *Cereb Cortex* **14**: 1164-1172.

Kastner, S. and Ungerleider, L., 2000, Mechanisms of visual attention in the human cortex, *Ann Rev Neurosci* **23**:315–341.

Khoe, W., Mitchell, J. F., Reynolds, J. H. and Hillyard, S. A., 2005, Exogenous attentional selection of transparent superimposed surfaces modulates early event-related potentials, *Vision Res* **45**(24):3004-14.

Kincade, J. M., Abrams, R. A., Astafiev, S. V., Shulman, G. L. and Corbetta, M., 2005, An event-related functional magnetic resonance imaging study of voluntary and stimulus-driven orienting of attention, *J Neurosci* **25**(18):4593-604.

Klein, R. M., 2000, Inhibition of return, *Trends Cogn Sci* **4**:138-147.

Knight, R. T., Grabowecky, M. F. and Scabini, D., 1995, Role of human prefrontal cortex in attention control, *Adv Neurol* **66**:21-34.

Krolak-Salmon, P., Henaff, M. A., Vighetto, A., Bertrand, O. and Mauguiere, F., 2004, Early amygdala reaction to fear spreading in occipital, temporal, and frontal cortex: a depth electrode ERP study in human, *Neuron* **42**(4):665-76.

LaBerge, D., 1997, Attention, awareness, and the triangular circuit, *Conscious Cogn* **6**: 149-181.

Luck, S. J., Hillyard, S. A., Mouloua, M., Woldorff, M. G., Clark, V. P. and Hawkins, H. L., 1994, Effects of spatial cuing on luminance detectability: psychophysical and electrophysiological evidence for early selection, *J Exp Psychol Human* **20**:887–904.

Luck, S. J., Chelazzi, L. Hillyard, S. A. and Desimone, R., 1997, Neural mechanisms of spatial selective attention in areas V1, V2 and V4 of macaque visual cortex, *J Neurophysiol* **77**:24-42.

Luck S. J., Woodman, G. F. and Vogel, E. K., 2000, Event-related potential studies of attention, *Trends Cogn Sci* **4**(11):432-440.

Mangun, G. R., 1995, Neural mechanisms of visual selective attention in humans, *Psychophysiology* **32**:4-18.

Marr, D., 1982, *Vision*, Freeman, New York.

Martinez, A., DiRusso, F., Anllo-Vento, L., Sereno, M. I., Buxton, R. B., and Hillyard, S. A., 2001. Putting spatial attention on the map: Timing and localization of stimulus selection processes in striate and extrastriate visual areas, *Vision Res* **41**:1437–1457.

Martinez, A., Anllo-Vento, L., Sereno, M. I., Frank, L. R., Buxton, R. B., Dubowitz, D. J., Wong, E. C., Heinze, H. J. and Hillyard, S. A.k, 1999, Involvement of striate and extrastriate visual cortical areas in spatial attention, *Nat Neurosci* **2**:364-369.

Maunsell, J. H., Cook, E. P., 2002, The role of attention in visual processing. *Philos Trans R Soc Lond B Biol Sci* **357**:1063-1072.

McAdams, C. J. and Maunsell, J. H., 1999, Effects of attention on the reliability of individual neurons in monkey visual cortex, *Neuron* **23**(4):765-73.

McAdams, C. and Reid, C., 2005, Attention modulates the responses of simple cells in monkey primary visual cortex, *J Neurosci* **25**(47):11023-33.

Mesulam, M. -M., 1981, A cortical network for directed attention and unilateral neglect, *Ann Neurology* **10**, 309-325.

Miller, E. K., 2000, The neural basis of top-down control of visual attention in the prefrontal cortex, in: Attention and Performance XVIII: *The Control Over Cognitive Processes* J. Driver, and S. Monsell, eds., MIT Press: Cambridge, Mass, pp. 511-534.

Moran, J. and Desimone, R., 1985, Selective attention gates visual processing in the extrastriate cortex, *Science* **229**:782-784.

Moore, T. and Fallah, M., 2004, Microstimulation of the frontal eye field and its effects on covert spatial attention, *J Neurophysiol* **91**:152-162.

Näätänen, R., 1975, Selective attention and evoked potentials in humans--a critical review, *Biol Psychol* **2**(4):237-307.

Nobre, A. C., Sebestyen, G. N. and Miniussi, C., 2000, The dynamics of shifting visuospatial attention revealed by event-related potentials, *Neuropsychologia* **38**(7):964-974

Noesselt, T., Hillyard, S. A., Woldorff, M. G., Shoenfeld, A., Hagner, T., Jancke, L., Tempelmann, C., Hinrichs, H., and Heinze., H- J., 2002, Delayed striate cortical activation during spatial attention, *Neuron* **35**:575-587.

Oakley, M. T., and Eason, R. G., 1990, The conjoint influence of spatial selective attention and motor set on very short latency VERs, *Neuropsychologia* **28**(5):487-97.

O'Connor, D. H., Fukui, M. M., Pinsk, M. A. and Kastner, S., 2002, Attention modulates responses in the human lateral geniculate nucleus, *Nat Neurosci* **5**:1203-1209.

Ollinger, J. M., Shulman, G. L. and Corbetta, M., 2001, Separating processes within a trial in event-related functional MRI, *Neuroimage* **13**(1):210-217.

Pashler, H., 1998, *The psychology of attention*, MIT Press (A Bradford Book), Cambridge, MASS.

Posner, M. I., 1978, *Chronometric Explorations of Mind*, Lawrence Erlbaum, Oxford, U.K..

Posner, M. I and Cohen, Y., 1984, Components of visual orienting, in: *Attention and Performance X*, H. Bouma and D. Bouwhis, eds., Erlbaum Associates, Hillsdale, N. J., pp. 531-556.

Posner, M. I., and Petersen, S. E., 1990, The attention system of the human brain, *Ann Rev Neurosci* **13**:25-42.

Pratt, H., Bleich, N. and Berliner, E., 1982, Short latency visual evoked potentials in man, *Electroencephalogr Clin Neurophysiol* **54**(1):55-62.

Recanzone, G. H., Wurtz, R. H. and Schwarz, U., 1993, Attentional modulation of neuronal responses in MT and MST of a macaque monkey performing a visual discrimination task, *Society for Neuroscience Abstract* **19**:973.

Reynolds, J. H., Chelazzi, L., and Desimone, R., 1999, Competitive mechanisms subserve attention in macaque areas V2 and V4, *J Neurosci* **19**:1736–1753.

Reynolds, J. H. and Chelazzi, L., 2004, Attentional modulation of visual processing, *Ann Rev Neurosci* **27**:611-647.

Rizzolatti, G., Riggio, L., Dascola, I. and Umilta, C., 1987, Reorienting attention across the horizontal and vertical meridians: evidence in favor of a premotor theory of attention, *Neuropsychologia* **25**(1A):31-40.

Robinson, D. L. and Petersen, S. E., 1992, The pulvinar and visual salience, *Trends Neurosci* **15**(4):127-32.

Rorden, C., Fruhmann-Berger, M. And Karnath, H. O., 2006, Disturbed line bisection is associated with posterior brain lesions, *Brain Res* **1080**(1):17-25.

Schroeder, C., Tenke, C. and Givre, S., 1992, Subcortical contributions to the surface-recorded flash-VEP in the awake macaque, *Electroencephalogr Clin Neurophysiol* **84**: 219-231.

Serences, J. T. and Yantis, S., 2006, Selective visual attention and perceptual coherence, *Trends Cogn Sci* **10**(1):38-45.

Serences, J. T., Shomstein, S., Leber, A. B., Golay, X., Egeth, H. E. and Yantis S., 2005, Coordination of voluntary and stimulus-driven attentional control in human cortex, *Psychol Sci* **16**(2):114-22.

Shepherd, M., Findlay, J. M., and Hockey, R. J., 1986, The relationship between eye movements and spatial attention, *QJ Exp Psychol A* **38**(3-A):475-491.

Sherman, S. M., 2004, Interneurons and triadic circuitry of the thalamus, *Trends Neurosci* **27**(11):670-5.

Sherman, S. M., and Guillery, R., 2002, The role of the thalamus in the flow of information to the cortex, *Philos Trans R Soc Lond B Biol Sci* **357**(1428):1695-708.

Shulman, G. L., d'Avossa, G., Tansy, A. P., and Corbetta, M., 2002, Two attentional processes in the parietal lobe, *Cereb Cortex* **12**(11):1124-1131.

Shulman, G. L., Ollinger, J. M., Akbudak, E., Conturo, T. E., Snyder, A. Z., Petersen, S. E., and Corbetta, M., 1999, Areas involved in encoding and applying directional expectations to moving objects, *J Neurosci* **19**(21):9480-96.

Spitzer, H. and Richmond, B. J., 1991, Task difficulty: ignoring, attending to, and discriminating a visual stimulus yield progressively more activity in inferior temporal neurons, *Exp Brain Res* **83**:340-8.

Treisman, A. M., 1964, Selective attention in man, *Br Med Bull* **20**:12-6.

Treisman, A., 1969, Strategies and models of selective attention, *Psychological Rev* **76**: 282- 299.

Treue, S., and Martínez Trujillo, J. C., 1999, Feature-based attention influences motion processing gain in macaque visual cortex, *Nature* **399**:575-579.

Treue, S., 2001, Neural correlates of attention in primate visual cortex, *Trends Neurosci* **24**:295-300.

Treue, S., 2003, Visual attention: the where, what, how and why of saliency, *Curr Opin Neurobiol* **13**:428-432.

Ungerleider, L. G. and Mishkin, M., 1982, Two cortical visual systems, in: *Analysis of Visual Behavior*, D. J. Ingle, M. A. Goodale, and R. J. W. Mansfield, eds., MIT Press, Cambridge, pp. 549-586.

Vanduffel, W., Tootell, R. B. H., and Orban, G. A., 2000, Attention-dependent suppression of metabolic activity in the early stages of the macaque visual system, *Cereb Cortex* **10**: 109-126.

Van Voorhis, S. and Hillyard, S. A., 1977, Visual evoked potentials and selective attention to points in space, *Percept Psychophys* **22**:54-62.

Vogel, E., Luck, S. and Shapiro, K., 1998, Electrophysiological evidence for a postperceptual locus of suppression during the attentional blink, *J Exp Psychol Human* **24**(6):1656-74.

Weissman, D. H., Mangun, G. R., and Woldorff, M. G., 2002a, A role for top-down attentional orienting during interference between global and local aspects of hierarchical stimuli, *Neuroimage* **17**(3):1266-1276.

Weissman, D. H., Woldorff, M. G., Hazlett, C. J. and Mangun, G. R., 2002b, Effects of practice on executive control investigated with fMRI, *Cognitive Brain Res* **15**(1):47-59.

Weissman, D. H., Giesbrecht, B., Song, A. W., Mangun, G. R., and Woldorff, M. G., 2003, Conflict monitoring in the human anterior cingulate cortex during selective attention to global and local object features, *Neuroimage* **19**:1361-1368.

Wilson, K. D., Woldorff, M. G., and Mangun, G. R., 2005, Control networks and hemispheric asymmetries in parietal cortex during attentional orienting in different spatial reference frames, *Neuroimage* **25**(3):668-683.

Wojciulik, E., and Kanwisher, N., 1999, The generality of parietal involvement in visual attention, *Neuron* **23**:747–764.

Woldorff, M. G., Hazlett, C. J., Fichtenholtz, H. M., Weissman, D. H., Dale, A. M., and Song, A. W., 2004, Functional parcellation of attentional control regions of the brain, *J Cognitive Neurosci* **16**(1):149–165.

Wright, M. J., 2005, Saliency predicts change detection in pictures of natural scenes, *Spat Vis* **18**:413-430.

Yingling, C. D. and Skinner, J. E., 1976, Selective regulation of thalamic sensory relay nuclei by nucleus reticularis thalami, *Electroencephalogr Clin Neurophysiol* **41**(5):476-82.

Acknowledgments

We gratefully acknowledge the support of the Swiss National Science Foundation (SNSF) und Deutsche Forschungsgemeinschaft (DFG). Without their support we would not have been able to carry out the preparatory steps for this project. We also would like to thank Armon Pfister for helping us to handle the manuscripts and to design the layout. Thanks also to all those researchers worldwide who inspired us with their most valuable work on spatial functions.

Index

Printed in the United States of America